Computational Methods of Linear Algebra

3rd Edition

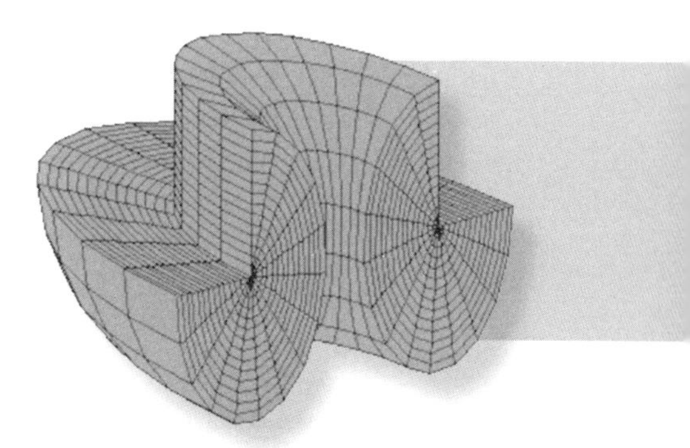

Computational Methods of Linear Algebra

3rd Edition

Granville Sewell
University of Texas El Paso, USA

NEW JERSEY • LONDON • SINGAPORE • BEIJING • SHANGHAI • HONG KONG • TAIPEI • CHENNAI

Published by

World Scientific Publishing Co. Pte. Ltd.
5 Toh Tuck Link, Singapore 596224
USA office: 27 Warren Street, Suite 401-402, Hackensack, NJ 07601
UK office: 57 Shelton Street, Covent Garden, London WC2H 9HE

Library of Congress Cataloging-in-Publication Data
Sewell, Granville, author.
Computational methods of linear algebra / by Granville Sewell (University of Texas El Paso, USA). -- 3rd edition.
pages cm
Includes bibliographical references and index.
ISBN 978-9814603850 (hardcover : alk. paper) -- ISBN 978-9814603867 (pbk : alk. paper)
1. Algebras, Linear--Textbooks. I. Title.
QA184.2.S44 2014
512'.5--dc23
2014016584

British Library Cataloguing-in-Publication Data
A catalogue record for this book is available from the British Library.

Printed in Singapore

To my son, Kevin

Contents

Preface

This text is appropriate for a course on the numerical solution of linear algebraic problems, designed for senior level undergraduate or beginning level graduate students. Although it will most likely be used for a second course in numerical analysis, the only prerequisite to using this text is a good course in linear or matrix algebra; however, such a course is an extremely important prerequisite.

Chapter 0 presents some basic definitions and results from linear algebra which are used in the later chapters. By no means can this short chapter be used to circumvent the linear algebra prerequisite mentioned above; it does not even contain a comprehensive review of the basic ideas of linear algebra. It is intended only to present some miscellaneous ideas, selected for inclusion because they are especially fundamental to later developments or may not be covered in a typical introductory linear algebra course.

Chapters 1–4 present and analyze methods for the solution of linear systems of equations (direct and iterative methods), linear least squares problems, linear eigenvalue problems, and linear programming problems; in short, we attack everything that begins with the word "linear". Truly "linear" numerical analysis problems have the common feature that they can be solved exactly, in a finite number of steps, if exact arithmetic is done. This means that all errors are due to roundoff; that is, they are attributable to the use of finite precision by the computer. (Iterative methods for linear systems and all methods for eigenvalue problems—which are not really linear—are exceptions.) Thus stability with respect to roundoff error must be a major consideration in the design of software for linear problems.

Chapter 5 discusses the fast Fourier transform. This is not a topic normally covered in texts on computational linear algebra. However, the fast Fourier transform is really just an efficient way of multiplying a special matrix times an arbitrary vector, and so it does not seem too out of place in a computational linear algebra text.

Chapter 6 contains a practical introduction for the student interested in writing computational linear algebra software that runs efficiently on today's vector and parallel supercomputers.

Double-precision FORTRAN90 subroutines, which solve each of the main problems covered using algorithms studied in the text, are presented and highlighted. A top priority in designing these subroutines was *readability*. Each subroutine is written in a well-documented, readable style, so that the student will be able to follow the program logic from start to finish. All loops are explicit and indented to make it easier for the student to analyze the computational complexity of the algorithms. Even though we have steadfastly resisted the temptation to make them slightly more efficient at the expense of readability, the subroutines that solve the truly linear problems are nearly state-of-the-art with regard to efficiency. The eigenvalue codes, on the other hand, are not state-of-the-art, but neither are they grossly inferior to the best programs available.

MATLAB® versions of the codes in Chapters 1–5 are listed in Appendix A. Machine-readable copies of the FORTRAN90 and MATLAB codes in the book can be downloaded from

http://www.math.utep.edu/Faculty/sewell/computational_methods

There is very little difference between the FORTRAN and MATLAB versions; they are almost line-by-line translations, so the student who is familiar with MATLAB (or any other programming language, for that matter) will have no trouble following the logic of the FORTRAN programs in the text. But students can do the computer problems using either FORTRAN or MATLAB, with the exception of the problems in Chapter 6. The problems in this chapter require a FORTRAN90 compiler and an MPI library.

Subroutines DEGNON, DPOWER, DFFT and NRFFT contain double-precision complex variables, typed COMPLEX*16, which is a nonstandard, but widely recognized, type. Otherwise, the subroutines conform to the FORTRAN90 standard and thus should be highly portable. In fact, the programs in Chapters 1–4 will run on most FORTRAN77 compilers.

For extensive surveys on other available mathematical software, including software for the problems studied in this text, the reader is referred to the books by Heath [2002; summary in each chapter] and Kincaid and Cheney [2004; Appendix B].

The author developed this text for a graduate course at the University of Texas El Paso and has also used it for a Texas A&M distance learning graduate course.

0

Reference Material

0.1 Miscellaneous Results from Linear Algebra

As mentioned in the preface, it is assumed that the reader has a good foundation in linear algebra; so we shall make no attempt here to present a comprehensive review of even the basic ideas of linear algebra. However, we have chosen a few results for review in this chapter, selected for inclusion either because they are especially important to later developments or else because they may not be covered in a typical introductory course in linear algebra.

Computational linear algebra deals primarily with matrices, rectangular arrays of real or complex numbers (unless otherwise stated, our matrices will consist only of real numbers). We shall consider vectors to be N by 1 matrices (i.e., column vectors), and we shall treat them differently from other matrices only in that they will be designated by italic boldface type, as $\boldsymbol{b}$, and their norms will be defined differently (Section 0.3).

If A is an N by M matrix with a_{ij} in row i, column j, then the **transpose** of A, denoted by A^{T}, is an M by N matrix with a_{ji} in row i, column j. For example, if

$$A = \begin{bmatrix} 3 & 2 \\ 1 & 4 \\ 1 & 5 \end{bmatrix}, \quad \text{then} \quad A^{\mathrm{T}} = \begin{bmatrix} 3 & 1 & 1 \\ 2 & 4 & 5 \end{bmatrix}.$$

If $\boldsymbol{x}$ and $\boldsymbol{y}$ are vectors, that is, N by 1 matrices, then $\boldsymbol{x}^{\mathrm{T}}$ is a 1 by N matrix, and $\boldsymbol{x}^{\mathrm{T}}\boldsymbol{y}$ is a 1 by 1 matrix, namely,

$$\boldsymbol{x}^{\mathrm{T}}\boldsymbol{y} = [x_1\, x_2\, \ldots\, x_N] \begin{bmatrix} y_1 \\ y_2 \\ \vdots \\ y_N \end{bmatrix} = x_1y_1 + x_2y_2 + \ldots + x_Ny_N.$$

Thus $\boldsymbol{x}^{\mathrm{T}}\boldsymbol{y}$ is the scalar product of vectors $\boldsymbol{x}$ and $\boldsymbol{y}$. $\boldsymbol{x}\boldsymbol{y}^{\mathrm{T}}$, on the other hand, is an N by N matrix.

We now present four miscellaneous theorems, which will be needed in later chapters.

Theorem 0.1.1. $(AB)^{\mathrm{T}} = B^{\mathrm{T}}A^{\mathrm{T}}$, *for arbitrary matrices* A *and* B, *provided that the product* AB *exists.*

Proof: If A is an N by K matrix with elements a_{ij}, and B is a K by M matrix with elements b_{ij}, then, using the definition of matrix multiplication, we have

$$\begin{aligned}(B^{\mathrm{T}}A^{\mathrm{T}})_{ij} &= \sum_{k=1}^{K}(B^{\mathrm{T}})_{ik}(A^{\mathrm{T}})_{kj} = \sum_{k=1}^{K} b_{ki}a_{jk} = \sum_{k=1}^{K} a_{jk}b_{ki} \\ &= (AB)_{ji} = [(AB)^{\mathrm{T}}]_{ij}. \qquad \blacksquare\end{aligned}$$

In other words, the transpose of a product is the product of the transposes, in reverse order.

Theorem 0.1.2.

(a) *The product (if it exists) of two upper triangular matrices is an upper triangular matrix (A is upper triangular if* $a_{ij} = 0$ *when* $j < i$*), and the product of two lower triangular matrices is lower triangular.*

(b) *The inverse (if it exists) of an upper triangular matrix is upper triangular, and the inverse (if it exists) of a lower triangular matrix is lower triangular.*

Proof:

(a) If A is an N by K upper triangular matrix with elements a_{ij}, and B is a K by M upper triangular matrix with elements b_{ij}, then $C \equiv AB$ has elements given by

$$c_{ij} = \sum_{k=1}^{K} a_{ik}b_{kj}.$$

Now, if $j < i$, it is not possible for $j \geq k$ and $k \geq i$ to hold simultaneously, for any k; otherwise we would have $j \geq i$. Therefore either $j < k$ or $k < i$, or both. In the former case, $b_{kj} = 0$ since B is upper triangular and, in the latter case, $a_{ik} = 0$ since A is upper triangular. Thus at least one of the factors in each term in the expression for c_{ij} is zero, and so $c_{ij} = 0$ if $j < i$. Therefore, $C = AB$ is upper triangular. If A and B are lower triangular, the proof that AB is lower triangular is similar.

(b) Suppose A is an N by N upper triangular matrix (we can assume that A is square; otherwise it has no inverse), and suppose B is an N by N matrix that is *not* upper triangular. Then let j be a column of B with nonzero elements below the diagonal, and let $i(i > j)$ be such that b_{ij} is the last nonzero element in that column. Then $C \equiv AB$ has elements given by

$$c_{ij} = \sum_{k=1}^{N} a_{ik} b_{kj}.$$

Now, since A is upper triangular, $a_{ik} = 0$ for $k < i$, and since b_{ij} is the last nonzero in column $j, b_{kj} = 0$ for $k > i$. Thus at least one of the factors in each term in the expression for c_{ij} is zero except the one with $k = i$. So $c_{ij} = a_{ii} b_{ij}$, but $b_{ij} \neq 0$ by assumption and, since the upper triangular matrix A is assumed to be nonsingular, $a_{ii} \neq 0$ also; therefore $c_{ij} \neq 0$. Since $i > j$, this means the matrix C cannot be diagonal, and thus AB cannot be equal to the identity matrix I. So we have shown that, if B is not upper triangular, it cannot be the inverse of A. The proof that the inverse of a lower triangular matrix is lower triangular is similar. ■

Theorem 0.1.3. *If $p \geq 0$, and N is a positive integer, then*

$$\frac{N^{p+1}}{p+1} \leq 1^p + 2^p + 3^p + \ldots + N^p \leq \frac{N^{p+1}}{p+1} \left(1 + \frac{p+1}{N}\right)$$

Proof: From calculus, we know that the area under the curve $y = x^p$ between $x = 0$ and $x = N$ is given by

$$\int_0^N x^p \, dx = \frac{N^{p+1}}{p+1}.$$

Since x^p is a nondecreasing function, we see in Figure 0.1.1 that the sum of the areas of the solid rectangles is less than the area under the curve $y = x^p$. Thus

$$0^p + 1^p + 2^p + \ldots + (N-1)^p \leq \frac{N^{p+1}}{p+1}.$$

Similarly, the sum of the areas of the dashed rectangles is greater than the area under the curve; so

$$1^p + 2^p + 3^p + \ldots + N^p \geq \frac{N^{p+1}}{p+1}.$$

The desired formula follows directly from these two bounds. ■

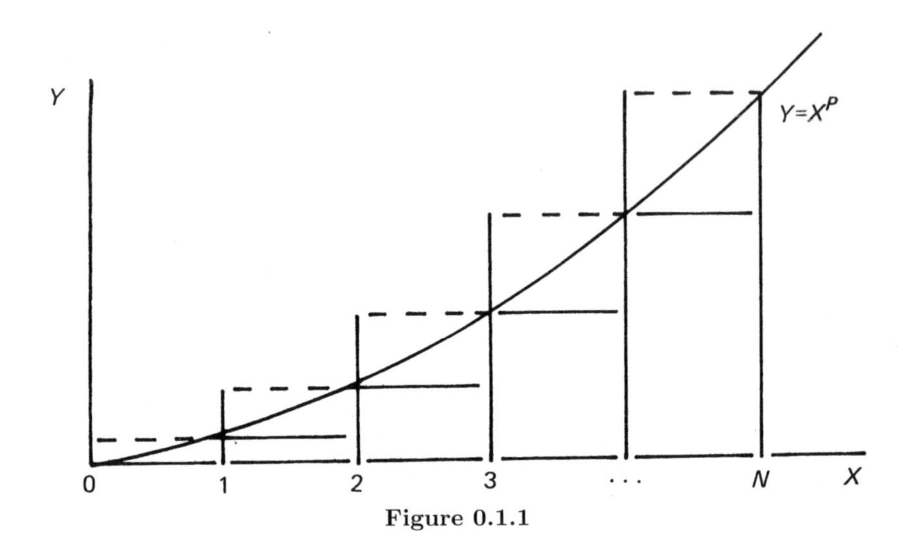

Figure 0.1.1

This theorem will be useful when we try to estimate the amount of work done by certain algorithms. When N is a large number, Theorem 0.1.3 implies that

$$\sum_{i=1}^{N} i^p \approx \frac{N^{p+1}}{p+1}. \tag{0.1.1}$$

Theorem 0.1.4. *If the matrices A and B have the block structure*

$$A = \overset{\quad m \qquad N-m}{\left[\begin{array}{c|c} A_{11} & A_{12} \\ \hline A_{21} & A_{22} \end{array}\right]} \begin{array}{l} m \\ N-m \end{array} \qquad B = \overset{\quad m \qquad N-m}{\left[\begin{array}{c|c} B_{11} & B_{12} \\ \hline B_{21} & B_{22} \end{array}\right]} \begin{array}{l} m \\ N-m \end{array}$$

where $A_{11}, \ldots, A_{22}$ and $B_{11}, \ldots, B_{22}$ are matrices with dimensions as indicated, then the product matrix will have the form

$$AB = \overset{\qquad\qquad m \qquad\qquad\qquad\qquad N-m}{\left[\begin{array}{c|c} A_{11}B_{11} + A_{12}B_{21} & A_{11}B_{12} + A_{12}B_{22} \\ \hline A_{21}B_{11} + A_{22}B_{21} & A_{21}B_{12} + A_{22}B_{22} \end{array}\right]} \begin{array}{l} m \\ N-m \end{array}$$

Proof: The proof involves simply carrying out the indicated multiplications and will be omitted. ■

Note that what Theorem 0.1.4 says is that block matrices can be multiplied as if the subblocks were just scalar elements.

0.2 Special Matrices

The algorithms that we shall study in later chapters involve a number of special matrix types. The more frequently used types are defined below.

The first few types are characterized by their nonzero structures. We shall be primarily interested in square matrices of the types listed below, even though the definitions themselves do not limit us to square matrices.

(1) A is **diagonal** if $a_{ij} = 0$ when $i \neq j$. An example is

$$\begin{bmatrix} 3 & 0 & 0 \\ 0 & 1 & 0 \\ 0 & 0 & -1 \end{bmatrix}.$$

(2) A is **tridiagonal** if $a_{ij} = 0$ when $|i - j| > 1$. An example is

$$\begin{bmatrix} 3 & 4 & 0 & 0 \\ 2 & 3 & -1 & 0 \\ 0 & 1 & 3 & 2 \\ 0 & 0 & 1 & 2 \end{bmatrix}.$$

(3) A is **upper triangular** if $a_{ij} = 0$ when $j < i$. An example is

$$\begin{bmatrix} 3 & 4 & 6 \\ 0 & 1 & 3 \\ 0 & 0 & 2 \end{bmatrix}.$$

A lower triangular matrix is defined similarly.

(4) A is **upper Hessenberg** if $a_{ij} = 0$ when $j < i - 1$. An example is

$$\begin{bmatrix} 3 & 4 & 5 & 5 \\ 2 & 3 & -1 & 2 \\ 0 & 1 & 3 & 2 \\ 0 & 0 & 1 & 2 \end{bmatrix}.$$

A lower Hessenberg matrix is defined similarly.

Note that each of these definitions specifies which positions must contain zeros and not which contain nonzero elements. Thus, for example, a diagonal matrix technically qualifies as a tridiagonal, an (upper or lower) triangular, and an (upper or lower) Hessenberg matrix, although we would not normally refer to it by any of these names.

The following matrix types are characterized by numerical properties, rather than by their nonzero structures. Each of the definitions below implicitly require A to be square, although in some cases this is not immediately obvious.

(5) A is **symmetric** if $a_{ij} = a_{ji}$, for each i, j. An example is

$$\begin{bmatrix} 3 & 4 & 6 \\ 4 & 1 & -1 \\ 6 & -1 & 0 \end{bmatrix}.$$

A symmetric matrix is thus one for which $A^{\mathrm{T}} = A$. An important feature of (real) symmetric matrices is that their eigenvalues are all real. To show this, suppose $A\boldsymbol{z} = \lambda\boldsymbol{z}$. Then (the overbar denotes complex conjugate)

$$\lambda\bar{\boldsymbol{z}}^{\mathrm{T}}\boldsymbol{z} = \bar{\boldsymbol{z}}^{\mathrm{T}}A\boldsymbol{z} = (\bar{\boldsymbol{z}}^{\mathrm{T}}A\boldsymbol{z})^{\mathrm{T}} = \boldsymbol{z}^{\mathrm{T}}A^{\mathrm{T}}\bar{\boldsymbol{z}} = \boldsymbol{z}^{\mathrm{T}}A\bar{\boldsymbol{z}} = \boldsymbol{z}^{\mathrm{T}}(\overline{A\boldsymbol{z}}) = \bar{\lambda}\boldsymbol{z}^{\mathrm{T}}\bar{\boldsymbol{z}},$$

and, since $\bar{\boldsymbol{z}}^{\mathrm{T}}\boldsymbol{z} = \boldsymbol{z}^{\mathrm{T}}\bar{\boldsymbol{z}} = |z_1|^2 + \ldots + |z_N|^2$ is real and positive, $\lambda = \bar{\lambda}$.

(6) A is **nonsingular** if a matrix A^{-1} exists such that $A^{-1}A = AA^{-1} = I$. (I is the identity matrix, a diagonal matrix with 1 elements along the diagonal.)

If A is not square, $A^{-1}A$ and AA^{-1} cannot be equal, as they will not even be of the same size. Thus only a square matrix can be nonsingular. If A is square, the following are true if and only if A is nonsingular:

(a) the rows of A are linearly independent;

(b) the columns of A are linearly independent;

(c) the determinant of A is nonzero;

(d) $A\boldsymbol{x} = \boldsymbol{b}$ has a unique solution (any $\boldsymbol{b}$).

(7) A is **positive-definite** if it is symmetric, and

$$\boldsymbol{x}^{\mathrm{T}}A\boldsymbol{x} > 0, \quad \text{for all } \boldsymbol{x} \neq \boldsymbol{0}.$$

Since a positive-definite matrix A is symmetric, its eigenvalues, and thus also its eigenvectors, are real. The eigenvalues are also positive for, if $A\boldsymbol{z} = \lambda\boldsymbol{z}$, then, since an eigenvector is by definition nonzero,

$$0 < \boldsymbol{z}^{\mathrm{T}}A\boldsymbol{z} = \lambda\boldsymbol{z}^{\mathrm{T}}\boldsymbol{z} = \lambda(z_1^2 + \ldots + z_N^2)$$

and so $\lambda > 0$. A is said to be positive-semidefinite if it is symmetric and $\boldsymbol{x}^{\mathrm{T}}A\boldsymbol{x} \geq 0$ for all $\boldsymbol{x}$. Clearly all eigenvalues of a positive-semidefinite matrix are nonnegative, but not necessarily positive.

(8) A is **orthogonal** if $A^{\mathrm{T}}A = AA^{\mathrm{T}} = I$.

In other words, A is orthogonal if its transpose is its inverse. Now

$$I_{ij} = (A^{\mathrm{T}}A)_{ij} = \sum_k (A^{\mathrm{T}})_{ik}(A)_{kj} = \sum_k a_{ki}a_{kj} = (\text{col. } i)^{\mathrm{T}}(\text{col. } j).$$

Since I_{ij} is 0 when $i \neq j$ and 1 when $i = j$, we see that the columns of an orthogonal matrix are mutually orthogonal and each column is normalized to have length one. The same can be said of its rows, since $AA^{\mathrm{T}} = I$ also.

Orthogonal matrices will be very important in Chapters 2 and 3.

0.3 Vector and Matrix Norms

Vector and matrix norms play important roles in later chapters. As seen below, although vector and matrix norms employ similar notations, they are defined quite differently.

For $p \geq 1$, the "p-norm" (or L_p-norm) of a vector $\boldsymbol{x} = (x_1, \ldots, x_N)$ is defined by

$$\|\boldsymbol{x}\|_p \equiv (|x_1|^p + \ldots + |x_N|^p)^{1/p}. \tag{0.3.1}$$

Thus the 1-norm of $\boldsymbol{x}$ is the sum of the absolute values of its components, and the 2-norm (also called the Euclidian norm) is the square root of the sum of squares of its components. Note that $\|\boldsymbol{x}\|_2^2 = \boldsymbol{x}^{\mathrm{T}}\boldsymbol{x}$.

Now suppose we hold the vector $\boldsymbol{x}$ fixed and let $p \to \infty$. If $M \equiv \max|x_i|$ and K denotes the number of components with $|x_i| = M$, then by (0.3.1)

$$\lim_{p\to\infty} \|\boldsymbol{x}\|_p = \lim_{p\to\infty} \left[M \left(\frac{|x_1|^p}{M^p} + \ldots + \frac{|x_N|^p}{M^p} \right)^{1/p} \right] = \lim_{p\to\infty} [M(K)^{1/p}] = M.$$

Thus it seems reasonable to define $\|\boldsymbol{x}\|_\infty \equiv M \equiv \max|x_i|$. The three norms $\|\boldsymbol{x}\|_1$, $\|\boldsymbol{x}\|_2$ and $\|\boldsymbol{x}\|_\infty$ are the only ones in common use. It is easy to verify that $\|\boldsymbol{x}\|_\infty \leq \|\boldsymbol{x}\|_p \leq N\|\boldsymbol{x}\|_\infty$ for any $p \geq 1$, so that these norms differ from each other by at most a factor of N.

For each vector norm, there is an associated matrix norm, defined by

$$\|A\|_p \equiv \max_{\boldsymbol{x}\neq\boldsymbol{0}} \left(\frac{\|A\boldsymbol{x}\|_p}{\|\boldsymbol{x}\|_p} \right). \tag{0.3.2}$$

Note that it follows immediately from (0.3.2) that $\|A\boldsymbol{x}\|_p \leq \|A\|_p\|\boldsymbol{x}\|_p$. However, it is not feasible to calculate a norm of a matrix using (0.3.2) directly. The following theorem provides formulas that can be used to calculate the three commonly used matrix norms. Although (0.3.2) does not require that A be square, in this book we shall only be interested in the norms of square matrices.

Theorem 0.3.1. *If the elements of the N by N matrix A are a_{ij}, then*

(a) $\|A\|_1 = \max_{1\le j\le N} \sum_{i=1}^N |a_{ij}|$,

(b) $\|A\|_\infty = \max_{1\le i\le N} \sum_{j=1}^N |a_{ij}|$,

(c) $\|A\|_2 = \mu_{\max}^{1/2}$,
where $\mu_{\max}$ is the largest eigenvalue of $A^{\mathrm{T}}A$.

Proof: For $p = 1, \infty$, and 2, we need to show first that with $\|A\|_p$ defined by formulas (a), (b), and (c), respectively, $\|A\boldsymbol{x}\|_p \le \|A\|_p\|\boldsymbol{x}\|_p$, for arbitrary $\boldsymbol{x}$. This verifies that our matrix norm formula is an upper bound on the ratios

$$\frac{\|A\boldsymbol{x}\|_p}{\|\boldsymbol{x}\|_p}. \tag{0.3.3}$$

Then, if we can find one nonzero vector $\boldsymbol{z}$ such that $\|A\boldsymbol{z}\|_p = \|A\|_p\|\boldsymbol{z}\|_p$, we have shown that our matrix norm formula is actually the least upper bound to the ratios (0.3.3), and hence it is really the matrix norm.

(a)

$$\begin{aligned}\|A\boldsymbol{x}\|_1 &= \sum_{i=1}^N |(A\boldsymbol{x})_i| = \sum_{i=1}^N \left|\sum_{j=1}^N a_{ij}x_j\right| \le \sum_{i=1}^N\sum_{j=1}^N |a_{ij}||x_j| \\ &= \sum_{j=1}^N \left(\sum_{i=1}^N |a_{ij}|\right)|x_j| \le \left(\max_{1\le j\le N}\sum_{i=1}^N |a_{ij}|\right)\sum_{j=1}^N |x_j| = \|A\|_1\|\boldsymbol{x}\|_1.\end{aligned}$$

If J is the value (or one of the values) of j for which the "column sum" $\sum_{i=1}^N |a_{ij}|$ is maximized, then we can choose $\boldsymbol{z}$ to be the vector that has $z_J = 1$ and all other components equal to zero. Then $\|\boldsymbol{z}\|_1 = 1$ and since $A\boldsymbol{z}$ is equal to the Jth column of A, $\|A\boldsymbol{z}\|_1 = \|A\|_1 = \|A\|_1\|\boldsymbol{z}\|_1$.

(b)

$$\begin{aligned}\|A\boldsymbol{x}\|_\infty &= \max_{1\le i\le N}|(A\boldsymbol{x})_i| = \max_{1\le i\le N}\left|\sum_{j=1}^N a_{ij}x_j\right| \le \max_{1\le i\le N}\left(\sum_{j=1}^N |a_{ij}||x_j|\right) \\ &\le \max_{1\le j\le N}|x_j|\left(\max_{1\le i\le N}\sum_{j=1}^N |a_{ij}|\right) = \|A\|_\infty\|\boldsymbol{x}\|_\infty.\end{aligned}$$

If I is the value (or one of the values) of i for which the "row sum" $\sum_{j=1}^{N} |a_{ij}|$ is maximized, then we can choose $\boldsymbol{z}$ to be the vector that has $z_j = 1$ when $a_{Ij} \geq 0$ and $z_j = -1$ when $a_{Ij} < 0$. Then $\|\boldsymbol{z}\|_\infty = 1$ and the largest component of $A\boldsymbol{z}$ will be component I, which will have the value $(A\boldsymbol{z})_I = a_{I1}z_1 + \ldots + a_{IN}z_N = |a_{I1}| + \ldots + |a_{IN}| = \|A\|_\infty$. Thus $\|A\boldsymbol{z}\|_\infty = \|A\|_\infty = \|A\|_\infty \|\boldsymbol{z}\|_\infty$.

(c) Even if A is not symmetric, the matrix $A^{\mathrm{T}}A$ is, since $(A^{\mathrm{T}}A)^{\mathrm{T}} = A^{\mathrm{T}}A$. It is also positive-semidefinite, since $\boldsymbol{x}^{\mathrm{T}}(A^{\mathrm{T}}A)\boldsymbol{x} = (A\boldsymbol{x})^{\mathrm{T}}(A\boldsymbol{x}) = \|A\boldsymbol{x}\|_2^2 \geq 0$; so the eigenvalues of $A^{\mathrm{T}}A$ are all real and nonnegative. Now by a basic theorem of linear algebra, since $A^{\mathrm{T}}A$ is symmetric, there exists an orthogonal matrix Q ($Q^{\mathrm{T}}Q = I$) such that $A^{\mathrm{T}}A = Q^{\mathrm{T}}DQ$, where D is a diagonal matrix containing the (nonnegative) eigenvalues of $A^{\mathrm{T}}A$. If we denote these eigenvalues by μ_i, and if the components of $Q\boldsymbol{x}$ are q_i, then

$$\begin{aligned}
\|A\boldsymbol{x}\|_2^2 &= (A\boldsymbol{x})^{\mathrm{T}}(A\boldsymbol{x}) = \boldsymbol{x}^{\mathrm{T}}(A^{\mathrm{T}}A)\boldsymbol{x} = \boldsymbol{x}^{\mathrm{T}}(Q^{\mathrm{T}}DQ)\boldsymbol{x} = (Q\boldsymbol{x})^{\mathrm{T}}D(Q\boldsymbol{x}) \\
&= \sum_{i=1}^{N} \mu_i q_i^2 \leq \mu_{\max} \sum_{i=1}^{N} q_i^2 = \mu_{\max}(Q\boldsymbol{x})^{\mathrm{T}}(Q\boldsymbol{x}) \\
&= \mu_{\max}\boldsymbol{x}^{\mathrm{T}}Q^{\mathrm{T}}Q\boldsymbol{x} = \mu_{\max}\|\boldsymbol{x}\|_2^2. \qquad (0.3.4)
\end{aligned}$$

Thus $\|A\boldsymbol{x}\|_2 \leq \mu_{\max}^{1/2}\|\boldsymbol{x}\|_2$. Now, if I is the value (or one of the values) of i that maximizes μ_i, we can choose $\boldsymbol{z}$ to be such that $Q\boldsymbol{z}$ has $q_I = 1$, and all other components equal to 0. Then clearly we can replace the only inequality in (0.3.4) by an equality, and $\|A\boldsymbol{z}\|_2 = \mu_{\max}^{1/2}\|\boldsymbol{z}\|_2$. ■

Note that, if A is symmetric, the eigenvalues μ_i of $A^{\mathrm{T}}A = AA$ are the squares of those (λ_i) of A, and so

$$\|A\|_2 = \mu_{\max}^{1/2} = |\lambda_{\max}|,$$

where $\lambda_{\max}$ is the eigenvalue of A of largest modulus.

As an example, consider the matrix

$$A = \begin{bmatrix} -3 & 4 & 1 \\ 2 & 0 & -1 \\ 3 & -2 & 5 \end{bmatrix}.$$

We calculate that $\|A\|_1 = 8$, since the sum of the absolute values of the three columns are 8, 6, and 7, and the 1-norm is the maximum of these. Also, $\|A\|_\infty = 10$, since the sum of the absolute values of the three rows are 8, 3, and 10, and the ∞-norm is the maximum of these. To calculate the 2-norm of A is much more difficult. First we form

$$A^{\mathrm{T}}A = \begin{bmatrix} 22 & -18 & 10 \\ -18 & 20 & -6 \\ 10 & -6 & 27 \end{bmatrix}.$$

Then, using a computer program to be presented in Section 3.2, we find the eigenvalues of $A^{\mathrm{T}}A$ to be 2.668, 20.400, and 45.931. Hence $\|A\|_2 = 45.931^{1/2} = 6.777$.

1

Systems of Linear Equations

1.1 Introduction

Consider the electrical circuit shown in Figure 1.1.1.

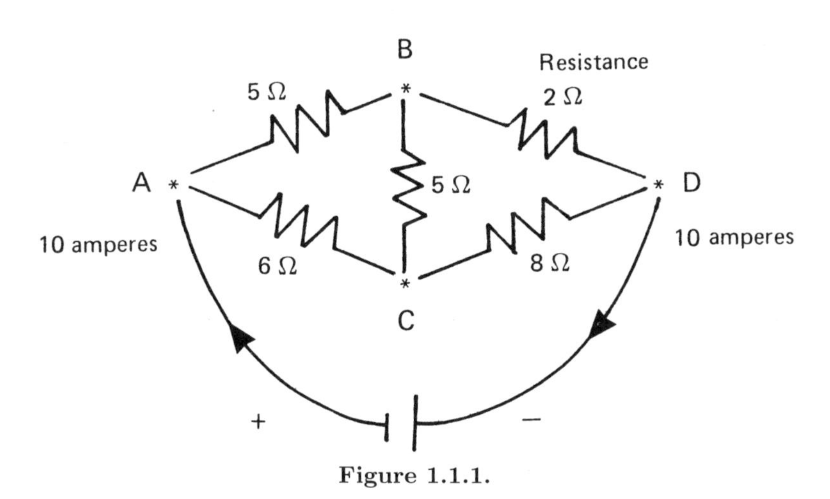

Figure 1.1.1.

We want to calculate the current I_{AB} flowing from node A to node B, and also the currents I_{BC}, I_{AC}, I_{BD} and I_{CD}. It is known that the net current flowing into any node is zero, and also that the voltage drop (current times resistance) around any closed loop is zero. From the first rule we deduce that

$$\begin{aligned} 10 - I_{AB} - I_{AC} &= 0, && (1.1.1a)\\ I_{AB} - I_{BC} - I_{BD} &= 0, && (1.1.1b)\\ I_{AC} + I_{BC} - I_{CD} &= 0, && (1.1.1c)\\ I_{BD} + I_{CD} - 10 &= 0. && (1.1.1d) \end{aligned}$$

The second rule gives

$$\begin{aligned} 5I_{AB} + 5I_{BC} - 6I_{AC} &= 0, && (1.1.2a)\\ 2I_{BD} - 8I_{CD} - 5I_{BC} &= 0, && (1.1.2b)\\ 5I_{AB} + 2I_{BD} - 8I_{CD} - 6I_{AC} &= 0. && (1.1.2c) \end{aligned}$$

It appears that we have seven linear equations for the five unknown currents. However, if we add equations 1.1.1a–1.1.1c together we get equation 1.1.1d; so the last equation is redundant and may be dropped. Similarly, if we add 1.1.2a and 1.1.2b, we get equation 1.1.2c, so that 1.1.2c can also be dropped, leaving us with the five equations

$$\begin{bmatrix} -1 & 0 & -1 & 0 & 0 \\ 1 & -1 & 0 & -1 & 0 \\ 0 & 1 & 1 & 0 & -1 \\ 5 & 5 & -6 & 0 & 0 \\ 0 & -5 & 0 & 2 & -8 \end{bmatrix} \begin{bmatrix} I_{AB} \\ I_{BC} \\ I_{AC} \\ I_{BD} \\ I_{CD} \end{bmatrix} = \begin{bmatrix} -10 \\ 0 \\ 0 \\ 0 \\ 0 \end{bmatrix}. \tag{1.1.3}$$

Systems of simultaneous linear equations, such as 1.1.3, have to be solved frequently by workers in almost every branch of science and engineering, and they arise during the solution of many of the other problems faced by numerical analysts, such as nonlinear systems, partial differential equations, and optimization problems. In this chapter we shall develop and analyze methods for the computer solution of linear systems.

1.2 Gaussian Elimination

The solution of linear systems is one of the few problems in numerical analysis where the first method taught in mathematics classes is still used by state-of-the-art software. The method of Gaussian elimination, whereby a system of linear equations is systematically reduced to an equivalent system of triangular form and then back solved for the unknowns, is the foundation for most modern algorithms—at least those designed for small dense systems. However, a few modifications are employed by sophisticated computer codes.

The basic tool used to reduce a general system of N linear equations in N unknowns to triangular form is the addition of a multiple of one equation to another. For example, the 2 by 2 system

$$\begin{aligned} 3x_1 + 2x_2 &= 5, \\ -6x_1 + 3x_2 &= -3 \end{aligned}$$

can be reduced to triangular form by adding twice the first row to the second row:

$$\begin{aligned} 3x_1 + 2x_2 &= 5, \\ 7x_2 &= 7. \end{aligned}$$

Now the last equation can be solved for $x_2 = 1$, and then the first equation can be solved for $x_1 = (5 - 2x_2)/3 = 1$.

For a general N by N system, $A\boldsymbol{x} = \boldsymbol{b}$:

$$\begin{bmatrix} a_{11} & a_{12} & a_{13} & \dots & a_{1N} \\ a_{21} & a_{22} & a_{23} & \dots & a_{2N} \\ a_{31} & a_{32} & a_{33} & \dots & a_{3N} \\ \vdots & \vdots & \vdots & \vdots & \vdots \\ a_{N1} & a_{N2} & a_{N3} & \dots & a_{NN} \end{bmatrix} \begin{bmatrix} x_1 \\ x_2 \\ x_3 \\ \vdots \\ x_N \end{bmatrix} = \begin{bmatrix} b_1 \\ b_2 \\ b_3 \\ \vdots \\ b_N \end{bmatrix}, \tag{1.2.1}$$

and the basic Gaussian elimination process is as follows. First, for $j = 2, \dots, N$ we take a multiple $-a_{j1}/a_{11}$ of the first row and add it to the jth row, to make $a_{j1} = 0$. Since we are really adding a multiple of the first equation to the jth equation, we must also add the same multiple of b_1 to b_j. Once we have knocked out all elements of A below the "pivot" element a_{11} in the first column, we have the system

$$\begin{bmatrix} a_{11} & a_{12} & a_{13} & \dots & a_{1N} \\ 0 & a'_{22} & a'_{23} & \dots & a'_{2N} \\ 0 & a'_{32} & a'_{33} & \dots & a'_{3N} \\ \vdots & \vdots & \vdots & \vdots & \vdots \\ 0 & a'_{N2} & a'_{N3} & \dots & a'_{NN} \end{bmatrix} \begin{bmatrix} x_1 \\ x_2 \\ x_3 \\ \vdots \\ x_N \end{bmatrix} = \begin{bmatrix} b_1 \\ b'_2 \\ b'_3 \\ \vdots \\ b'_N \end{bmatrix},$$

which is equivalent to 1.2.1.

Now, for $j = 3, \dots, N$, we take a multiple $-a_{j2}/a_{22}$ of the second row and add it to the jth row. When we have finished this, all subdiagonal elements in the second column are zero, and we are ready to process the third column. Applying this process to columns $i = 1, \dots, N - 1$ (there are no subdiagonal elements to knock out in the Nth column) completes the forward elimination stage, and the coefficient matrix A has been reduced to upper triangular form (see 1.2.7).

$$A = \begin{array}{c} \\ \\ \\ \\ \text{row } i \\ \\ \\ \text{row } j \\ \\ \\ \end{array} \begin{array}{c} \text{column} \\ i \end{array}\left[\begin{array}{ccccccccc}
X & X & X & X & X & X & X & X & X \\
0 & X & X & X & X & X & X & X & X \\
0 & 0 & X & X & X & X & X & X & X \\
0 & 0 & 0 & a_{ii} & X & X & X & X & X \\
0 & 0 & 0 & X & X & X & X & X & X \\
0 & 0 & 0 & X & X & X & X & X & X \\
0 & 0 & 0 & a_{ji} & X & X & X & X & X \\
0 & 0 & 0 & X & X & X & X & X & X \\
0 & 0 & 0 & X & X & X & X & X & X
\end{array}\right]. \tag{1.2.2}$$

It is important to note that, because we eliminate subdiagonal elements column by column, beginning with the first column, the zeroed elements remain zero; while we are adding a multiple $-a_{ji}/a_{ii}$ of row i to row j ($j > i$), to knock out a_{ji}, we are just adding multiples of zero to zero in columns 1 to $i-1$ (see 1.2.2). In fact, we can skip over these unnecessary calculations when we program this process.

Obviously, if one of the diagonal pivots a_{ii} is zero, we cannot use a_{ii} to knock out the elements below it; we cannot change a_{ji} by adding *any* multiple of $a_{ii} = 0$ to it. We must switch row i with another row l below it (switching with a row above i would destroy some of the zeros introduced earlier), which contains a nonzero element a_{li} in the ith column. Again, since we cannot switch the left-hand sides of two equations without interchanging the right-hand sides, we have to switch the values of b_i and b_l also. Now the new pivot a_{ii} is not zero, and we can continue the Gaussian elimination scheme as usual.

If $a_{li} = 0$ for $l = i, \ldots, N$, then it will not be satisfactory to switch row i with any of the rows below it, as all the potential pivots are zero. It is time to give up, because the reduced matrix (and thus also the original matrix) is singular. To see this, note that the last $N - i + 1$ rows have nonzero elements only in the last $N - i$ columns (see 1.2.2); so they all belong to the same $(N - i)$-dimensional subspace of R^N. However, $N - i + 1$ vectors in an $(N - i)$-dimensional space are necessarily linearly dependent, which means that the matrix is singular. If A is square and singular, $A\boldsymbol{x} = \boldsymbol{b}$ may (or may not) still have solutions, but it will never have a unique solution. In Chapter 2 we give an algorithm that can be used to return a solution if solutions exist, or a near-solution if solutions do not exist, but in this chapter we are only interested in systems with unique solutions, and we shall simply give up if A is found to be singular.

Now, as long as all arithmetic is done exactly, the above-described forward-elimination process will, unless A is singular, produce an exactly equivalent triangular system, from which the unknowns can be solved in reverse order. However, when we solve linear systems on a computer—which does not do arithmetic exactly—we must modify the algorithm somewhat. As long as we do arithmetic exactly, there is no need to switch rows until a pivot is found that is *exactly* zero; when computations are done on a computer we find that a *nearly* zero pivot is also to be avoided, as it can lead to serious inaccuracies in the final answer.

Perhaps the best way to see that a nearly zero pivot poses a threat to accuracy is to consider a linear system such as

$$\begin{bmatrix} 1 & \frac{1}{3} & 0 \\ 3 & 1 & 1 \\ 3 & 2 & 1 \end{bmatrix} \begin{bmatrix} x \\ y \\ z \end{bmatrix} = \begin{bmatrix} \frac{4}{3} \\ 5 \\ 6 \end{bmatrix}. \tag{1.2.3}$$

If exact arithmetic is done, we could take -3 times the first equation and add it to the second and third equations, reducing 1.2.3 to

$$\begin{bmatrix} 1 & \frac{1}{3} & 0 \\ 0 & 0 & 1 \\ 0 & 1 & 1 \end{bmatrix} \begin{bmatrix} x \\ y \\ z \end{bmatrix} = \begin{bmatrix} \frac{4}{3} \\ 1 \\ 2 \end{bmatrix}. \tag{1.2.4}$$

We now have to switch rows 2 and 3, to bring a nonzero up to the pivot (a_{22}) position. Then back substitution produces $z = y = x = 1$, as can easily be verified.

On the other hand, if exact arithmetic is *not* done, a_{22} in 1.2.4 will not in general be exactly zero, but some number ϵ that is essentially a small random number:

$$\begin{bmatrix} 1 & \frac{1}{3} & 0 \\ 0 & \epsilon & 1 \\ 0 & 1 & 1 \end{bmatrix} \begin{bmatrix} x \\ y \\ z \end{bmatrix} = \begin{bmatrix} \frac{4}{3} \\ 1 \\ 2 \end{bmatrix}. \tag{1.2.5}$$

We may argue that "since a_{22} is not zero, there is no need to switch rows" and proceed to add $-1/\epsilon$ times the second equation to the third equation, and we may then back solve for the unknowns. However, we are taking as significant a number that is essentially garbage (ϵ is as likely to be positive as negative), and the result on a real computer is that our solution will probably also be garbage. In fact, when we solved 1.2.3 using single-precision arithmetic, with *no row interchanges*, the result was $x = -4, y = 16, z = 1$, which is nowhere close to the true solution $x = y = z = 1$. On the same computer, when rows 2 and 3 were switched, the answer was accurate to six decimal places in each unknown.

Thus the most important modification to the classical elimination scheme that must be made to produce a good computer algorithm is this: We interchange rows whenever $|a_{ii}|$ is small, and not only when it is zero.

But how small is "small"? Several strategies are available for deciding when a pivot is too small to use. We could, for example, switch rows when $|a_{ii}|$ is less than a certain arbitrary threshold. However, we shall see when we program Gaussian elimination (Figure 1.2.1) that switching rows requires a negligible amount of work compared with the actual elimination calculations; so it is not unreasonable to use the following strategy, called partial pivoting: For each column i, we *always* switch row i with row l, where a_{li} is the largest (in absolute value) of all the potential pivots $a_{ii}, a_{i+1,i}, \ldots, a_{Ni}$.

The partial pivoting strategy avoids the problem of choosing an arbitrary threshold, by using the largest pivot available each time. Although it is possible to construct example systems that fool this algorithm into making a poor choice of pivot (see Section 1.8), partial pivoting is what is used by most of the popular Gaussian elimination computer codes, and on "real-world" problems it is very unlikely that we can find another pivoting strategy that will significantly improve on the accuracy produced by partial pivoting.

A competing strategy is "complete pivoting" in which the entire lower right hand $N-i+1$ by $N-i+1$ submatrix is searched for the element that is largest in absolute value. By row *and* column interchanges (row interchanges are accompanied by interchanges in $\boldsymbol{b}$, while column interchanges must be accompanied by interchanges in $\boldsymbol{x}$) this element is brought to the pivot position a_{ii} before proceeding. Most experts agree, however, that the gain in numerical stability offered by complete pivoting is not usually worth the extra effort.

Now let us summarize Gaussian elimination with partial pivoting (see 1.2.6).

$$\begin{bmatrix} a_{11} & \cdots & a_{1i} & \cdots & a_{1N} \\ \vdots & \ddots & \vdots & & \vdots \\ 0 & \cdots & a_{ii} & \cdots & a_{iN} \\ \cdot & & \cdot & & \cdot \\ 0 & \cdots & a_{ji} & \cdots & a_{jN} \\ \cdot & & \cdot & & \cdot \\ 0 & \cdots & a_{Ni} & \cdots & a_{NN} \end{bmatrix} \begin{bmatrix} x_1 \\ \vdots \\ x_i \\ \cdot \\ \cdot \\ \cdot \\ x_N \end{bmatrix} = \begin{bmatrix} b_1 \\ \vdots \\ b_i \\ \cdot \\ b_j \\ \cdot \\ b_N \end{bmatrix}. \tag{1.2.6}$$

For a given column i ($i = 1, \ldots, N-1$) we first search the potential pivots a_{ii}, $a_{i+1,i}, \ldots, a_{Ni}$ for the one that has largest absolute value (the method used to break ties is unimportant). If the largest one lies in row l (a_{li}), we switch rows i and l, and also switch b_i and b_l. If all the potential pivots are zero, we give up. Then, for $j = i+1, \ldots, N$, we knock out a_{ji} by adding $-a_{ji}/a_{ii}$ times the ith row to the jth row, and $-a_{ji}/a_{ii}$ times b_i to b_j.

Once we have finished this forward-elimination phase, we begin the back substitution phase, to solve the reduced triangular system

$$\begin{bmatrix} a_{11} & \cdots & a_{1i} & \cdots & a_{1N} \\ & \ddots & \vdots & & \vdots \\ & & a_{ii} & \cdots & a_{iN} \\ & & & \ddots & \vdots \\ & & & & a_{NN} \end{bmatrix} \begin{bmatrix} x_1 \\ \vdots \\ x_i \\ \vdots \\ x_N \end{bmatrix} = \begin{bmatrix} b_1 \\ \vdots \\ b_i \\ \vdots \\ b_N \end{bmatrix}. \tag{1.2.7}$$

At this point we know that the diagonal elements (the pivots) are all nonzero; otherwise we would have given up earlier (actually $a_{NN} \neq 0$ must be checked separately, since we have only processed columns 1 to $N-1$). Thus we can solve the last equation $a_{NN}x_N = b_N$ for $x_N = b_N/a_{NN}$ and then solve for the other unknowns in reverse order; for each i ($i = N-1$ to 1) we have

$$a_{ii}x_i + \sum_{j=i+1}^{N} a_{ij}x_j = b_i$$

or, since x_{i+1} to x_N are by now known,

$$x_i = \frac{1}{a_{ii}}\left(b_i - \sum_{j=i+1}^{N} a_{ij}x_j\right).$$

A FORTRAN program that implements Gaussian elimination with partial pivoting is shown in Figure 1.2.1. Note that the program considers the matrix to be singular only if the largest potential pivot at any stage is exactly zero. Alternatively, it would reasonable to give up when the largest potential pivot is less than some small threshold, in absolute value. A code that uses that strategy may occasionally give up prematurely, when it could still have produced a solution with some significant digits. Our code, on the other hand, runs a risk of returning garbage without a warning.

There are two features of DLINEQ that may be confusing to the reader at this point, but whose purposes will be clarified in the next section. One is the use of the permutations vector IPERM and the other is statement 30, where A(J, I) is set to LJI, when it clearly should be set to zero (after all, the whole aim of loop 30 is to make A(J, I) = 0!). For now, we can simply consider IPERM to be an output vector containing useless information, and it should be noticed that subdiagonal element A(J, I) is never used again after it is zeroed; so, as long as we remember that it is really zero, we can store whatever information we like in that piece of memory.

```
      SUBROUTINE DLINEQ(A,N,X,B,IPERM)
      IMPLICIT DOUBLE PRECISION (A-H,O-Z)
C                                DECLARATIONS FOR ARGUMENTS
      DOUBLE PRECISION A(N,N),X(N),B(N)
      INTEGER N,IPERM(N)
C                                DECLARATIONS FOR LOCAL VARIABLES
      DOUBLE PRECISION B_(N),LJI
C
C  SUBROUTINE DLINEQ SOLVES THE LINEAR SYSTEM A*X=B
C
C  ARGUMENTS
C
C             ON INPUT                                  ON OUTPUT
```

```
C            --------                               ---------
C
C    A      - THE N BY N COEFFICIENT MATRIX.        THE DIAGONAL AND UPPER
C                                                   TRIANGLE OF A CONTAINS U
C                                                   AND THE LOWER TRIANGLE
C                                                   OF A CONTAINS THE LOWER
C                                                   TRIANGLE OF L, WHERE
C                                                   PA = LU, P BEING THE
C                                                   PERMUTATION MATRIX
C                                                   DEFINED BY IPERM.
C
C    N      - THE SIZE OF MATRIX A.
C
C    X      -                                       AN N-VECTOR CONTAINING
C                                                   THE SOLUTION.
C
C    B      - THE RIGHT HAND SIDE N-VECTOR.
C
C    IPERM  -                                       AN N-VECTOR CONTAINING
C                                                   A RECORD OF THE ROW
C                                                   INTERCHANGES MADE.  IF
C                                                   J = IPERM(K), THEN ROW
C                                                   J ENDED UP AS THE K-TH
C                                                   ROW.
C
C-----------------------------------------------------------------------
      DO 10 K=1,N
C                                INITIALIZE IPERM = (1,2,3,...,N)
         IPERM(K) = K
C                                COPY B TO B_, SO B WILL NOT BE ALTERED
         B_(K) = B(K)
   10 CONTINUE
C                                BEGIN FORWARD ELIMINATION
      DO 35 I=1,N-1
C                                SEARCH FROM A(I,I) ON DOWN FOR
C                                LARGEST POTENTIAL PIVOT, A(L,I)
         BIG = ABS(A(I,I))
         L = I
         DO 15 J=I+1,N
            IF (ABS(A(J,I)).GT.BIG) THEN
               BIG = ABS(A(J,I))
               L = J
            ENDIF
   15    CONTINUE
C                                IF LARGEST POTENTIAL PIVOT IS ZERO,
C                                MATRIX IS SINGULAR
         IF (BIG.EQ.0.0) GO TO 50
C                                SWITCH ROW I WITH ROW L, TO BRING
```

```
C                                 UP LARGEST PIVOT
         DO 20 K=1,N
            TEMP = A(L,K)
            A(L,K) = A(I,K)
            A(I,K) = TEMP
   20    CONTINUE
C                                 SWITCH B_(I) AND B_(L)
         TEMP = B_(L)
         B_(L) = B_(I)
         B_(I) = TEMP
C                                 SWITCH IPERM(I) AND IPERM(L)
         ITEMP = IPERM(L)
         IPERM(L) = IPERM(I)
         IPERM(I) = ITEMP
         DO 30 J=I+1,N
C                                 CHOOSE MULTIPLIER TO ZERO A(J,I)
            LJI = A(J,I)/A(I,I)
            IF (LJI.NE.0.0) THEN
C                                 SUBTRACT LJI TIMES ROW I FROM ROW J
               DO 25 K=I+1,N
                  A(J,K) = A(J,K) - LJI*A(I,K)
   25          CONTINUE
C                                 SUBTRACT LJI TIMES B_(I) FROM B_(J)
               B_(J) = B_(J) - LJI*B_(I)
            ENDIF
C                                 SAVE LJI IN A(J,I).  IT IS UNDERSTOOD,
C                                 HOWEVER, THAT A(J,I) IS REALLY ZERO.
   30       A(J,I) = LJI
   35 CONTINUE
      IF (A(N,N).EQ.0.0) GO TO 50
C                                 SOLVE U*X = B_ USING BACK SUBSTITUTION.
      X(N) = B_(N)/A(N,N)
      DO 45 I=N-1,1,-1
         SUM = 0.0
         DO 40 J=I+1,N
            SUM = SUM + A(I,J)*X(J)
   40    CONTINUE
         X(I) = (B_(I)-SUM)/A(I,I)
   45 CONTINUE
      RETURN
C                                 MATRIX IS NUMERICALLY SINGULAR.
   50 PRINT 55
   55 FORMAT (' ***** THE MATRIX IS SINGULAR *****')
      RETURN
      END
```

Figure 1.2.1

When DLINEQ was used to solve 1.1.3 for the currents in Figure 1.1.1, the result was $I_{AB} = 6.05, I_{BC} = -1.30, I_{AC} = 3.95, I_{BD} = 7.35$, and $I_{CD} = 2.65$. Note that I_{BC} is negative, which means the current really flows from C to B. When we try, using DLINEQ, to solve naively the first five of the seven equations (1.1.1–1.1.2), we get an error message from DLINEQ indicating that the matrix is singular (owing to the redundancies mentioned earlier).

A superficial analysis of subroutine DLINEQ shows that loop 25, where a multiple of row i is subtracted from row j, is where nearly all the computer time is spent, when N is large (note that this is the only triply nested DO loop in the program). Thus, for large N, the total amount of work done by DLINEQ is nearly equal to the amount done by the abbreviated program

```
      DO 35 I=1,N-1
         DO 30 J=I+1,N
            DO 25 K=I+1,N
               A(J,K) = A(J,K)- LJI*A(I,K)
25          CONTINUE
30       CONTINUE
35    CONTINUE
```

The number of multiplications done by this program is (see 0.1.1)

$$\sum_{i=1}^{N-1}(N-i)^2 = \sum_{i=1}^{N-1}(N^2 - 2Ni + i^2) \approx N^3 - N^3 + \frac{N^3}{3} = \frac{1}{3}N^3.$$

Thus the total work done by DLINEQ to solve an N by N linear system is about $\frac{1}{3}N^3$ multiplications. Note that the work to carry out the back substitution is only $O(N^2)$ (it uses only double nested loops); most of the computer time is spent in the forward-elimination phase.

1.3 Solving Several Systems with the Same Matrix

Frequently in applications we have to solve several systems with the same coefficient matrix but different right-hand side vectors. As an example, consider the shifted inverse power iteration 3.5.3, where we solve a sequence of linear systems with identical coefficient matrices. The electrical circuit of Figure 1.1.1 suggests another example. If we connect the battery leads to different nodes or vary the battery voltage, only the right-hand side of 1.1.3 will change.

One alternative in such situations is obviously to calculate A^{-1}. Once we have an inverse it only takes N^2 multiplications to solve $A\boldsymbol{x} = \boldsymbol{b}$ by multiplying A^{-1} times $\boldsymbol{b}$. However, there is a more efficient way to accomplish the same thing; we shall see how to calculate something that serves the same purpose as an inverse but that falls out "for free" when the first linear system is solved.

If $\boldsymbol{b}$ and $\boldsymbol{c}$ are simultaneously available, we can of course solve two linear systems $A\boldsymbol{x} = \boldsymbol{b}$ and $A\boldsymbol{x} = \boldsymbol{c}$ for essentially the price of one, by modifying DLINEQ to process $\boldsymbol{c}$ along with $\boldsymbol{b}$ (switching c_i and c_l each time that b_i and b_l are switched, etc.), but we shall not use this approach, because often it is necessary to solve the first system *before* solving the second; for example, in the inverse power iteration 3.5.3 we must solve $(A - pI)\boldsymbol{v}_{n+1} = \boldsymbol{v}_n$ before we know what the next right-hand side is.

Subroutine DRESLV (Figure 1.3.1) allows us to solve a second system $A\boldsymbol{x} = \boldsymbol{c}$ almost for free (in $O(N^2)$ operations) *after* we have finished solving the first system $A\boldsymbol{x} = \boldsymbol{b}$ the hard way (using DLINEQ). The idea is simply to keep up with what was done to $\boldsymbol{b}$ when $A\boldsymbol{x} = \boldsymbol{b}$ was solved and then to do the same things to $\boldsymbol{c}$. To explain how DRESLV works, let us first assume that no row interchanges are done by either DLINEQ or DRESLV. Later we shall take the row interchanges into account.

```
      SUBROUTINE DRESLV(A,N,X,C,IPERM)
      IMPLICIT DOUBLE PRECISION (A-H,O-Z)
C                                  DECLARATIONS FOR ARGUMENTS
      DOUBLE PRECISION A(N,N),X(N),C(N)
      INTEGER N,IPERM(N)
C                                  DECLARATIONS FOR LOCAL VARIABLES
      DOUBLE PRECISION C_(N),LJI
C
C  SUBROUTINE DRESLV SOLVES THE LINEAR SYSTEM A*X=C IN O(N**2) TIME,
C    AFTER DLINEQ HAS PRODUCED AN LU DECOMPOSITION OF PA.
C
C  ARGUMENTS
C
C                ON INPUT                          ON OUTPUT
C                --------                          ---------
C
C    A       - THE N BY N COEFFICIENT MATRIX
C              AFTER PROCESSING BY DLINEQ.
C              AS OUTPUT BY DLINEQ, A CONTAINS
C              AN LU DECOMPOSITION OF PA.
C
C    N       - THE SIZE OF MATRIX A.
C
C    X       -                                     AN N-VECTOR CONTAINING
C                                                  THE SOLUTION.
C
C    C       - THE RIGHT HAND SIDE N-VECTOR.
C
C    IPERM   - THE PERMUTATION VECTOR OF
C              LENGTH N OUTPUT BY DLINEQ.
C
C-----------------------------------------------------------------------
```

```
C                                  CALCULATE C_=P*C, WHERE P IS PERMUTATION
C                                  MATRIX DEFINED BY IPERM.
      DO 10 K=1,N
         J = IPERM(K)
         C_(K) = C(J)
   10 CONTINUE
C                                  BEGIN FORWARD ELIMINATION, TO CALCULATE
C                                  C_ = L**(-1)*C_
      DO 20 I=1,N-1
         DO 15 J=I+1,N
C                                  RETRIEVE MULTIPLIER SAVED IN A(J,I)
            LJI = A(J,I)
C                                  SUBTRACT LJI TIMES C_(I) FROM C_(J)
            C_(J) = C_(J) - LJI*C_(I)
   15    CONTINUE
   20 CONTINUE
C                                  SOLVE U*X = C_ USING BACK SUBSTITUTION.
      X(N) = C_(N)/A(N,N)
      DO 30 I=N-1,1,-1
         SUM = 0.0
         DO 25 J=I+1,N
            SUM = SUM + A(I,J)*X(J)
   25    CONTINUE
         X(I) = (C_(I)-SUM)/A(I,I)
   30 CONTINUE
      RETURN
      END
```

Figure 1.3.1

When DLINEQ subtracts a multiple l_{ji} of row i from row j (and the same multiple of b_i from b_j), a_{ji} is overwritten by l_{ji}, even though it is understood that a_{ji} is now really zero. This does not cause DLINEQ any problems, since $A(j,i)$ is never used again by this subroutine, in the rest of the forward elimination or in the back substitution. When DLINEQ has finished, A has the form

$$\begin{bmatrix} u_{11} & u_{12} & u_{13} & \dots & u_{1N} \\ (l_{21}) & u_{22} & u_{23} & \dots & u_{2N} \\ (l_{31}) & (l_{32}) & u_{33} & \dots & u_{3N} \\ \vdots & \vdots & \vdots & & \vdots \\ (l_{N1}) & (l_{N2}) & (l_{N3}) & \dots & u_{NN} \end{bmatrix}. \tag{1.3.1}$$

A is really, at this point, in its reduced triangular form U. The elements below the diagonal are enclosed in parentheses to indicate that the corresponding elements of U are really zero, but we are using $A(j,i)(j > i)$ to store the multiple (l_{ji}) of row i, which was subtracted from row j to eliminate a_{ji}.

Thus, when DRESLV is used to solve $A\boldsymbol{x} = \boldsymbol{c}$, A as output by DLINEQ contains not only the final upper triangular form of A but also a record of the multiples used to reduce A to this triangular form. $A(j,i)$ tells us what multiple of row i we must subtract from row j, and what multiple of c_i we must subtract from c_j, in order to reduce A to upper triangular form again. However, this time we can skip the operations on the rows of A, and only perform them on $\boldsymbol{c}$, since we already know what these multiples are going to be. This allows us to skip the expensive loop 25, where a multiple of row i is subtracted from row j. The forward-elimination phase (loop 20 in Figure 1.3.1) now requires only about $\frac{1}{2}N^2$ multiplications, since (see 0.1.1)

$$\sum_{i=1}^{N-1}(N-i) \approx N^2 - \frac{1}{2}N^2 = \frac{1}{2}N^2.$$

Since the upper triangle of A, as output by DLINEQ, already contains the final (reduced) triangular form of A, we can now back solve for the solution of $A\boldsymbol{x} = \boldsymbol{c}$. This back-substitution phase (loop 30 in Figure 1.3.1) requires another $\frac{1}{2}N^2$ multiplications, since

$$\sum_{i=N-1}^{1}(N-i) \approx \frac{1}{2}N^2.$$

So DRESLV does a total of about N^2 multiplications to solve $A\boldsymbol{x} = \boldsymbol{c}$, the same number as required by the multiplication $\boldsymbol{x} = A^{-1}\boldsymbol{c}$, but note that it cost DLINEQ virtually nothing extra to save the multipliers used by DRESLV, whereas computing the inverse of A is an expensive process (see Problem 2).

So far we have assumed that no row interchanges are done by DLINEQ, but usually this will not be the case. However, DLINEQ also provides DRESLV with a record of the interchanges it makes in the permutation vector IPERM, so that it can make the same interchanges when it solves $A\boldsymbol{x} = \boldsymbol{c}$. IPERM is set to $(1,2,3,\ldots,N)$ initially by DLINEQ; then each time that rows i and l are switched, IPERM(i) and IPERM(l) are also interchanged. After DLINEQ is finished, IPERM(k) contains the number of the original row (equation), which ended up as the kth row (equation).

Now, if we had known before we began to solve $A\boldsymbol{x} = \boldsymbol{b}$ in what order the equations would end up, we could have (in DLINEQ) reordered these equations (i.e., rows of A and elements of $\boldsymbol{b}$) *before* we started, and then no row interchanges would need to be done during the elimination; at each step the largest potential pivot would already be in the pivot position. Then the final A (stored multipliers and upper triangle) and the final $\boldsymbol{b}$ would be exactly the same as if the interchanges were done one by one during the elimination, as prescribed by the partial pivoting formula. Although DLINEQ in fact has to do the interchanges one by one (it cannot predict the future), since the final A and $\boldsymbol{b}$ are the same in either case, we can pretend that the rows of A and the elements of $\boldsymbol{b}$ are permuted at the beginning only. Note that in DLINEQ when we switch rows i and l (loop 20), we switch not only the nonzero parts in columns i to N but also the elements of these rows in columns 1 to $i-1$, which contain a record of the multiples used to zero elements in the interchanged rows. If these elements were actually storing zeros, loop 20 could run from $K = I$ to $K = N$.

Now DRESLV, because it has IPERM available to it, *does* know the final equation order before it starts, and so it does actually permute the elements of $\boldsymbol{c}$ into their final order then, just as (let's pretend!) DLINEQ does with $\boldsymbol{b}$ and the rows of A. If $j = \text{IPERM}(k)$ $(k = 1, \ldots, N)$, c_j is destined to end up as the kth element of $\boldsymbol{c}$; so we move it there to begin with. From there the forward elimination proceeds exactly as described above, with no further interchanges.

1.4 The *LU* Decomposition

The two row operations used to reduce A to upper triangular form can be thought of as resulting from premultiplications by certain elementary matrices. Subtracting a multiple l_{ji} of row i from row j $(j > i)$ is the result of premultiplying A by

$$M_{ij} = \begin{matrix} \\ \\ \text{row } i \\ \\ \\ \text{row } j \\ \\ \\ \end{matrix} \begin{bmatrix} 1 & & & & & & & \\ & 1 & & & & & & \\ & & 1 & & & & & \\ & & & 1 & & & & \\ & & & & 1 & & & \\ & & -l_{ji} & & & 1 & & \\ & & & & & & 1 & \\ & & & & & & & 1 \end{bmatrix}, \qquad (1.4.1)$$

(with column i and column j labelled above the matrix)

as is easy to verify.

Also, any row interchanges can be performed by premultiplying with an appropriate permutation matrix P, where P is equal to the identity matrix with its rows permuted in a similar manner. For example, if the three rows of P are the third, first, and second rows of the identity,

$$PA = \begin{bmatrix} 0 & 0 & 1 \\ 1 & 0 & 0 \\ 0 & 1 & 0 \end{bmatrix} \begin{bmatrix} a_{11} & a_{12} & a_{13} \\ a_{21} & a_{22} & a_{23} \\ a_{31} & a_{32} & a_{33} \end{bmatrix} = \begin{bmatrix} a_{31} & a_{32} & a_{33} \\ a_{11} & a_{12} & a_{13} \\ a_{21} & a_{22} & a_{23} \end{bmatrix}$$

and the three rows of PA are the third, first, and second rows of A.

Now we can describe the processes of the last two sections in terms of premultiplications by P and M matrices. The analysis is much easier if we pretend, again, that in solving $A\boldsymbol{x} = \boldsymbol{b}$ using Gaussian elimination with partial pivoting we somehow have the foresight to permute the equations (rows of A and elements of $\boldsymbol{b}$) into their final order at the start, rather than doing the switches one by one as the need becomes apparent.

After permuting the equations, our linear system now has the form

$$PA\boldsymbol{x} = P\boldsymbol{b}, \tag{1.4.2}$$

where P is some appropriate permutation matrix. No further interchanges will be required.

Now to zero a subdiagonal element a_{ji} we can premultiply both sides of 1.4.2 by M_{ij}, where M_{ij} is as shown in 1.4.1, with l_{ji} chosen appropriately. If we zero the elements in the proper (column by column) order, we have

$$M_{N-1,N} \ldots M_{12}PA\boldsymbol{x} = M_{N-1,N} \ldots M_{12}P\boldsymbol{b}. \tag{1.4.3}$$

These premultiplications are designed so that

$$M_{N-1,N} \ldots M_{12}PA = U, \tag{1.4.4}$$

where U is upper triangular. If we call the right-hand side of 1.4.3 $\boldsymbol{y}$, then 1.4.3 can be written in the form

$$\begin{aligned} \boldsymbol{y} &= M_{N-1,N} \ldots M_{12}P\boldsymbol{b}, \\ U\boldsymbol{x} &= \boldsymbol{y}. \end{aligned}$$

The upper triangular system $U\boldsymbol{x} = \boldsymbol{y}$ is solved by back substitution.

Now, if we have a new system $A\boldsymbol{x} = \boldsymbol{c}$ with the same coefficient matrix, DRESLV (Figure 1.3.1) solves this system similarly, by forming

$$\boldsymbol{y} = M_{N-1,N} \ldots M_{12}P\boldsymbol{c} \tag{1.4.5a}$$

and then back solving

$$U\boldsymbol{x} = \boldsymbol{y}. \tag{1.4.5b}$$

That is, DRESLV first premultiplies $\boldsymbol{c}$ by P (using the information saved in IPERM), then premultiplies by the M_{ij} matrices (M_{ij} is of the form 1.4.1 where l_{ji} is saved in subdiagonal position (j, i) of A (see 1.3.1)) and finally solves 1.4.5b by back substitution (U is stored in the diagonal and upper triangle of A).

Now from 1.4.4 we have

$$PA = M_{12}^{-1} \ldots M_{N-1,N}^{-1}U. \tag{1.4.6}$$

The inverse of M_{ij} (see 1.4.1) is easily seen to be

$$M_{ij}^{-1} = \begin{array}{r} \\ \\ \text{row } i \\ \\ \text{row } j \\ \\ \\ \end{array} \begin{bmatrix} 1 & & & & & & & \\ & 1 & & & & & & \\ & & 1 & & & & & \\ & & & 1 & & & & \\ & & & & 1 & & & \\ & & l_{ji} & & & 1 & & \\ & & & & & & 1 & \\ & & & & & & & 1 \end{bmatrix}.$$

(with column i and column j labelled above the matrix)

Since the product of lower triangular matrices is lower triangular (Theorem 0.1.2), from 1.4.6 we have

$$PA = LU, \tag{1.4.7}$$

where $L \equiv M_{12}^{-1} \ldots M_{N-1,N}^{-1}$ is a lower triangular matrix, and U is upper triangular. In fact, it can be verified (Problem 3) that

$$L = M_{12}^{-1} \ldots M_{N-1,N}^{-1} = \begin{bmatrix} 1 & & & & \\ l_{21} & 1 & & & \\ l_{31} & l_{32} & 1 & & \\ \vdots & \vdots & \vdots & \ddots & \\ l_{N1} & l_{N2} & l_{N3} & \ldots & 1 \end{bmatrix}. \tag{1.4.8}$$

Thus the matrix A as output by DLINEQ (see 1.3.1) contains U on its diagonal and upper triangle and the lower triangular part of L in its lower triangle. In other words, DLINEQ calculates the LU decomposition 1.4.7 not of A but of a row-wise permutation PA of A and overwrites A with L and U (the diagonal elements of L are not saved, but we can remember that they are all ones).

The traditional way to use this LU decomposition to solve a new system $A\boldsymbol{x} = \boldsymbol{c}$ is to premultiply both sides by P and then to rewrite $PA\boldsymbol{x} = P\boldsymbol{c}$ as $LU\boldsymbol{x} = P\boldsymbol{c}$; finally, this system is solved in the form

$$\begin{aligned} L\boldsymbol{y} &= P\boldsymbol{c}, \\ U\boldsymbol{x} &= \boldsymbol{y}. \end{aligned}$$

Recalling the definition (see 1.4.8) of L, we see that the equations 1.4.5 used by DRESLV to solve a new system $A\boldsymbol{x} = \boldsymbol{c}$ can be written in the form

$$\boldsymbol{y} = L^{-1}P\boldsymbol{c}, \tag{1.4.9a}$$
$$U\boldsymbol{x} = \boldsymbol{y}. \tag{1.4.9b}$$

Now 1.4.9 is essentially equivalent to the traditional approach, except that $L\boldsymbol{y} = P\boldsymbol{c}$ is normally solved by forward substitution (the unknowns are solved for in forward order).

If A is positive-definite, the diagonal elements are all positive initially, since $a_{ii} = \boldsymbol{e}_i^{\mathrm{T}} A \boldsymbol{e}_i > 0$, where $\boldsymbol{e}_i$ is the vector with 1 in position i and zeros elsewhere. Now it can be shown [Sewell 2005, Section 0.2] that, if no pivoting is done, the diagonal elements will remain positive throughout the Gaussian elimination process. Thus, when the LU decomposition of A is formed, the diagonal portion D of U consists of positive elements. It is shown in Problem 4b that (since A is symmetric) $A = PA = LU = LDL^{\mathrm{T}}$. Now, since D consists of positive elements, $D^{1/2}$ is real, and

$$A = LDL^{\mathrm{T}} = (LD^{1/2})(D^{1/2}L^{\mathrm{T}}) = (LD^{1/2})(LD^{1/2})^{\mathrm{T}} \equiv L_1 L_1^{\mathrm{T}}. \quad (1.4.10)$$

Equation 1.4.10 is called the Cholesky decomposition of a positive-definite matrix, and it will be useful later (Section 3.6). Note that $L_1 \equiv LD^{1/2}$ is lower triangular and, since it has positive diagonal elements (the square roots of the elements of D), it is nonsingular. Now

$$\boldsymbol{x}^{\mathrm{T}} A \boldsymbol{x} = \boldsymbol{x}^{\mathrm{T}} (L_1 L_1^{\mathrm{T}}) \boldsymbol{x} = (L_1^{\mathrm{T}} \boldsymbol{x})^{\mathrm{T}} (L_1^{\mathrm{T}} \boldsymbol{x}) = \| L_1^{\mathrm{T}} \boldsymbol{x} \|_2^2 \geq 0$$

and, since L_1^{T} is nonsingular, this can be zero only when $\boldsymbol{x} = \boldsymbol{0}$. Thus we see that A has a Cholesky decomposition if *and only if* it is positive-definite.

1.5 Banded Systems

The large linear systems that arise in applications are usually sparse; that is, most of the matrix coefficients are zero. Many of these systems can, by properly ordering the equations and unknowns, be put into "banded" form, where all elements of the coefficient matrix are zero outside some relatively small band around the main diagonal. For example, the huge linear systems that arise when differential equations are solved using finite difference or finite element methods are nearly always banded. A typical band matrix is exhibited in Figure 1.5.1. N_{LD} is the number of diagonals below the main diagonal which contain some nonzero elements and N_{UD} is the number of diagonals above the main diagonal. For the matrix shown in Figure 1.5.1, $N_{\mathrm{LD}} = 2$ and $N_{\mathrm{UD}} = 3$. Thus the total number of diagonals, called the band width, is $N_{\mathrm{LD}} + N_{\mathrm{UD}} + 1$.

When Gaussian elimination is applied to a linear system with a banded coefficient matrix, if we take advantage of the fact that the elements outside the band are zero, we can solve such a system much more rapidly than a similar size system with a full coefficient matrix.

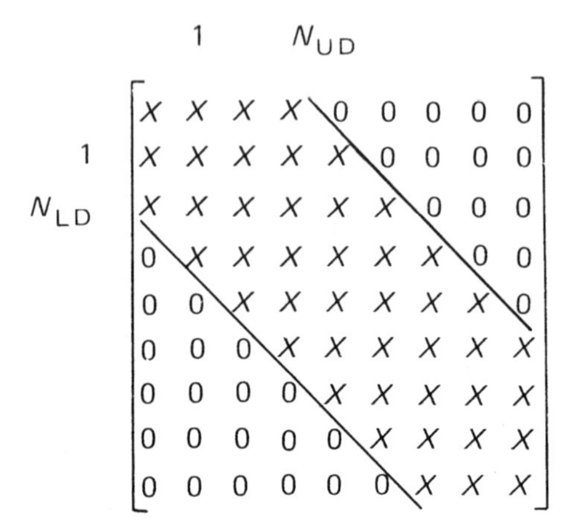

Figure 1.5.1
Band Matrix
X, not necessarily zero.

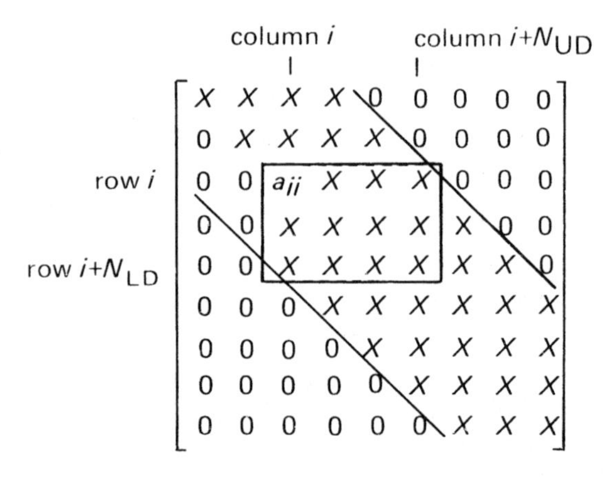

Figure 1.5.2
Partially Processed Band Matrix (No Pivoting)

In fact, if no row interchanges are done, all elements outside the band remain zero throughout the forward elimination. This is clear when we look at Figure 1.5.2, which shows a partially processed band matrix. When the pivot element a_{ii} is used to knock out the N_{LD} nonzero subdiagonal elements, only the elements in the $N_{\mathrm{LD}}+1$ by $N_{\mathrm{UD}}+1$ rectangle shown will change, as multiples of the pivot row i are subtracted from rows $i+1$ to $i+N_{\mathrm{LD}}(\min(i+N_{\mathrm{LD}}, N)$ actually).

Furthermore, since the LU decomposition of A is formed by simply overwriting the zeroed elements a_{ji} with the multipliers l_{ji}, we see that L and

U (see 1.3.1) will necessarily be band matrices also, and neither will extend outside the original band of A. The inverse of a band matrix is generally a full matrix; so we see that for banded systems the LU decomposition is *much* more economical to form and use than the inverse (and it serves the same purpose).

If partial pivoting is used, as it normally should be, the N_{LD} diagonals immediately above the band may "fill in" and become nonzero during the Gaussian elimination, but all elements below the band will remain zero. Figure 1.5.3 illustrates how these diagonals may fill in. In the worst case, row i may have to be interchanged with row $i+N_{\mathrm{LD}}$, in which case row i will then be nonzero out to column $i+N_{\mathrm{UD}}+N_{\mathrm{LD}}$ and, when multiples of row i are added to rows $i+1$ to $i+N_{\mathrm{LD}}$, everything inside the rectangle shown may fill in.

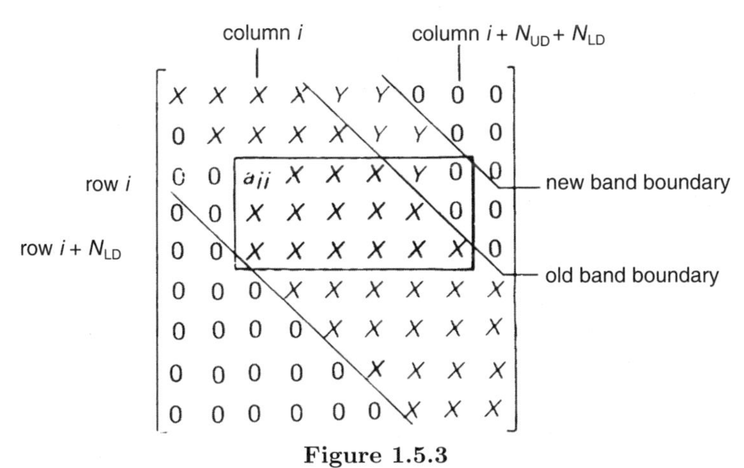

Figure 1.5.3
Partially Processed Band Matrix (Pivoting Allowed)
X, nonzero originally; Y, nonzero owing to previous fill-in.

FORTRAN subroutine DBAND (Figure 1.5.4) solves a banded linear system $A\boldsymbol{x} = \boldsymbol{b}$ using Gaussian elimination with partial pivoting. The program follows closely the pattern of DLINEQ, but the limits of most of the DO loops are designed to take advantage of the fact that only the first N_{LD} subdiagonals, the main diagonal, and the first $N_{\mathrm{UD}}+N_{\mathrm{LD}}$ superdiagonals will ever contain nonzero elements. Furthermore, only these $2N_{\mathrm{LD}}+N_{\mathrm{UD}}+1$ diagonals are actually stored; we do not waste computer memory storing elements that will always be zero. The columns of the FORTRAN array A hold the diagonals of the matrix A, as shown in Figure 1.5.5. The first subscript of array A represents the row number, while the second represents the diagonal number, which can vary from $-N_{\mathrm{LD}}$ to $N_{\mathrm{UD}}+N_{\mathrm{LD}}$. Diagonal number 0 (column 0 of array A) is the main diagonal. In general we see that array element $A(i, j-i)$ holds matrix element (i, j). There will be some elements of the array A (denoted by Z in Figure 1.5.5) that represent positions outside the matrix, and these will never be used, but they represent only a small amount of wasted storage.

Problem 8b shows how DBAND could be modified to save the multipliers, for the efficient solution of several systems with the same band matrix. The approach is similar to that used by DLINEQ and DRESLV, but the permutation must be saved in a different form, if pivoting is done.

```
      SUBROUTINE DBAND(A,N,NLD,NUD,X,B)
      IMPLICIT DOUBLE PRECISION (A-H,O-Z)
C                                  DECLARATIONS FOR ARGUMENTS
      DOUBLE PRECISION A(N,-NLD:NUD+NLD),X(N),B(N)
      INTEGER N,NLD,NUD
C                                  DECLARATIONS FOR LOCAL VARIABLES
      DOUBLE PRECISION B_(N),LJI
C
C  SUBROUTINE DBAND SOLVES THE LINEAR SYSTEM A*X=B, WHERE A IS A
C    BAND MATRIX.
C
C  ARGUMENTS
C
C              ON INPUT                          ON OUTPUT
C              --------                          ---------
C
C    A       - THE BAND MATRIX OF SIZE N,        DESTROYED.
C              DIMENSIONED A(N,-NLD:NUD+NLD)
C              IN THE MAIN PROGRAM.  COLUMNS
C              -NLD THROUGH NUD OF A CONTAIN
C              THE NONZERO DIAGONALS OF A.  THE
C              LAST NLD COLUMNS ARE USED AS
C              WORKSPACE (TO HOLD THE FILL-IN
C              IN THE NLD DIAGONALS DIRECTLY
C              ABOVE A).
C
C    N       - THE SIZE OF MATRIX A.
C
C    NLD     - NUMBER OF NONZERO LOWER DIAGONALS
C              IN A, I.E., NUMBER OF DIAGONALS
C              BELOW THE MAIN DIAGONAL.
C
C    NUD     - NUMBER OF NONZERO UPPER DIAGONALS
C              IN A, I.E., NUMBER OF DIAGONALS
C              ABOVE THE MAIN DIAGONAL.
C
C    X       -                                   AN N-VECTOR CONTAINING
C                                                THE SOLUTION.
C
C    B       - THE RIGHT HAND SIDE N-VECTOR.
C
```

```
C
C-----------------------------------------------------------------------
      DO 10 I=1,N
C                             COPY B TO B_, SO B WILL NOT BE ALTERED
         B_(I) = B(I)
         DO 5 J=NUD+1,NUD+NLD
C                             ZERO TOP NLD DIAGONALS (WORKSPACE)
            A(I,J) = 0.0
    5    CONTINUE
   10 CONTINUE
C                             BEGIN FORWARD ELIMINATION
      DO 35 I=1,N-1
C                             SEARCH FROM AII ON DOWN FOR
C                             LARGEST POTENTIAL PIVOT, ALI
         BIG = ABS(A(I,0))
         L = I
         DO 15 J=I+1,MIN(I+NLD,N)
            IF (ABS(A(J,I-J)).GT.BIG) THEN
               BIG = ABS(A(J,I-J))
               L = J
            ENDIF
   15    CONTINUE
C                             IF LARGEST POTENTIAL PIVOT IS ZERO,
C                             MATRIX IS SINGULAR
         IF (BIG.EQ.0.0) GO TO 50
C                             SWITCH ROW I WITH ROW L, TO BRING
C                             UP LARGEST PIVOT
         DO 20 K=I,MIN(I+NUD+NLD,N)
            TEMP = A(L,K-L)
            A(L,K-L) = A(I,K-I)
            A(I,K-I) = TEMP
   20    CONTINUE
C                             SWITCH B_(I) AND B_(L)
         TEMP = B_(L)
         B_(L) = B_(I)
         B_(I) = TEMP
         DO 30 J=I+1,MIN(I+NLD,N)
C                             CHOOSE MULTIPLIER TO ZERO AJI
            LJI = A(J,I-J)/A(I,0)
            IF (LJI.NE.0.0) THEN
C                             SUBTRACT LJI TIMES ROW I FROM ROW J
               DO 25 K=I,MIN(I+NUD+NLD,N)
                  A(J,K-J) = A(J,K-J) - LJI*A(I,K-I)
   25          CONTINUE
C                             SUBTRACT LJI TIMES B_(I) FROM B_(J)
               B_(J) = B_(J) - LJI*B_(I)
            ENDIF
   30    CONTINUE
```

```
   35 CONTINUE
      IF (A(N,0).EQ.0.0) GO TO 50
C                                  SOLVE U*X = B_ USING BACK SUBSTITUTION.
      X(N) = B_(N)/A(N,0)
      DO 45 I=N-1,1,-1
         SUM = 0.0
         DO 40 J=I+1,MIN(I+NUD+NLD,N)
            SUM = SUM + A(I,J-I)*X(J)
   40    CONTINUE
         X(I) = (B_(I)-SUM)/A(I,0)
   45 CONTINUE
      RETURN
C                                  MATRIX IS NUMERICALLY SINGULAR.
   50 PRINT 55
   55 FORMAT (' ***** THE MATRIX IS SINGULAR *****')
      RETURN
      END
```

Figure 1.5.4

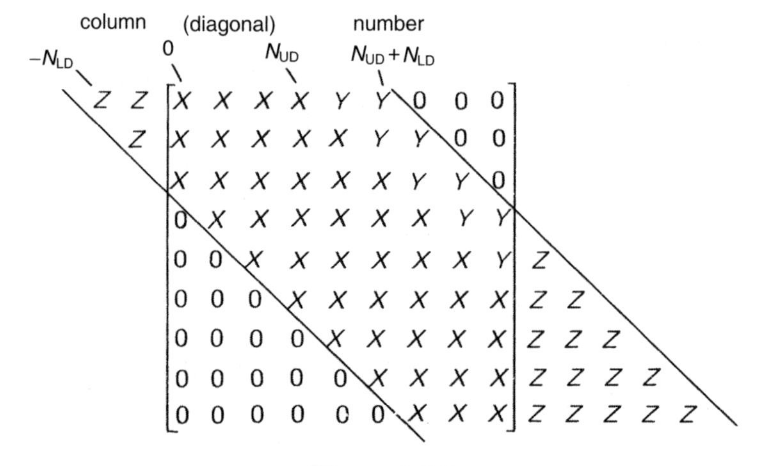

Figure 1.5.5
Diagonals of Matrix Stored in Columns of FORTRAN Array
X, nonzero originally; Y, nonzero owing to fill-in;
Z, unaccessed elements outside A.

Loop 25 is (as it was for DLINEQ) where most of the computer time is spent, if $N_{\rm LD}$ and $N_{\rm UD}$ are large compared with 1. Then the following abbreviated program does approximately the same quantity of work as DBAND:

```
      DO 35  I=1,N-1
         DO 30  J=I+1,MIN(I+NLD,N)
```

```
            DO 25  K=I,MIN(I+NUD+NLD,N)
               A(J,K-J)=A(J,K-J)-LJI*A(I,K-I)
   25       CONTINUE
   30    CONTINUE
   35 CONTINUE
```

If we also assume that, even though they are large compared with 1, N_{LD} and N_{UD} are small compared with N, then "usually" $\min(i + N_{\text{LD}}, N) = i + N_{\text{LD}}$ and $\min(i + N_{\text{UD}} + N_{\text{LD}}, N) = i + N_{\text{UD}} + N_{\text{LD}}$ and the total number of multiplications done by this program is about

$$(N-1)N_{\text{LD}}(N_{\text{UD}} + N_{\text{LD}} + 1) \approx N N_{\text{LD}}(N_{\text{UD}} + N_{\text{LD}}).$$

Hence this also approximates the work done by DBAND.

If N_{LD} and N_{UD} are constant (e.g., tridiagonal matrices have $N_{\text{LD}} = 1$ and $N_{\text{UD}} = 1$), the total work required to solve a banded linear system is $O(N)$, which compares *very* favorably with the $O(N^3)$ work required by DLINEQ to solve a full N by N system. The amount of computer memory required is $N(2N_{\text{LD}} + N_{\text{UD}} + 1)$, which also compares very favorably with the $O(N^2)$ memory used by DLINEQ.

1.6 Sparse Systems

Although the simplest way to take advantage of sparsity during Gaussian elimination is to order the unknowns and equations so that the matrix is banded, there are often other orderings which may produce even less fill-in, and thus use less memory and presumably less computer time. Algorithms which use such orderings are generally complicated to implement efficiently and the details are beyond the scope of this book.

However, the importance of ordering the unknowns and equations wisely, for efficient Gaussian elimination, is illustrated by Figures 1.6.1a-b. Elements that are nonzero in the original matrix are marked by (X) and elements that are originally zero but become nonzero due to fill-in are marked by (+).

```
X X X X X X X X
X X + + + + + +
X + X + + + + +
X + + X + + + +
X + + + X + + +
X + + + + X + +
X + + + + + X +
X + + + + + + X
```

Figure 1.6.1a

```
X             X
  X           X
    X         X
      X       X
        X     X
          X   X
            X X
X X X X X X X X
```

Figure 1.6.1b

In Figure 1.6.1a, the nonzero elements on the top row cause the entire matrix to fill-in (no pivoting is assumed), while when the first and last rows, and first and last columns, are switched (Figure 1.6.1b), there is no fill-in at all. In general, rows with many nonzeros near the bottom of the matrix do not have "time" to generate much further fill-in.

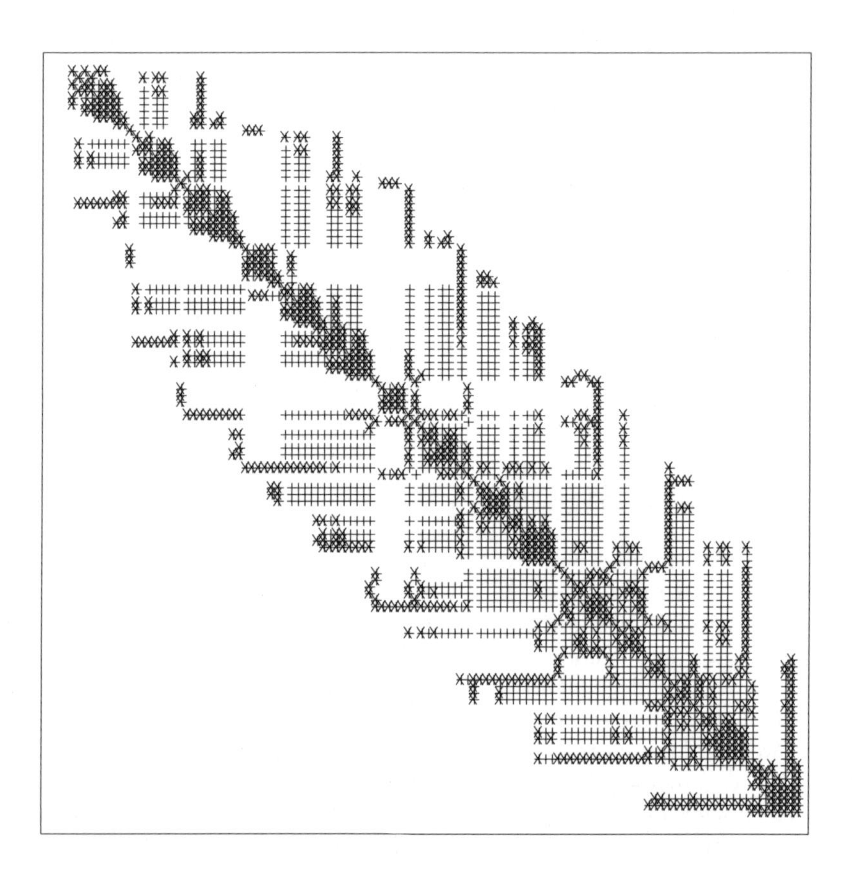

Figure 1.6.2
Fill-in for Band Ordering

Figure 1.6.2 shows how the matrix fills in when a band ordering is used, with no pivoting done, for a linear system generated by the finite element program PDE2D [Sewell 2005] (www.pde2d.com). There are $N = 113$ unknowns in this linear system, with 1041 nonzero elements (again marked by (X)) in the original matrix, and 1508 additional originally-zero elements (again marked by (+)) fill in during the elimination. (Remember, however, that a band solver must actually store everything inside the band, which for this matrix means $N(N_{LD}+N_{UD}+1) = 113(35+35+1) = 8023$ elements.) Figure 1.6.3 shows how the same matrix fills in when the ordering generated by Harwell Library routine MA27 [Duff and Reed 1983] is used. MA27 uses a variant of the "minimal degree" algorithm, which orders the unknowns so that the next unknown eliminated (x_k) is always the one which, of all those ($x_k, ..., x_N$) which have not yet been eliminated, is currently connected to the fewest other uneliminated unknowns. Thus unknowns which would generate more fill-in are eliminated "later," when they will do less damage (notice that the fill-in is concentrated toward the lower and right portions of the matrix). Now only 450 zero elements fill in, even though the bandwidth is very large. When the reverse of the MA27 ordering is used, 5936 zero elements fill in!

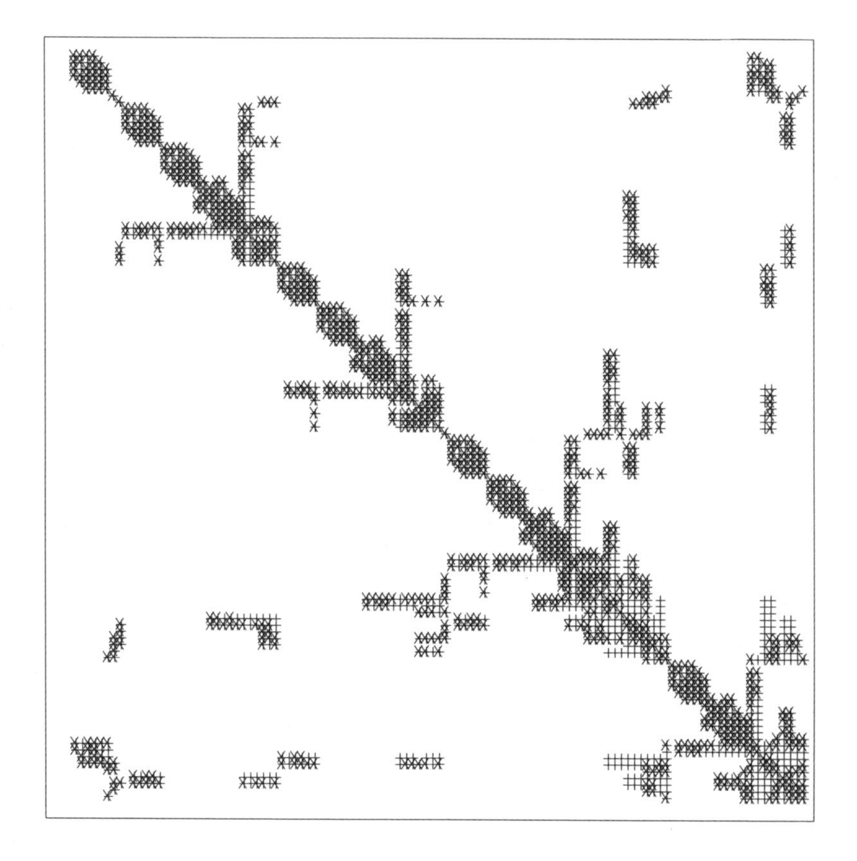

Figure 1.6.3
Fill-in for Minimal Degree Ordering

Useful references for "sparse direct" methods are Duff and Reid [1986], and George and Liu [1981]. Problem 7b gives a simple example of a linear system that is sparse but not banded, and that can be solved efficiently using Gaussian elimination.

1.7 Application: Cubic Spline Interpolation

Many of the algorithms used to solve other numerical analysis problems have as one of their steps the solution of a system of linear equations. If you look, for example, at a subroutine in the IMSL or NAG library which solves partial differential equations, integral equations, systems of nonlinear equations, or multivariate optimization problems, you will often find that it calls a linear system solver; in fact, for most of these applications, the bulk of the computer time is spent in the linear system solver.

Interpolation and smoothing of data are other problem areas where the solution of a linear system is often required. Currently the functions that are most widely used to interpolate between accurate data points, or to smooth a set of inaccurate data points, are cubic splines. A cubic spline is a function $s(x)$ that is a cubic polynomial in each subinterval (x_i, x_{i+1}) of the range of data, with $s(x), s'(x)$, and $s''(x)$ continuous at the knots x_i. Thus for x in (x_i, x_{i+1}) we have

$$s(x) = a_i + b_i(x - x_i) + c_i(x - x_i)^2 + d_i(x - x_i)^3. \tag{1.7.1}$$

Suppose we want to determine the coefficients of a cubic spline $s(x)$ that interpolates to the data points $(x_i, y_i), i = 1, \ldots, N$. The requirement that $s(x_i) = y_i$ and $s(x_{i+1}) = y_{i+1}$ gives us two equations that must be satisfied by the four coefficients a_i, b_i, c_i, and d_i that determine the form of $s(x)$ in (x_i, x_{i+1}). If we also specify the second derivative $s''(x_i) = \sigma_i$ at each data point, setting $s''(x_i) = \sigma_i$ and $s''(x_{i+1}) = \sigma_{i+1}$ will give us two more equations, and the coefficients of $s(x)$ in the subinterval (x_i, x_{i+1}) are uniquely determined, namely,

$$\begin{aligned}
a_i &= y_i, \\
b_i &= \frac{y_{i+1} - y_i}{h_i} - \frac{h_i(2\sigma_i + \sigma_{i+1})}{6}, \\
c_i &= \frac{\sigma_i}{2}, \\
d_i &= \frac{\sigma_{i+1} - \sigma_i}{6h_i},
\end{aligned} \tag{1.7.2}$$

where $h_i \equiv x_{i+1} - x_i$. It is easy to verify by direct calculation that $s(x)$ as expressed in 1.7.1 with coefficients given by 1.7.2, does satisfy $s(x_i) = y_i$, $s(x_{i+1}) = y_{i+1}$, $s''(x_i) = \sigma_i$, and $s''(x_{i+1}) = \sigma_{i+1}$.

Now how should the second derivatives $\sigma_1, \ldots, \sigma_N$ be chosen? We have already ensured that the piecewise cubic polynomial $s(x)$ will interpolate to the data points (x_i, y_i); so all that remains is to make sure that $s(x)$ is a cubic spline, that is, that s, s', and s'' are continuous. However, no matter what the values of the y-coordinates $y_1, \ldots, y_N$ are, $s(x)$ will be continuous at the knots, since the cubic polynomials on each side of a knot have to match the same value (y_i). Similarly, no matter how we choose the values $\sigma_1, \ldots, \sigma_N$ of the second derivatives, $s''(x)$ will be continuous at the knots, since the cubic polynomials on each side of a knot have the same second derivative σ_i. So our only consideration in choosing the parameters σ_i is that $s'(x)$ must also be continuous at each knot.

From 1.7.1 we see that, in (x_i, x_{i+1}),

$$\begin{aligned} s'(x) &= b_i + 2c_i(x - x_i) + 3d_i(x - x_i)^2, \\ s'(x_{i+1}) &= b_i + 2c_i h_i + 3d_i h_i^2, \end{aligned}$$

while, in (x_{i+1}, x_{i+2}),

$$\begin{aligned} s'(x) &= b_{i+1} + 2c_{i+1}(x - x_{i+1}) + 3d_{i+1}(x - x_{i+1})^2, \\ s'(x_{i+1}) &= b_{i+1}. \end{aligned}$$

Thus for continuity of $s'(x)$ across the knot x_{i+1}, we must have

$$b_i + 2c_i h_i + 3d_i h_i^2 = b_{i+1}.$$

Using equations 1.7.2, this equation simplifies to

$$\frac{h_i}{6}\sigma_i + \frac{h_i + h_{i+1}}{3}\sigma_{i+1} + \frac{h_{i+1}}{6}\sigma_{i+2} = \frac{y_{i+2} - y_{i+1}}{h_{i+1}} - \frac{y_{i+1} - y_i}{h_i} \qquad (i = 1, \ldots, N-2). \tag{1.7.3}$$

Equation 1.7.3 must hold at each knot x_{i+1} that is not an end point (there are no continuity requirements at the end points); that is, 1.7.3 must hold for $i = 1, \ldots, N - 2$.

$$\begin{bmatrix} \frac{h_1+h_2}{3} & \frac{h_2}{6} & & & \\ & \ddots & & & \\ & \frac{h_i}{6} & \frac{h_i+h_{i+1}}{3} & \frac{h_{i+1}}{6} & \\ & & & \ddots & \\ & & & \frac{h_{N-2}}{6} & \frac{h_{N-2}+h_{N-1}}{3} \end{bmatrix} \begin{bmatrix} \sigma_2 \\ \sigma_3 \\ \vdots \\ \sigma_{i+1} \\ \sigma_{i+2} \\ \vdots \\ \sigma_{N-2} \\ \sigma_{N-1} \end{bmatrix} = \begin{bmatrix} r_1 - \frac{h_1}{6}\sigma_1 \\ \vdots \\ r_i \\ \vdots \\ r_{N-2} - \frac{h_{N-1}}{6}\sigma_N \end{bmatrix}$$

Figure 1.7.1
Tridiagonal Linear System for Spline Second Derivatives

This gives us $N-2$ linear equations for the N unknowns $\sigma_1,\ldots,\sigma_N$. If we arbitrarily specify values for σ_1 and σ_N, then we have the same number of equations as unknowns, and the linear system 1.7.3 has the form shown in Figure 1.7.1. (There, $r_i = (y_{i+2}-y_{i+1})/h_{i+1} - (y_{i+1}-y_i)/h_i$)

In Figure 1.7.2 a FORTRAN subroutine is given which calculates the cubic spline that interpolates to a set of data points $(x_i, y_i), i = 1,\ldots,N$. The user specifies the second derivatives σ_1 and σ_N at the end points, and the other second derivatives $\sigma_2,\ldots,\sigma_{N-1}$ are determined by solving the tridiagonal linear system in Figure 1.7.1, using the banded system solver DBAND in Figure 1.5.4. The number of lower diagonals and the number of upper diagonals are $N_{\mathrm{LD}} = N_{\mathrm{UD}} = 1$ for a tridiagonal system; so the total work to solve this N by N linear system is only $O(N)$. Once we have $\sigma_1,\ldots,\sigma_N$, the coefficients of the cubic spline in (x_i, x_{i+1}) are determined by equations 1.7.2, and then this spline is evaluated at user-specified output points, using equation 1.7.1.

How should the user specify the second derivatives $\sigma_1 = s''(x_i)$ and $\sigma_N = s''(x_N)$? One popular choice is simply to set $\sigma_1 = \sigma_N = 0$. The resulting spline is called a "natural" cubic spline. The popularity of the natural cubic spline interpolant derives from the following theorem.

```
      SUBROUTINE DSPLN(X,Y,N,YXX1,YXXN,XOUT,YOUT,NOUT)
      IMPLICIT DOUBLE PRECISION (A-H,O-Z)
C                                 DECLARATIONS FOR ARGUMENTS
      DOUBLE PRECISION X(N),Y(N),YXX1,YXXN,XOUT(NOUT),YOUT(NOUT)
      INTEGER N,NOUT
C                                 DECLARATIONS FOR LOCAL VARIABLES
      DOUBLE PRECISION A(N-2,-1:2),COEFF(4,N),SIG(N),R(N)
C
C  SUBROUTINE DSPLN FITS AN INTERPOLATORY CUBIC SPLINE THROUGH THE
C    POINTS (X(I),Y(I)), I=1,...,N, WITH SPECIFIED SECOND DERIVATIVES
C    AT THE END POINTS, AND EVALUATES THIS SPLINE AT THE OUTPUT POINTS
C    XOUT(1),...,XOUT(NOUT).
C
C  ARGUMENTS
C
C              ON INPUT                          ON OUTPUT
C              --------                          ---------
C
C    X       - A VECTOR OF LENGTH N CONTAINING
C              THE X-COORDINATES OF THE DATA
C              POINTS.
C
C    Y       - A VECTOR OF LENGTH N CONTAINING
C              THE Y-COORDINATES OF THE DATA
C              POINTS.
C
C    N       - THE NUMBER OF DATA POINTS
C              (N.GE.3).
```

```
C
C     YXX1    - THE SECOND DERIVATIVE OF THE
C                CUBIC SPLINE AT X(1).
C
C     YXXN    - THE SECOND DERIVATIVE OF THE
C                CUBIC SPLINE AT X(N).  (YXX1=0
C                AND YXXN=0 GIVES A NATURAL
C                CUBIC SPLINE)
C
C     XOUT    - A VECTOR OF LENGTH NOUT CONTAINING
C                THE X-COORDINATES AT WHICH THE
C                CUBIC SPLINE IS EVALUATED.  THE
C                ELEMENTS OF XOUT MUST BE IN
C                ASCENDING ORDER.
C
C     YOUT    -                                  A VECTOR OF LENGTH NOUT.
C                                                YOUT(I) CONTAINS THE
C                                                VALUE OF THE SPLINE
C                                                AT XOUT(I).
C
C     NOUT    - THE NUMBER OF OUTPUT POINTS.
C
C-----------------------------------------------------------------------
      SIG(1) = YXX1
      SIG(N) = YXXN
C                                SET UP TRIDIAGONAL SYSTEM SATISFIED
C                                BY SECOND DERIVATIVES (SIG(I)=SECOND
C                                DERIVATIVE AT X(I)).
      DO 5 I=1,N-2
         HI   = X(I+1)-X(I)
         HIP1 = X(I+2)-X(I+1)
         R(I) = (Y(I+2)-Y(I+1))/HIP1 - (Y(I+1)-Y(I))/HI
         A(I,-1) = HI/6.0
         A(I, 0) = (HI + HIP1)/3.0
         A(I, 1) = HIP1/6.0
         IF (I.EQ.  1) R(1)   = R(1)   - HI/  6.0*SIG(1)
         IF (I.EQ.N-2) R(N-2) = R(N-2) - HIP1/6.0*SIG(N)
    5 CONTINUE
C                                CALL DBAND TO SOLVE TRIDIAGONAL SYSTEM
      NLD = 1
      NUD = 1
      CALL DBAND(A,N-2,NLD,NUD,SIG(2),R)
C                                CALCULATE COEFFICIENTS OF CUBIC SPLINE
C                                IN EACH SUBINTERVAL
      DO 10 I=1,N-1
         HI = X(I+1)-X(I)
         COEFF(1,I) = Y(I)
         COEFF(2,I) = (Y(I+1)-Y(I))/HI - HI/6.0*(2*SIG(I)+SIG(I+1))
```

```
         COEFF(3,I) = SIG(I)/2.0
         COEFF(4,I) = (SIG(I+1)-SIG(I))/(6.0*HI)
   10 CONTINUE
      L = 1
      DO 25 I=1,NOUT
C                               FIND FIRST VALUE OF J FOR WHICH X(J+1) IS
C                               GREATER THAN OR EQUAL TO XOUT(I).  SINCE
C                               ELEMENTS OF XOUT ARE IN ASCENDING ORDER,
C                               WE ONLY NEED CHECK THE KNOTS X(L+1)...X(N)
C                               WHICH ARE GREATER THAN OR EQUAL TO
C                               XOUT(I-1).
         DO 15 J=L,N-1
            JSAVE = J
            IF (X(J+1).GE.XOUT(I)) GO TO 20
   15    CONTINUE
   20    L = JSAVE
C                               EVALUATE CUBIC SPLINE IN INTERVAL
C                               (X(L),X(L+1))
         P = XOUT(I)-X(L)
         YOUT(I) = COEFF(1,L)      + COEFF(2,L)*P
     &           + COEFF(3,L)*P*P + COEFF(4,L)*P*P*P
   25 CONTINUE
      RETURN
      END
```

Figure 1.7.2

Theorem 1.7.1. *Among all functions that are continuous, with continuous first and second derivatives, which interpolate to the data points* $(x_i, y_i), i = 1, \ldots, N$, *the natural cubic spline* $s(x)$ *interpolant minimizes*

$$\int_{x_1}^{x_N} [s''(x)]^2 dx.$$

Proof: Let $u(x)$ be any other function that is continuous, with continuous first and second derivatives, which also interpolates to the data points. Then define $e(x) \equiv u(x) - s(x)$ and

$$\begin{aligned} \int_{x_1}^{x_N} [u''(x)]^2 dx &= \int_{x_1}^{x_N} [s''(x)]^2 dx + \int_{x_1}^{x_N} [e''(x)]^2 dx \\ &\quad + 2\int_{x_1}^{x_N} s''(x)e''(x) dx. \end{aligned}$$

If we can show that the last integral (of $s''(x)e''(x)$) is zero, we are finished, since then the integral of $s''(x)^2$ is less than or equal to the integral of $u''(x)^2$. Now

$$\int_{x_1}^{x_N} s''e''dx = \sum_{i=1}^{N-1} \int_{x_i}^{x_{i+1}} s''e''dx = \sum_{i=1}^{N-1} \int_{x_i}^{x_{i+1}} (s''e')' - (s'''e)' + s^{\text{iv}}e \; dx.$$

Since $s(x)$ is a polynomial of degree three in (x_i, x_{i+1}), $s^{\text{iv}} \equiv 0$; so the integral of $s''e''$ becomes

$$\sum_{i=1}^{N-1} s''(x_{i+1})e'(x_{i+1}) - s''(x_i)e'(x_i) - s'''(x_{i+1})e(x_{i+1}) + s'''(x_i)e(x_i).$$

Since $s(x)$ and $u(x)$ interpolate the same values at the knots, $e(x_i) = e(x_{i+1}) = 0$. Finally, since $s''(x)e'(x)$ is continuous across the knots, the sums involving $s''e'$ telescope and, since $s'' = 0$ at the end points, we have

$$\int_{x_1}^{x_N} s''(x)e''(x)dx = s''(x_N)e'(x_N) - s''(x_1)e'(x_1) = 0. \qquad \blacksquare$$

Since the L_2-norm of the second derivative is a measure of the curvature of a function, we can state Theorem 1.7.1 in the following words: Of all functions of continuous curvature that can be drawn through a given set of data points, the natural cubic spline has the least curvature. If we take a semirigid rod and force it (using nails?) to pass through our data points, it will very nearly take on the form of the natural cubic spline interpolant. This gives some insight into the nature of the natural cubic spline interpolant; it remains as "straight" as it can, given that it has to pass through the data points.

It is remarkable that Theorem 1.7.1 does not say that the natural cubic spline interpolant is the piecewise cubic of minimum curvature, or even the piecewise polynomial of minimum curvature, which interpolates to the data, but the (smooth) *function* of minimum curvature.

If (x_i, y_i) represent points on a smooth curve $y = f(x)$ (which may or may not be known explicitly), there is another obvious choice for the user-specified end conditions: We can set σ_1 and σ_N equal to estimates of $f''(x_1)$ and $f''(x_N)$. If we know (or can at least estimate) the second derivative of f, we would expect $s(x)$ to be a better approximation to $f(x)$ if we set $s''(x_1) = f''(x_1)$ and $s''(x_N) = f''(x_N)$, rather than arbitrarily setting the second derivatives to zero.

If we set $s''(x)$ equal to $f''(x)$ at the end points, the cubic spline error is $O(h_{max}^4)$. It can be shown [de Boor 1978, Chapter 4] that setting $s'' = 0$ at the end points knocks the order of accuracy down to $O(h_{max}^2)$, unless, of course, f'' happens to be zero at the end points.

In summary, if the data points lie on a smooth curve $y = f(x)$, and if $f''(x)$ can be estimated, we shall get a more accurate representation of $f(x)$ if we set $s''(x_1) \approx f''(x_1)$ and $s''(x_N) \approx f''(x_N)$. However, if we simply want to draw a smooth, visually pleasing curve through a set of given data points, the natural cubic spline end conditions are recommended.

Figure 1.7.3 shows the natural cubic spline, calculated by DSPLN, which interpolates to the points $(1,1)$, $(2,3)$, $(3,2)$, $(4,3)$, and $(5,4)$. In Section 2.4 we shall see cubic splines used in another application, one designed to illustrate linear least squares applications.

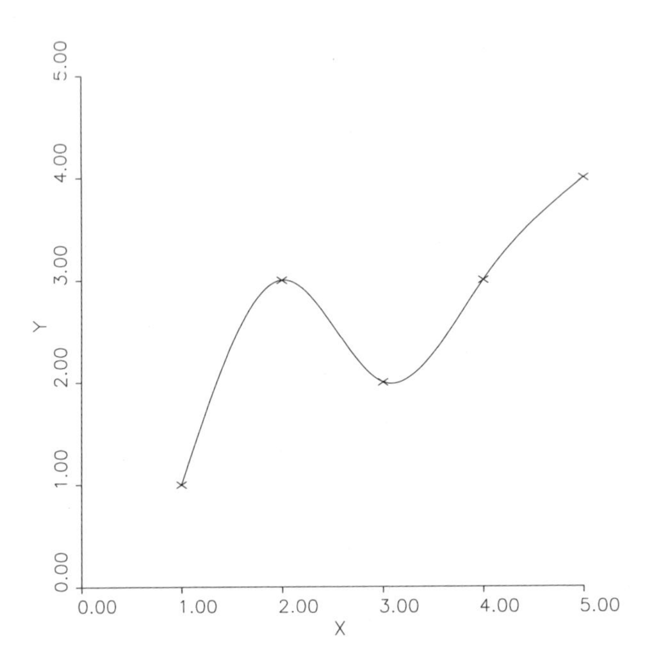

Figure 1.7.3
Natural Cubic Spline Interpolation

1.8 Roundoff Error

The errors committed by numerical methods can be grouped into two categories. Those that would still exist even if the arithmetic were done exactly are called truncation errors, and those that go away when exact arithmetic is done are called roundoff errors. Since Gaussian elimination with partial pivoting gives exact answers when the arithmetic is exact (as do most methods applied to linear problems of any sort), there is no truncation error to worry about; so we have to worry about roundoff errors in this section. Unfortunately, while the analysis of truncation errors is a science, the analysis of roundoff errors is an art, and the best that we can hope to do is to find some qualitative rules that allow us to predict (usually) the order of magnitude of the roundoff errors. As evidence that we cannot hope for more, recall that, even when we solve problems that are inherently susceptible to roundoff, it is possible for the roundoff error to be zero, owing to fortuitous cancellations.

Let us direct our attention first to 2 by 2 linear systems, since in this case we can analyze the situation geometrically. When we solve two linear equations in two unknowns, we are really just finding the intersection of two straight lines. If the lines $a_{11}x + a_{12}y = b_1$ and $a_{21}x + a_{22}y = b_2$ are parallel, the matrix

$$A = \begin{bmatrix} a_{11} & a_{12} \\ a_{21} & a_{22} \end{bmatrix}$$

is singular, and there are either many solutions (if the two lines coincide) or no solutions (if they do not coincide). If the two lines are nearly parallel, we say that the system $A\boldsymbol{x} = \boldsymbol{b}$ is "ill-conditioned" and we should expect to have problems, because A is nearly singular. In fact, it is clear that, if the two lines are nearly parallel, the minute that we store b_1 in our computer memory we are in trouble, as chopping it off to fit our floating point word length changes b_1 slightly, effectively translating the first line slightly (Figure 1.8.1).

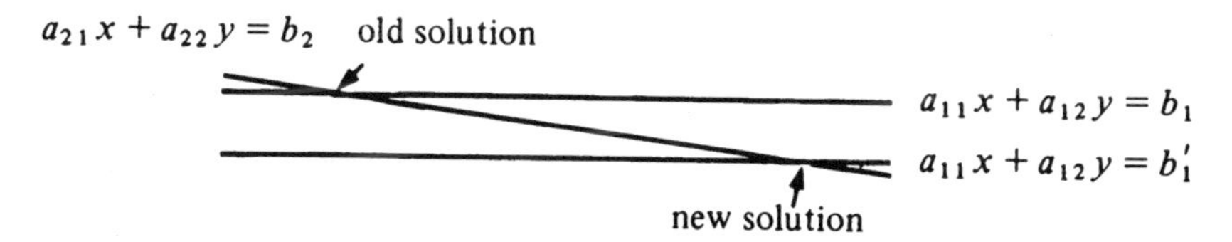

Figure 1.8.1
The Intersection of Nearly Parallel Lines

If the two lines are nearly parallel, this small change in b_1 will make a large change in the solution, even if we could solve the new system exactly. Further errors are introduced when we store the other coefficients $b_2, a_{11}, a_{12}, a_{21}, a_{22}$ and—of course—when we do the Gaussian elimination arithmetic itself. Note that, since we cannot generally even store the coefficients of a linear system exactly, no modification to our solution algorithm (e.g., to employ a better pivoting strategy) can save us from large errors if the system is ill-conditioned. The only thing we can do is to use higher-precision arithmetic.

If there is no cure for ill-conditioning, we would at least like to develop some tools to help us to diagnose this disease. This is not quite as easy as it sounds. Consider, for example, the following four linear systems:

$$B\boldsymbol{x} = \boldsymbol{b}, \quad C\boldsymbol{x} = \boldsymbol{c}, \quad D\boldsymbol{x} = \boldsymbol{d}, \quad E\boldsymbol{x} = \boldsymbol{e} \tag{1.8.1}$$

where

$$B = \begin{bmatrix} 10^{-10} & 10^{-20} \\ 10^{-20} & 10^{-10} \end{bmatrix}, \qquad C = \begin{bmatrix} 1 & 10^{-10} \\ 10^{-20} & 10^{-10} \end{bmatrix},$$

$$D = \begin{bmatrix} 1 & 10^{-10} \\ 10^{-10} & 1 \end{bmatrix}, \qquad E = \begin{bmatrix} 1.0001 & 1 \\ 1 & 1 \end{bmatrix},$$

and the right-hand side vectors $\boldsymbol{b}, \boldsymbol{c}, \boldsymbol{d}, \boldsymbol{e}$ are chosen so that the exact solution in each case is $(1, 1)$.

If we plot the two lines for each system, it is clear that $E\boldsymbol{x} = \boldsymbol{e}$ is the most ill-conditioned system, since the first three involve two nearly perpendicular lines, while the last involves two nearly parallel lines. This conclusion was confirmed when we actually solved the four systems, using single precision, with partial pivoting (equivalent to no pivoting, in each case). For the first three systems the calculated solution was accurate to machine precision, but the last system produced a 1% error in each component. Thus, for systems of two equations, the angle θ between the two lines appears to be a good measure of conditioning, but we would like a measure that generalizes to N by N systems.

Since a singular matrix has zero determinant, we might naively take the size of the determinant as a measure of how ill-conditioned a matrix is. The determinants of the above four matrices are

$$\det(B) \approx 10^{-20}, \quad \det(C) \approx 10^{-10}, \quad \det(D) \approx 1, \quad \det(E) \approx 10^{-4}.$$

The determinant incorrectly points to $B\boldsymbol{x} = \boldsymbol{b}$ as the most poorly conditioned system, because it is scale sensitive: $\det(\alpha A) = \alpha^N \det A$. The systems $B\boldsymbol{x} = \boldsymbol{b}$ and $D\boldsymbol{x} = \boldsymbol{d}$ represent the exact same pairs of lines but, since $B = 10^{-10}D$, B has a smaller determinant.

A more commonly used measure of the condition of a matrix is

$$c(A) \equiv \|A\|\|A^{-1}\|,$$

where any matrix norm may be used. This measure has the advantage that it is not scale sensitive, since $c(\alpha A) = \|\alpha A\|\, \|(\alpha A)^{-1}\| = c(A)$, and the following theorem accounts for the popularity of this condition number.

Theorem 1.8.1. *If the right-hand side of $A\boldsymbol{x} = \boldsymbol{b}$ is changed by $\Delta\boldsymbol{b}$, and the coefficient matrix is changed by ΔA, the change $\Delta\boldsymbol{x}$ in the solution satisfies*

$$(1 - \|A^{-1}\Delta A\|)\frac{\|\Delta\boldsymbol{x}\|}{\|\boldsymbol{x}\|} \le \|A\|\|A^{-1}\| \left(\frac{\|\Delta\boldsymbol{b}\|}{\|\boldsymbol{b}\|} + \frac{\|\Delta A\|}{\|A\|} \right),$$

where any associated vector–matrix norm pair can be used.

Proof: If $A\boldsymbol{x} = \boldsymbol{b}$ and $(A + \Delta A)(\boldsymbol{x} + \Delta\boldsymbol{x}) = \boldsymbol{b} + \Delta\boldsymbol{b}$, then

$$\begin{aligned}
A\,\Delta\boldsymbol{x} &= \Delta\boldsymbol{b} - \Delta A\,\boldsymbol{x} - \Delta A\,\Delta\boldsymbol{x} \\
\Delta\boldsymbol{x} &= A^{-1}\Delta\boldsymbol{b} - A^{-1}\Delta A\,\boldsymbol{x} - A^{-1}\Delta A\,\Delta\boldsymbol{x} \\
\|\Delta\boldsymbol{x}\| &\le \|A^{-1}\|\|\Delta\boldsymbol{b}\| + \|A^{-1}\|\|\Delta A\|\|\boldsymbol{x}\| + \|A^{-1}\Delta A\|\|\Delta\boldsymbol{x}\|
\end{aligned}$$

$$\begin{aligned}
(1 - \|A^{-1}\Delta A\|)\frac{\|\Delta\boldsymbol{x}\|}{\|\boldsymbol{x}\|} &\le \|A^{-1}\| \left(\frac{\|\Delta\boldsymbol{b}\|}{\|\boldsymbol{x}\|} + \|\Delta A\| \right) \\
&= \|A\|\|A^{-1}\| \left(\frac{\|\Delta\boldsymbol{b}\|}{\|A\|\|\boldsymbol{x}\|} + \frac{\|\Delta A\|}{\|A\|} \right).
\end{aligned}$$

Since $\boldsymbol{b} = A\boldsymbol{x}$ and so $\|\boldsymbol{b}\| \leq \|A\|\|\boldsymbol{x}\|$, the result of Theorem 1.8.1 follows immediately. ■

When ΔA is small, the quantity in parentheses on the left-hand side is nearly equal to one; so in this case Theorem 1.8.1 says that the relative change in the solution is bounded by the relative change in $\boldsymbol{b}$ plus the relative change in A, all multiplied by $c(A)$.

As mentioned earlier, just chopping off the elements of A and $\boldsymbol{b}$ to fit them into memory causes A and $\boldsymbol{b}$ to change. In fact, the relative changes in A and $\boldsymbol{b}$ should be no more than about ϵ, the machine relative precision. Thus Theorem 1.8.1 suggests that an upper bound on the relative error in the solution, owing to the inaccuracies in A and $\boldsymbol{b}$, should be around $c(A)\epsilon$.

Of course, additional errors arise when we actually do the arithmetic of Gaussian elimination. However, it has been shown [Stewart 1973, Section 3.5] that, when Gaussian elimination with partial pivoting is done, the resulting approximate solution $\boldsymbol{x}_1$ is always the exact solution to another linear system $(A+\Delta A)\boldsymbol{x}_1 = \boldsymbol{b}$, where $\|\Delta A\|/\|A\| \approx \epsilon$. Thus the errors resulting from doing inexact arithmetic during Gaussian elimination can be thought of as resulting from making further changes to A, changes that are comparable with those made when we chop the coefficients off to fit them into memory. So Theorem 1.8.1 effectively gives a bound on the overall error in $\boldsymbol{x}$.

Note that $c(A) = \|A\|\|A^{-1}\| \geq \|AA^{-1}\| = \|I\| = 1$ (in any matrix norm), so that the condition number is always greater than or equal to one. The condition number is essentially the factor by which the small errors in the coefficients are amplified when $A\boldsymbol{x} = \boldsymbol{b}$ is solved.

Now the condition number $c(A)$ appears to be the roundoff error predictor for which we were looking. If $c(A) = 10^m$, we can use Theorem 1.8.1 to predict that the relative error in our solution will be at most about $10^m\epsilon$; that is, we expect to lose at most m significant decimal digits in our answer. Unfortunately, Theorem 1.8.1 provides only an upper bound, and this bound can be so pessimistic as to be very misleading. For example, let us calculate the condition numbers of the four systems 1.8.1, using the matrix infinity norm:

$$c(B) \approx 1, \quad c(C) \approx 10^{10}, \quad c(D) \approx 1, \quad c(E) \approx 4 \times 10^4.$$

We see that the condition number incorrectly points to $C\boldsymbol{x} = \boldsymbol{c}$ as the most poorly conditioned system. It predicts that we may lose up to ten significant decimal digits in our answer, whereas we have already mentioned that essentially no significant digits are lost in the actual computer solution of this system.

Although the condition number $c(A)$ is not fooled (as is the determinant) when we multiply all equations by a scalar, and it is able to recognize that $B\boldsymbol{x} = \boldsymbol{b}$ and $D\boldsymbol{x} = \boldsymbol{d}$ are equally well-conditioned, it is fooled (for reasons explored in Problem 9) when we multiply a single equation through by a scalar, and it is not able to recognize that $C\boldsymbol{x} = \boldsymbol{c}$ is essentially equivalent to $D\boldsymbol{x} = \boldsymbol{d}$, as it involves the same pair of lines. When, in the Gaussian elimination process, we add a multiple of one row to another, the scaling of the rows affects the size of the multiple used, and nothing else.

However, this does not mean that $c(A)$ is useless as a predictor of roundoff, for consider the 2 by 2 system

$$\begin{bmatrix} a_{11} & a_{12} \\ a_{21} & a_{22} \end{bmatrix} \begin{bmatrix} x \\ y \end{bmatrix} = \begin{bmatrix} b_1 \\ b_2 \end{bmatrix}.$$

It can be shown (Problem 10) that

$$\frac{M^2}{|\det(A)|} \leq \|A\|_\infty \|A^{-1}\|_\infty \leq \frac{4M^2}{|\det(A)|}, \tag{1.8.2}$$

where M is the largest element of A, in absolute value. It can also be shown (Problem 10) that

$$\frac{M_1 M_2}{|\det(A)|} \leq \frac{1}{|\sin(\theta)|} \leq \frac{2M_1 M_2}{|\det(A)|}, \tag{1.8.3}$$

where θ is the angle of intersection of the two lines, and M_i is the largest element in row i, in absolute value. From 1.8.2 and 1.8.3 we see that, if the rows of A are scaled so that $M_1 \approx M_2 \approx M, c(A)$ is approximately equal to $1/|\sin(\theta)|$. Since we have taken θ as our definitive measure of the condition of a system of two equations, we conclude that $c(A) = \|A\|_\infty \|A^{-1}\|_\infty$ is a good measure of conditioning, *provided that the rows of A are comparably scaled.* Although we have only analyzed the situation for $N = 2$, the above conclusion is more generally valid.

In the above discussion, we have assumed that the norm (any norm) of $\Delta\boldsymbol{x}$ is a reasonable way to measure roundoff error. However, if the components of $\boldsymbol{x}$ are scaled very differently—for example, if (x_1, x_2) represent the coordinates of a ship, with x_1 measured in microns and x_2 measured in kilometers—$\|\Delta x\|$ may be an unreasonable way to measure error. For the example just given, a norm of $(10^{-6}\Delta x_1, 10^3 \Delta x_2)$ would be a better way to measure error. In general, if the unknowns are poorly scaled, we should use $\|D\ \Delta\boldsymbol{x}\|$ to measure roundoff, where D is a diagonal matrix of appropriately chosen weights. Then, if we write $A\boldsymbol{x} = \boldsymbol{b}$ in the form $(AD^{-1})(D\boldsymbol{x}) = \boldsymbol{b}$, we can see that $c(AD^{-1})$ is a better estimate of the true condition of the system than is $c(A)$, since it predicts the error $D\ \Delta\boldsymbol{x}$, rather than $\Delta\boldsymbol{x}$. Since postmultiplying by D^{-1} multiplies column i of A by $1/d_{ii}$, we are really just rescaling the columns of A, when we calculate AD^{-1}.

In summary, the size of $c(A)$ is a reasonable predictor of the accuracy in the solution to be expected when $A\boldsymbol{x} = \boldsymbol{b}$ is solved, *provided that the rows of A are comparably scaled.* Fortunately, in most "real-world" applications, the rows (and columns) are already comparably scaled, since usually the unknowns represent similar quantities, and the equations are generated by similar considerations. However, if the equations (rows) are not similarly scaled, we must rescale the rows before calculating $c(A)$ if we want to estimate the loss of accuracy in $\boldsymbol{x}$ and, if we want to estimate the loss of accuracy in $D\boldsymbol{x}$, we must rescale the columns (postmultiply A by D^{-1}) also.

Now rescaling the rows and columns of A is not important only for predicting roundoff error. Rescaling the rows is also important for the proper functioning of our algorithm itself. Consider the linear system $D\boldsymbol{x} = \boldsymbol{d}$ exhibited in 1.8.1. The partial pivoting strategy correctly tells us not to switch rows. Switching rows would bring the 10^{-10} to the pivot position with disastrous results, despite the fact that this is a very well-conditioned system and large roundoff errors are by no means inevitable. Now suppose we multiply the second equation of $D\boldsymbol{x} = \boldsymbol{d}$ through by 10^{11}:

$$\begin{bmatrix} 1 & 10^{-10} \\ 10 & 10^{11} \end{bmatrix} \begin{bmatrix} x \\ y \end{bmatrix} = \begin{bmatrix} 1+10^{-10} \\ 10+10^{11} \end{bmatrix}. \tag{1.8.4}$$

Now, when Gaussian elimination with partial pivoting is applied to the linear system 1.8.4, the rows *will* be switched, since 10 is larger than 1. The result is still a disaster: The calculated solution (using single precision) is $(0, 1)$, whereas the exact solution is $(1, 1)$. The vastly different norms of the two rows fools our partial pivoting algorithm into switching rows; it does not realize that $a_{21} = 10$ is really a very small number, while $a_{11} = 1$ is really a large number and should be used as the pivot. So we see that scaling the rows comparably may be important not only for predicting the error but also for the proper functioning of our algorithm itself.

A popular alternative to actually rescaling the equations is "scaled partial pivoting". In this approach, we compare $|a_{ii}|/s_i, \ldots, |a_{Ni}|/s_N$, rather than $|a_{ii}|, \ldots, |a_{Ni}|$, to determine which element to use as the pivot for column i, where s_i is the magnitude of the largest element in row i of the original matrix. The resulting pivoting strategy is exactly the same as results from partial pivoting applied to the matrix after each row has been rescaled to have the largest element of magnitude one. Scaled partial pivoting, or actually rescaling the equations, is only necessary, however, when the equations are grossly out of scale. Modifying DLINEQ (Figure 1.2.1) to do scaled partial pivoting would no doubt make it more robust, but differences of a few orders of magnitude in the scaling factors $s_1, \ldots, s_N$ can be tolerated by DLINEQ as written.

Let us now perform a numerical experiment. The Hilbert matrix H_N, whose elements are given by the formula

$$H_N(i,j) = \frac{1}{i+j-1}, \qquad i,j = 1, \ldots, N,$$

is known to be very ill-conditioned, for large N. Note that H_N has reasonably well-scaled rows and columns. We solve several systems of the form $H_N\boldsymbol{x} = \boldsymbol{b}$, where $\boldsymbol{b}$ is rigged so that the exact solution is $x^* = (1, 1, \ldots, 1)$, for different values of N. The results using DLINEQ are displayed in Table 1.8.1.

Note that, for large N, the relative error in $\boldsymbol{x}$ is large and approximately equal to $c(H_N)\epsilon$, as expected, where $\epsilon \approx 10^{-16}$ since double-precision arithmetic was used. However, even when the approximate solution appears to be pure garbage ($N = 15$), it still satisfies $H_N\boldsymbol{x} = \boldsymbol{b}$ very closely—see Problem 11 for an explanation.

Table 1.8.1

Hilbert Matrix Example

N	$c(H_N) = \|H_N\|_\infty \|H_N^{-1}\|_\infty$	$\|\boldsymbol{x} - \boldsymbol{x}^*\|_\infty / \|\boldsymbol{x}^*\|_\infty$	$\|H_N\boldsymbol{x} - \boldsymbol{b}\|_\infty$
3	7.5×10^2	0.1×10^{-13}	0.3×10^{-15}
6	2.9×10^7	0.1×10^{-9}	0.4×10^{-15}
9	1.1×10^{12}	0.3×10^{-5}	0.6×10^{-15}
12	4.2×10^{16}	0.4×10^{-1}	0.6×10^{-15}
15	3.3×10^{20}	223	0.2×10^{-14}

At this point we may well ask, if we have to calculate A^{-1} in order to compute the condition number, of what practical use is $c(A)$? The answer is that it *is* possible, if we are clever, to estimate $\|A^{-1}\|$ without actually calculating A^{-1} [Hagar 1988, Section 3.4]. The usual approach is to solve several systems $A\boldsymbol{x}_i = \boldsymbol{b}_i$, where the right-hand sides are chosen cleverly (or even randomly), and to estimate the norm of A^{-1} by

$$\|A^{-1}\| \approx \max_i \left(\frac{\|A^{-1}\boldsymbol{b}_i\|}{\|\boldsymbol{b}_i\|} \right) = \max_i \left(\frac{\|\boldsymbol{x}_i\|}{\|\boldsymbol{b}_i\|} \right).$$

As discussed in Section 1.4, the systems $A\boldsymbol{x}_i = \boldsymbol{b}_i$ can be solved inexpensively once an LU decomposition of A is available.

Some sophisticated codes [Forsythe et al. 1977, Dongarra et al. 1979] for solving $A\boldsymbol{x} = \boldsymbol{b}$ return an estimate of $c(A) = \|A\|\|A^{-1}\|$ along with the solution, so that the user has some idea of the accuracy of the returned solution.

1.9 Iterative Methods: Jacobi, Gauss-Seidel and SOR

If direct methods for linear systems, which find the exact answer (to within roundoff error) in a finite number of steps, are available to us, why would anyone bother to develop iterative methods? The answer is that iterative methods are attractive only for the linear systems that arise in certain types of applications. We cannot do better than Gaussian elimination if our linear system has an arbitrary full coefficient matrix, but the huge sparse systems that arise in certain applications often can be solved more efficiently by iterative methods. The iterative methods that we shall discuss were designed primarily to solve the systems that arise when partial differential equations are solved by finite difference or finite element methods, but they are useful in some other applications also.

The iterative methods that we shall introduce in this section are based on the following idea. We split A into two parts, B and $A-B$, and write $A\boldsymbol{x} = \boldsymbol{b}$ in the form

$$B\boldsymbol{x} + (A - B)\boldsymbol{x} = \boldsymbol{b}$$

or

$$B\boldsymbol{x} = (B - A)\boldsymbol{x} + \boldsymbol{b}. \tag{1.9.1}$$

Now we define an iterative method based on this formula:

$$B\boldsymbol{x}_{n+1} = (B - A)\boldsymbol{x}_n + \boldsymbol{b}. \tag{1.9.2}$$

First, it is clear that we must choose B so that we can easily solve 1.9.2; otherwise each iteration may require as much work as solving the original system $A\boldsymbol{x} = \boldsymbol{b}$. Now, when we compare 1.9.1 and 1.9.2, we see that, if $\boldsymbol{x}_n$ converges to a vector $\boldsymbol{x}_\infty$, then $\boldsymbol{x}_\infty$ will satisfy 1.9.1, and thus will be a solution to $A\boldsymbol{x} = \boldsymbol{b}$.

To determine when this iteration will converge, we subtract 1.9.1 from 1.9.2 and get

$$B\boldsymbol{e}_{n+1} = (B - A)\boldsymbol{e}_n,$$

where $\boldsymbol{e}_n \equiv \boldsymbol{x}_n - \boldsymbol{x}$ is the error after n iterations. Then

$$\boldsymbol{e}_{n+1} = (I - B^{-1}A)\boldsymbol{e}_n = (I - B^{-1}A)^{n+1}\boldsymbol{e}_0. \tag{1.9.3}$$

The question of convergence reduces to whether or not the powers of the matrix $I - B^{-1}A$ converge to zero. To answer this question we need the following theorem.

Theorem 1.9.1. *H^n converges to zero as $n \to \infty$, if and only if all eigenvalues of H are less than one in absolute value.*

Proof: The proof is almost trivial if H is diagonalizable, but not every matrix is similar to a diagonal matrix. However, for any matrix H there exists a matrix J of Jordan canonical form such that $H = SJS^{-1}$ and J has the nearly diagonal form [Wilkinson 1965]

$$J = \begin{bmatrix} J_1 & & & \\ & J_2 & & \\ & & \ddots & \\ & & & J_m \end{bmatrix},$$

where each J_i is an α_i by α_i square block of the form

$$J_i = \begin{bmatrix} \lambda_i & 1 & & & \\ & \lambda_i & 1 & & \\ & & \ddots & \ddots & \\ & & & \lambda_i & 1 \\ & & & & \lambda_i \end{bmatrix}.$$

The diagonal elements λ_i of J_i are eigenvalues of the upper triangular matrix J, and thus also of H. Now $H^n = (SJS^{-1})(SJS^{-1})\ldots(SJS^{-1}) = SJ^nS^{-1}$; so (since $J^n = S^{-1}H^nS$ also) $H^n \to 0$ if and only if $J^n \to 0$. Now (see Theorem 0.1.4)

$$J^n = \begin{bmatrix} J_1^n & & & \\ & J_2^n & & \\ & & \ddots & \\ & & & J_m^n \end{bmatrix}.$$

Let us look at an individual block J_i^n. If J_i is 2 by 2, J_i^n has the form (Problem 15)

$$J_i^n = \begin{bmatrix} \lambda_i & 1 \\ 0 & \lambda_i \end{bmatrix}^n = \begin{bmatrix} \lambda_i^n & n\lambda_i^{n-1} \\ 0 & \lambda_i^n \end{bmatrix},$$

and if J_i is 3 by 3, J_i^n has the form

$$J_i^n = \begin{bmatrix} \lambda_i & 1 & 0 \\ 0 & \lambda_i & 1 \\ 0 & 0 & \lambda_i \end{bmatrix}^n = \begin{bmatrix} \lambda_i^n & n\lambda_i^{n-1} & \frac{1}{2}n(n-1)\lambda_i^{n-2} \\ 0 & \lambda_i^n & n\lambda_i^{n-1} \\ 0 & 0 & \lambda_i^n \end{bmatrix}.$$

In general, we see that, if J_i is α_i by α_i, the components of J_i^n will all be polynomials in n of order less than α_i, times λ_i^n. Now clearly, if $|\lambda_i| \geq 1$, the nonzero elements of J_i^n do not converge to zero as $n \to \infty$, and it is well known (and easy to prove) that, if $|\lambda_i| < 1$, λ_i^n times any polynomial in n goes to zero as $n \to \infty$. Thus, in summary, if all eigenvalues λ_i of H are less than one in absolute value, each J_i^n, and thus J^n, converges to the zero matrix while, if any eigenvalue is greater than or equal to one in absolute value, J^n does not converge to the zero matrix. ■

Now we can give a convergence theorem.

Theorem 1.9.2. *If A and B are both nonsingular, and if $B\boldsymbol{x}_{n+1} = (B - A)\boldsymbol{x}_n + \boldsymbol{b}$, then $\boldsymbol{x}_n$ converges to the solution $\boldsymbol{x}$ of $A\boldsymbol{x} = \boldsymbol{b}$, for all initial guesses $\boldsymbol{x}_0 \neq \boldsymbol{x}$, if and only if all eigenvalues of $I - B^{-1}A$ are less than one in absolute value.*

Proof: The result follows immediately by applying Theorem 1.9.1 to equation 1.9.3, with $H \equiv I - B^{-1}A$. ■

Note that neither the right-hand side vector $\boldsymbol{b}$ nor the initial guess $\boldsymbol{x}_0$ has any influence on whether or not the iteration converges, although choosing $\boldsymbol{x}_0$ close to the true solution will clearly cause convergence to occur earlier, if the iteration does converge.

How then should we choose B? As mentioned earlier, we must choose B so that $B\boldsymbol{x}_{n+1} = (B-A)\boldsymbol{x}_n + \boldsymbol{b}$ is easily solved for $\boldsymbol{x}_{n+1}$; otherwise the iteration is not of much use. However, from the proof of Theorem 1.9.1 it is obvious that the smaller the maximum eigenvalue of $I - B^{-1}A = B^{-1}(B - A)$, the faster the iteration will converge; so we also want to choose B as close to A (in some sense) as we can, so that the eigenvalues of $B^{-1}(B-A)$ are as small as possible, and certainly less than one in absolute value.

One obvious choice for B is D, the diagonal matrix that is equal to A on the diagonal; it is easy to solve a linear system with a diagonal coefficient matrix. This choice defines the Jacobi iteration

$$D\boldsymbol{x}_{n+1} = (D - A)\boldsymbol{x}_n + \boldsymbol{b}$$

or

$$(x_i)_{n+1} = \frac{1}{a_{ii}}\left(b_i - \sum_{j\neq i} a_{ij}(x_j)_n\right), \qquad i = 1, \ldots, N. \tag{1.9.4}$$

Intuitively, we expect that the Jacobi method will converge if A is in some sense well approximated by its diagonal portion, and this is confirmed by the following theorem.

Theorem 1.9.3. *If A is diagonal-dominant, that is, if for each i*

$$|a_{ii}| > \sum_{j\neq i} |a_{ij}|,$$

then the Jacobi iteration 1.9.4 will converge.

Proof: The matrix $I - B^{-1}A = I - D^{-1}A$ has the form

$$\begin{bmatrix} 0 & -a_{12}/a_{11} & -a_{13}/a_{11} & \ldots & -a_{1N}/a_{11} \\ -a_{21}/a_{22} & 0 & -a_{23}/a_{22} & \ldots & -a_{2N}/a_{22} \\ -a_{31}/a_{33} & -a_{32}/a_{33} & 0 & \ldots & -a_{3N}/a_{33} \\ \vdots & \vdots & \vdots & & \vdots \\ -a_{N1}/a_{NN} & -a_{N2}/a_{NN} & -a_{N3}/a_{NN} & \ldots & 0 \end{bmatrix}.$$

Since A is diagonal-dominant, the sum of the absolute values of the elements of any row of the above matrix is less than one; hence the infinity norm of $I - D^{-1}A$ is less than one. Now we shall show that, for any matrix H, all eigenvalues are less (in absolute value) than any matrix norm $\|H\|$. If λ is an eigenvalue of H, with eigenvector $\boldsymbol{z}$, then

$$|\lambda|\|\boldsymbol{z}\| = \|H\boldsymbol{z}\| \leq \|H\|\|\boldsymbol{z}\|.$$

Since an eigenvector cannot be zero, we see that $|\lambda| \leq \|H\|$, and so any matrix norm of H is an upper bound on the absolute values of the eigenvalues of H. We have found a matrix norm of $I - D^{-1}A$ which is less than one, and therefore all eigenvalues of $I - D^{-1}A$ must be less than one, and the Jacobi method converges, by Theorem 1.9.2. ■

Note that this theorem does not say that the Jacobi method will converge *only* if A is diagonal-dominant. In fact, the matrix

$$A = \begin{bmatrix} 1 & 2 \\ 0 & 1 \end{bmatrix}$$

is not diagonal-dominant, and yet

$$I - D^{-1}A = \begin{bmatrix} 0 & -2 \\ 0 & 0 \end{bmatrix}$$

has eigenvalues $\lambda_1 = \lambda_2 = 0$; so the Jacobi method will converge (in only two iterations!) when applied to a system $A\boldsymbol{x} = \boldsymbol{b}$. Intuitively, however, for the Jacobi method to converge, the diagonal elements must be large in *some* sense; indeed we cannot even solve for $\boldsymbol{x}_{n+1}$ if any of the diagonal elements a_{ii} are zero.

Another reasonable choice for B is $B = L + D$, where L is defined to be the matrix that is equal to A below the diagonal and equal to 0 on and above the diagonal, and D is the diagonal matrix that has the same diagonal elements as A. Since B is lower triangular, it is easy to solve for $\boldsymbol{x}_{n+1}$ each iteration, and $L + D$ would seem to be a better approximation to A than D alone. This choice for B gives the Gauss–Seidel iteration. If we further define U to be the matrix that is equal to A above the diagonal and equal to 0 on and below the diagonal, $A = L + D + U$, and the Gauss–Seidel iteration can be written in matrix-vector form as

$$(L + D)\boldsymbol{x}_{n+1} = -U\boldsymbol{x}_n + \boldsymbol{b}$$

or

$$\boldsymbol{x}_{n+1} = D^{-1}(\boldsymbol{b} - L\boldsymbol{x}_{n+1} - U\boldsymbol{x}_n),$$

or in component form as

$$(x_i)_{n+1} = \frac{1}{a_{ii}}\left(b_i - \sum_{j<i} a_{ij}(x_j)_{n+1} - \sum_{j>i} a_{ij}(x_j)_n\right). \tag{1.9.5}$$

Although 1.9.5 appears to be implicit, note that, if we calculate new unknowns in the natural order $i = 1, \ldots, N$, then, by the time that we calculate a new value for x_i, new values for $x_1, \ldots, x_{i-1}$ are already available, so that everything on the right-hand side is known. In fact, the only difference between iterations 1.9.4 and 1.9.5 is that the Jacobi method insists that all variables on the right-hand side be from the previous iteration vector $\boldsymbol{x}_n$, while the Gauss–Seidel method uses the latest available values for the variables on the right-hand side.

The Gauss–Seidel iteration is also guaranteed to converge when A is diagonal-dominant (see Problem 16).

A third popular iterative method, called successive overrelaxation (SOR) [Young 1971], is defined by the choice $B = L + D/\omega$, where ω is a parameter whose value will be discussed shortly. The SOR iteration can be written in matrix-vector form as

$$\left(L + \frac{D}{\omega}\right)\boldsymbol{x}_{n+1} = \left(\frac{D}{\omega} - D - U\right)\boldsymbol{x}_n + \boldsymbol{b}$$

or

$$\boldsymbol{x}_{n+1} = \omega D^{-1}(\boldsymbol{b} - L\boldsymbol{x}_{n+1} - U\boldsymbol{x}_n) + (1-\omega)\boldsymbol{x}_n,$$

or in component form as

$$(x_i)_{n+1} = \frac{\omega}{a_{ii}}\left(b_i - \sum_{j<i} a_{ij}(x_j)_{n+1} - \sum_{j>i} a_{ij}(x_j)_n\right) + (1-\omega)(x_i)_n. \quad (1.9.6)$$

Note that the SOR formula assigns a value to the new x_i, which is a weighted average of the old value of x_i and the new value assigned by the Gauss–Seidel formula. When $\omega = 1$, SOR reduces to the Gauss–Seidel method.

We shall now show that, if A is positive-definite and $0 < \omega < 2$, the SOR method (and thus also the Gauss–Seidel method) will converge to the solution of $A\boldsymbol{x} = \boldsymbol{b}$.

Theorem 1.9.4. *If A is positive-definite and $0 < \omega < 2$, the eigenvalues of $I - B^{-1}A$ are all less than one in absolute value, when $B = L + D/\omega$. Therefore, the SOR iteration 1.9.6 will converge.*

Proof: Suppose

$$(I - B^{-1}A)\boldsymbol{z} = \lambda\boldsymbol{z}.$$

Since $I - B^{-1}A$ is not necessarily symmetric (even though A is), we have to assume that the eigenvalue λ and eigenvector $\boldsymbol{z}$ may be complex. Then

$$\begin{aligned}(B - A)\boldsymbol{z} &= \lambda B\boldsymbol{z},\\ (D - \omega D - \omega U)\boldsymbol{z} &= \lambda(D + \omega L)\boldsymbol{z}.\end{aligned}$$

Since A is symmetric, $U = L^{\mathrm{T}}$. Now we premultiply both sides by $\bar{\boldsymbol{z}}^{\mathrm{T}}$:

$$(1-\omega)\bar{\boldsymbol{z}}^{\mathrm{T}}D\boldsymbol{z} - \omega\bar{\boldsymbol{z}}^{\mathrm{T}}L^{\mathrm{T}}\boldsymbol{z} = \lambda\bar{\boldsymbol{z}}^{\mathrm{T}}D\boldsymbol{z} + \lambda\omega\bar{\boldsymbol{z}}^{\mathrm{T}}L\boldsymbol{z}. \tag{1.9.7}$$

If we define $d \equiv \bar{\boldsymbol{z}}^{\mathrm{T}}D\boldsymbol{z}$ and $l \equiv \bar{\boldsymbol{z}}^{\mathrm{T}}L\boldsymbol{z}$, we have

$$\bar{l} = \boldsymbol{z}^{\mathrm{T}}L\bar{\boldsymbol{z}} = (\boldsymbol{z}^{\mathrm{T}}L\bar{\boldsymbol{z}})^{\mathrm{T}} = \bar{\boldsymbol{z}}^{\mathrm{T}}L^{\mathrm{T}}\boldsymbol{z}.$$

Then 1.9.7 can be written

$$(1-\omega)d - \omega\bar{l} = \lambda d + \lambda\omega l. \tag{1.9.8}$$

Since $A = L + D + L^{\mathrm{T}}$, we also have

$$\bar{\boldsymbol{z}}^{\mathrm{T}}A\boldsymbol{z} = \bar{\boldsymbol{z}}^{\mathrm{T}}L\boldsymbol{z} + \bar{\boldsymbol{z}}^{\mathrm{T}}D\boldsymbol{z} + \bar{\boldsymbol{z}}^{\mathrm{T}}L^{\mathrm{T}}\boldsymbol{z}.$$

If we further define $a \equiv \bar{\boldsymbol{z}}^{\mathrm{T}}A\boldsymbol{z}$, we can write this as

$$a = l + d + \bar{l}. \tag{1.9.9}$$

Since A is positive-definite, both a and d must be real and positive (see Problem 12); l may be complex, however. Now replace $\bar{l}$ in 1.9.8 by $\bar{l} = a - l - d$, then solve 1.9.8 for l, giving $l = -d/\omega + a/(1-\lambda)$. Finally, replacing l and its conjugate by this formula in 1.9.9 gives

$$1 - |\lambda|^2 = \frac{d(2-\omega)|1-\lambda|^2}{a\omega}.$$

Since $a, d, 2-\omega$, and ω are positive, the right-hand side is positive, unless λ is exactly equal to 1. However, if λ were equal to 1, then we would have $(I - B^{-1}A)\boldsymbol{z} = \boldsymbol{z}$, or $A\boldsymbol{z} = \boldsymbol{0}$, which is impossible for nonzero $\boldsymbol{z}$, since A is positive-definite and therefore nonsingular. Thus the left-hand side must be positive, and so $|\lambda|^2 < 1$, for an arbitrary eigenvalue λ of $I - B^{-1}A$. ■

Since these three iterative methods were originally designed to solve the types of linear systems that arise when partial differential equations are solved, and they are not very useful for solving arbitrary full systems, to appreciate them fully we must study them in their native habitat. So let us consider the partial differential equation

$$-U_{xx} - U_{yy} - U_{zz} = 1 \qquad \text{in } 0 \le x \le 1, \quad 0 \le y \le 1, \quad 0 \le z \le 1$$

with

$$U = 0 \quad \text{on the boundary.}$$

We can approximate the differential equation and boundary conditions using finite differences as follows:

$$-\frac{U_{i+1,j,k} - 2U_{i,j,k} + U_{i-1,j,k}}{h^2} - \frac{U_{i,j+1,k} - 2U_{i,j,k} + U_{i,j-1,k}}{h^2}$$
$$-\frac{U_{i,j,k+1} - 2U_{i,j,k} + U_{i,j,k-1}}{h^2} = 1, \qquad \text{for } i,j,k = 1,\ldots,M-1,$$

or

$$6U_{i,j,k} - U_{i+1,j,k} - U_{i-1,j,k} - U_{i,j+1,k} - U_{i,j-1,k} - U_{i,j,k+1} \\ -U_{i,j,k-1} = h^2, \qquad \text{for } i,j,k = 1,\ldots,M-1, \tag{1.9.10a}$$

with

$$U_{i,j,k} = 0 \tag{1.9.10b}$$

when

$$i = 0, \quad i = M, \quad j = 0, \quad j = M, \quad k = 0, \quad k = M,$$

where $h = 1/M$ and $U_{i,j,k}$ approximates $U(i/M, j/M, k/M)$.

Now 1.9.10 represents a system of $N = (M-1)^3$ linear equations for the $(M-1)^3$ unknowns $U_{i,j,k}, i,j,k = 1,\ldots,M-1$. The coefficient matrix has the following form: In the row corresponding to equation (i,j,k) (which row this is depends on the ordering that we assign to the equations) there will be a 6 in the diagonal position (corresponding to unknown (i,j,k)) and at most six -1's scattered about. Most rows will have exactly six -1's but those adjacent to the boundary will have fewer than six, since $U_{i,j,k}$ can be replaced by 0 when i, j, or k is equal to 0 or M. Thus our coefficient matrix is almost, but not quite, diagonal-dominant. It can be shown that the matrix is positive-definite, however (see Problem 13); so, by Theorem 1.9.4, both Gauss–Seidel and SOR (with $0 < \omega < 2$) are guaranteed to converge.

Now explicitly forming this coefficient matrix is somewhat involved, and we first would have to decide on how to order the $(M-1)^3$ equations and unknowns. Fortunately, our three iterative methods can be programmed to solve this problem without ever even forming the matrix. For the Jacobi method 1.9.4, all we do is solve the ith equation for the ith unknown, then input old values on the right-hand side and calculate new values on the left-hand side. In our case we should say, solve the (i,j,k)th equation for the (i,j,k)th unknown, since we never explicitly decided on an ordering for the equations and unknowns, nor do we need to do this. The Jacobi method applied to the system 1.9.10 is then

$$U_{i,j,k}^{n+1} = \frac{h^2}{6} + \frac{U_{i+1,j,k}^n + U_{i-1,j,k}^n + U_{i,j+1,k}^n + U_{i,j-1,k}^n + U_{i,j,k+1}^n + U_{i,j,k-1}^n}{6}.$$

Now some of the "unknowns" on the right-hand side are really known boundary values. When $i = 1$, for example, $U_{i-1,j,k}$ is known to be zero. This causes us no confusion when we program the method, however, if we store all values of $U_{i,j,k}$—unknowns and knowns—together in one $M+1$ by $M+1$ by $M+1$ array. Figure 1.9.1 shows a FORTRAN program that solves the system 1.9.10 using the Jacobi iteration, with an initial guess of $\boldsymbol{x}_0 = \mathbf{0}$. The iteration halts when the relative residual $\|A\boldsymbol{x}_n - \boldsymbol{b}\|_\infty / \|\boldsymbol{b}\|_\infty$ ($\|\boldsymbol{b}\|_\infty = h^2$ for the system 1.9.10) is less than 10^{-10}. Note that we only check for convergence every ten iterations; calculating the residual every iteration would slow the program down appreciably.

To use the Gauss–Seidel iteration 1.9.5 we also solve the ith equation ((i,j,k)th equation, in our case) for the ith ((i,j,k)th) unknown, and iterate with this formula. However, while the Jacobi method requires that we use all old (nth iteration) values on the right-hand side, the Gauss–Seidel formula uses the latest available values for each unknown on the right-hand side. This is even easier to program than the Jacobi method, as now we only need one array to hold U. Figure 1.9.2 gives a FORTRAN program that solves the linear system 1.9.10 using the Gauss–Seidel method.

This program is exceptionally simple; if we use only one array to store U, the latest available value of each unknown on the right-hand side will automatically be used, since the latest value is the only one that has been saved. We do not even have to think about which of the unknowns on the right-hand side are new and which are old.

Finally, the SOR iteration 1.9.6 can be programmed by making a trivial modification to Figure 1.9.2. We simply replace statement 10 by

```
10     U(I,J,K) = W*GAUSS + (1-W)*U(I,J,K)
```

where W represents the parameter ω.

Table 1.9.1 shows the number of iterations which were required by the Jacobi, Gauss–Seidel, and SOR methods to solve 1.9.10. Several values of ω were tried for the SOR method. The same convergence criterion (relative residual less than 10^{-10}) was used in each test.

The results are typical for this type of problem. The Gauss–Seidel method converged about twice as fast as the Jacobi method, and the SOR method did even better, for $\omega > 1$. Typically the best value of ω for the SOR method is between 1 and 2.

If the system 1.9.10 is solved using Gaussian elimination, the amount of work is $O(N^3)$, where $N = (M-1)^3$ is the number of unknowns. If this system is put into band form and a band solver is used, the amount of work will be $O(M^7) = O(N^{2.333})$. From the results in Table 1.9.1 it would appear that the Jacobi method converges in $O(M^2)$ iterations, since the number of iterations quadruples as M is doubled. In Problem 14 this experimental estimate is confirmed theoretically. Now, since the amount of work per iteration is $O(M^3)$ (see Figure 1.9.1), this means that the total work required by the Jacobi method is $O(M^5) = O(N^{1.667})$. The results in Table 1.9.1 also suggest that the number of iterations required by the Gauss–Seidel and SOR methods (with ω chosen optimally) appear to be $O(M^2)$ and $O(M)$, respectively, for this problem. These estimates are, in fact, confirmed by theoretical calculations. Thus the total work to solve this linear system is $O(N^{1.667})$ for the Gauss–Seidel method, and $O(N^{1.333})$ for the SOR method, with ω chosen optimally. When N is very large, then, these iterative methods are much more efficient than even the band solver. Note that the storage requirements for the iterative

Table 1.9.1
Performance of Jacobi, Gauss–Seidel, and SOR Iterative Methods

Method	Number of Iterations		
	$M = 10$ $N = 729$	$M = 20$ $N = 6859$	$M = 40$ $N = 59319$
Jacobi	480	1920	7700
SOR ($\omega = 1.0$) ($\equiv$ Gauss–Seidel)	240	960	3850
SOR ($\omega = 1.1$)	200	790	3150
SOR ($\omega = 1.2$)	160	640	2570
SOR ($\omega = 1.3$)	130	520	2070
SOR ($\omega = 1.4$)	100	410	1640
SOR ($\omega = 1.5$)	70	310	1270
SOR ($\omega = 1.6$)	60*	220	950
SOR ($\omega = 1.7$)	80	140	660
SOR ($\omega = 1.8$)	130	130*	380
SOR ($\omega = 1.9$)	250	260	280*

*, Optimum value

methods are very low also, since the coefficient matrix is never actually stored, even in band form.

```
C                                  JACOBI METHOD
      PARAMETER (M=10)
      IMPLICIT DOUBLE PRECISION (A-H,O-Z)
      DOUBLE PRECISION UOLD(0:M,0:M,0:M),UNEW(0:M,0:M,0:M)
      H = 1.D0/M
C                                  SET BOUNDARY KNOWNS TO ZERO PERMANENTLY
C                                  AND INTERIOR UNKNOWNS TO ZERO TEMPORARILY
      DO 5 I=0,M
      DO 5 J=0,M
      DO 5 K=0,M
    5 UOLD(I,J,K) = 0.0
C                                  BEGIN JACOBI ITERATION
      NITER = (M-1)**3
      DO 25 ITER = 1,NITER
C                                  UPDATE UNKNOWNS ONLY
         DO 10 I=1,M-1
         DO 10 J=1,M-1
         DO 10 K=1,M-1
   10    UNEW(I,J,K) = H**2/6.0 + ( UOLD(I+1,J,K) + UOLD(I-1,J,K)
     &                              + UOLD(I,J+1,K) + UOLD(I,J-1,K)
     &                              + UOLD(I,J,K+1) + UOLD(I,J,K-1))/6.0
C                                  COPY UNEW ONTO UOLD
         DO 15 I=1,M-1
```

```
      DO 15 J=1,M-1
      DO 15 K=1,M-1
   15 UOLD(I,J,K) = UNEW(I,J,K)
C                         EVERY 10 ITERATIONS CALCULATE MAXIMUM
C                         RESIDUAL AND CHECK FOR CONVERGENCE
      IF (MOD(ITER,10).NE.0) GO TO 25
      RMAX = 0.0
      DO 20 I=1,M-1
      DO 20 J=1,M-1
      DO 20 K=1,M-1
         RESID = 6*UOLD(I,J,K) - UOLD(I+1,J,K) - UOLD(I-1,J,K)
     &                         - UOLD(I,J+1,K) - UOLD(I,J-1,K)
     &                         - UOLD(I,J,K+1) - UOLD(I,J,K-1) - H**2
         RMAX = MAX(RMAX,ABS(RESID))
   20 CONTINUE
      RMAX = RMAX/H**2
      PRINT *, ITER, RMAX
      IF (RMAX.LE.1.D-10) STOP
   25 CONTINUE
      STOP
      END
```

Figure 1.9.1

```
C                         GAUSS-SEIDEL METHOD
      PARAMETER (M=10)
      IMPLICIT DOUBLE PRECISION (A-H,O-Z)
      DOUBLE PRECISION U(0:M,0:M,0:M)
      H = 1.D0/M
C                         SET BOUNDARY KNOWNS TO ZERO PERMANENTLY
C                         AND INTERIOR UNKNOWNS TO ZERO TEMPORARILY
      DO 5 I=0,M
      DO 5 J=0,M
      DO 5 K=0,M
    5 U(I,J,K) = 0.0
C                         BEGIN GAUSS-SEIDEL ITERATION
      NITER = (M-1)**3
      DO 20 ITER = 1,NITER
C                         UPDATE UNKNOWNS ONLY
         DO 10 I=1,M-1
         DO 10 J=1,M-1
         DO 10 K=1,M-1
            GAUSS = H**2/6.0 + ( U(I+1,J,K) + U(I-1,J,K)
     &                         + U(I,J+1,K) + U(I,J-1,K)
     &                         + U(I,J,K+1) + U(I,J,K-1))/6.0
   10    U(I,J,K) = GAUSS
C                         EVERY 10 ITERATIONS CALCULATE MAXIMUM
C                         RESIDUAL AND CHECK FOR CONVERGENCE
```

```
        IF (MOD(ITER,10).NE.0) GO TO 20
        RMAX = 0.0
        DO 15 I=1,M-1
        DO 15 J=1,M-1
        DO 15 K=1,M-1
           RESID = 6*U(I,J,K) - U(I+1,J,K) - U(I-1,J,K)
     &                        - U(I,J+1,K) - U(I,J-1,K)
     &                        - U(I,J,K+1) - U(I,J,K-1) - H**2
           RMAX = MAX(RMAX,ABS(RESID))
   15   CONTINUE
        RMAX = RMAX/H**2
        PRINT *, ITER, RMAX
        IF (RMAX.LE.1.D-10) STOP
   20 CONTINUE
      STOP
      END
```

Figure 1.9.2

1.10 The Conjugate Gradient Method

A more modern iterative method, for symmetric systems, which is not based on a splitting (1.9.2) of A, is the conjugate-gradient method, which can be defined by:

$$\begin{aligned}
\mathbf{x_0} &= \text{starting guess}, \\
\mathbf{r_0} &= \mathbf{b} - A\mathbf{x_0}, \\
\mathbf{p_0} &= \mathbf{r_0}, \\
\lambda_n &= (\mathbf{r_n}^T\mathbf{p_n})/(\mathbf{p_n}^T A\mathbf{p_n}), \qquad n = 0, 1, 2..., \qquad (1.10.1)\\
\mathbf{x_{n+1}} &= \mathbf{x_n} + \lambda_n \mathbf{p_n}, \\
\mathbf{r_{n+1}} &= \mathbf{r_n} - \lambda_n A\mathbf{p_n}, \\
\alpha_n &= -(\mathbf{r_{n+1}}^T A\mathbf{p_n})/(\mathbf{p_n}^T A\mathbf{p_n}), \\
\mathbf{p_{n+1}} &= \mathbf{r_{n+1}} + \alpha_n \mathbf{p_n},
\end{aligned}$$

A more complete analysis of the convergence of this iterative method is given in [Sewell 2005, Section 4.8] and in many other references, here we will only show that if A is not only symmetric, but also positive definite, the residual $\boldsymbol{b} - A\boldsymbol{x}_n$ is guaranteed nonincreasing, in a certain norm.

First, note that $\boldsymbol{r}_n$ is the residual $\boldsymbol{b} - A\boldsymbol{x}_n$, as can be proved by induction. For $n = 0$, this is true by the definition of $\boldsymbol{r}_0$. Assume that it is true for n, that is, $\boldsymbol{r}_n = \boldsymbol{b} - A\boldsymbol{x}_n$, and we verify that

$$\boldsymbol{r}_{n+1} = \boldsymbol{r}_n - \lambda_n A\boldsymbol{p}_n = \boldsymbol{b} - A\boldsymbol{x}_n - A\lambda_n\boldsymbol{p}_n = \boldsymbol{b} - A\boldsymbol{x}_{n+1}.$$

If A, and therefore also A^{-1}, is positive-definite, the following defines a legitimate norm since it is always nonnegative, and zero only when $\boldsymbol{e} = \mathbf{0}$:

$$\|\boldsymbol{e}\| \equiv (\boldsymbol{e}^T A^{-1} \boldsymbol{e})^{1/2}$$

Then in this norm,

$$\begin{aligned}
\|\boldsymbol{r}_{n+1}\|^2 &= (\boldsymbol{r}_n - \lambda_n A\boldsymbol{p}_n)^T A^{-1} (\boldsymbol{r}_n - \lambda_n A\boldsymbol{p}_n) \\
&= \boldsymbol{r}_n^T A^{-1} \boldsymbol{r}_n - 2\lambda_n \boldsymbol{r}_n^T \boldsymbol{p}_n + \lambda_n^2 \boldsymbol{p}_n^T A \boldsymbol{p}_n \\
&= \|\boldsymbol{r}_n\|^2 - 2\lambda_n \boldsymbol{r}_n^T \boldsymbol{p}_n + \lambda_n \boldsymbol{r}_n^T \boldsymbol{p}_n \\
&= \|\boldsymbol{r}_n\|^2 - (\boldsymbol{r}_n^T \boldsymbol{p}_n)^2 / (\boldsymbol{p}_n^T A \boldsymbol{p}_n)
\end{aligned}$$

Since $\boldsymbol{p}_n^T A \boldsymbol{p}_n \geq 0$, this shows that the residual is nonincreasing, in this norm. (It can also be shown that $\boldsymbol{p}_n$ cannot be zero unless the residual is zero, and thus $\boldsymbol{p}_n^T A \boldsymbol{p}_n > 0$ unless we already have the exact solution.)

It can be shown that a lower condition number for the matrix means faster convergence for the conjugate-gradient method, which explains why there is much interest in "preconditioning" the matrix, that is, premultiplying $A\boldsymbol{x} = \boldsymbol{b}$ by the inverse of a positive definite matrix B such that $B^{-1}A$ has a lower condition number than A, before applying the conjugate-gradient iteration 1.10.1. Since $B^{-1}A$ will generally no longer be symmetric, we need to use the "preconditioned" algorithm:

$$\begin{aligned}
\mathbf{X_0} &= \text{starting guess,} \\
\mathbf{R_0} &= B^{-1}(\mathbf{b} - A\mathbf{X_0}), \\
\mathbf{P_0} &= \mathbf{R_0}, \\
\lambda_n &= (\mathbf{R_n}^T B\mathbf{P_n})/(\mathbf{P_n}^T A\mathbf{P_n}), \qquad n = 0, 1, 2..., \qquad (1.10.2) \\
\mathbf{X_{n+1}} &= \mathbf{X_n} + \lambda_n \mathbf{P_n}, \\
\mathbf{R_{n+1}} &= \mathbf{R_n} - \lambda_n B^{-1} A\mathbf{P_n}, \\
\alpha_n &= -(\mathbf{R_{n+1}}^T A\mathbf{P_n})/(\mathbf{P_n}^T A\mathbf{P_n}), \\
\mathbf{P_{n+1}} &= \mathbf{R_{n+1}} + \alpha_n \mathbf{P_n},
\end{aligned}$$

If we make the substitutions $\mathbf{x_n} = L^T\mathbf{X_n}, \mathbf{r_n} = L^T\mathbf{R_n}, \mathbf{p_n} = L^T\mathbf{P_n}$ where $B = LL^T$ is the Cholesky decomposition (see Section 1.4) of the positive definite matrix B, the preconditioned algorithm 1.10.2 reduces to the usual conjugate-gradient algorithm 1.10.1 applied to the linear system $H\mathbf{x} = L^{-1}\mathbf{b}$, where $H = L^{-1}AL^{-T}$ (note that this is equivalent to $A\mathbf{X} = \mathbf{b}$). It is easy to verify that, assuming A and B are positive definite, H is also symmetric and positive definite, and therefore the preconditioned algorithm will still converge. Also, since $L^{-T}HL^T = B^{-1}A$, H and $B^{-1}A$ have the same (positive) eigenvalues and therefore the same spectral condition number (ratio of largest to smallest eigenvalue).

If B is chosen so that $B^{-1}A$ (and thus H) has a smaller condition number than A, the preconditioned conjugate-gradient method above should converge faster than the unmodified version. Choosing $B = A$ would make $H = I$, which has the lowest possible condition number (one), but since B^{-1} appears in the iteration, we want to choose B close to A in some sense, but such that it is easy to solve systems with matrix B. The simplest choice is $B =$ diagonal(A).

Figure 1.10.1 shows a subroutine DCG which solves a linear system using the preconditioned conjugate-gradient algorithm 1.10.2, with a diagonal preconditioning matrix D. If $D(I) = 1$, DCG will be the unmodified conjugate-gradient algorithm, but normally D should be the diagonal part of A, in which case this is called the "Jacobi" conjugate-gradient method. The matrix A is stored in sparse matrix format, that is, with only the NZ nonzeros stored in $A(I), I = 1, ..., NZ$, and with arrays $IROW$ and $JCOL$ to identify the row and column numbers of the nonzeros.

```
      SUBROUTINE DCG(A,IROW,JCOL,NZ,X,B,N,D)
      IMPLICIT DOUBLE PRECISION (A-H,O-Z)
C                                        DECLARATIONS FOR ARGUMENTS
      DOUBLE PRECISION A(NZ),B(N),X(N),D(N)
      INTEGER IROW(NZ),JCOL(NZ)
C                                        DECLARATIONS FOR LOCAL VARIABLES
      DOUBLE PRECISION R(N),P(N),AP(N),LAMBDA
C
C  SUBROUTINE DCG SOLVES THE SYMMETRIC LINEAR SYSTEM A*X=B, USING THE
C     CONJUGATE GRADIENT ITERATIVE METHOD.  THE NON-ZEROS OF A ARE STORED
C     IN SPARSE FORMAT.
C
C  ARGUMENTS
C
C             ON INPUT                          ON OUTPUT
C             --------                          ---------
C
C    A      - A(IZ) IS THE MATRIX ELEMENT IN
C             ROW IROW(IZ), COLUMN JCOL(IZ),
C             FOR IZ=1,...,NZ.
C
C    IROW   - (SEE A).
C
C    JCOL   - (SEE A).
C
C    NZ     - NUMBER OF NONZEROS.
C
C    X      -                                   AN N-VECTOR CONTAINING
C                                               THE SOLUTION.
C
C    B      - THE RIGHT HAND SIDE N-VECTOR.
C
C    N      - SIZE OF MATRIX A.
```

```
C
C    D       - VECTOR HOLDING A DIAGONAL
C              PRECONDITIONING MATRIX.
C              D = DIAGONAL(A) IS RECOMMENDED.
C              D(I) = 1 FOR NO PRECONDITIONING
C
C-----------------------------------------------------------------------
C                                    X0 = 0
C                                    R0 = D**(-1)*B
C                                    P0 = R0
      R0MAX = 0
      DO 10 I=1,N
         X(I) = 0
         R(I) = B(I)/D(I)
         R0MAX = MAX(R0MAX,ABS(R(I)))
         P(I) = R(I)
   10 CONTINUE
C                                    NITER = MAX NUMBER OF ITERATIONS
      NITER = 3*N
      DO 90 ITER=1,NITER
C                                    AP = A*P
         DO 20 I=1,N
            AP(I) = 0
   20    CONTINUE
         DO 30 IZ=1,NZ
            I = IROW(IZ)
            J = JCOL(IZ)
            AP(I) = AP(I) + A(IZ)*P(J)
   30    CONTINUE
C                                    PAP = (P,AP)
C                                    RP = (R,D*P)
         PAP = 0.0
         RP = 0.0
         DO 40 I=1,N
            PAP = PAP + P(I)*AP(I)
            RP = RP + R(I)*D(I)*P(I)
   40    CONTINUE
C                                    LAMBDA = (R,D*P)/(P,AP)
         LAMBDA = RP/PAP
C                                    X = X + LAMBDA*P
C                                    R = R - LAMBDA*D**(-1)*AP
         DO 50 I=1,N
            X(I) = X(I) + LAMBDA*P(I)
            R(I) = R(I) - LAMBDA*AP(I)/D(I)
   50    CONTINUE
C                                    RAP = (R,AP)
         RAP = 0.0
         DO 60 I=1,N
```

```
            RAP = RAP + R(I)*AP(I)
   60     CONTINUE
C                                          ALPHA = -(R,AP)/(P,AP)
          ALPHA = -RAP/PAP
C                                          P = R + ALPHA*P
          DO 70 I=1,N
             P(I) = R(I) + ALPHA*P(I)
   70     CONTINUE
C                                          RMAX = MAX OF RESIDUAL (R)
          RMAX = 0
          DO 80 I=1,N
             RMAX = MAX(RMAX,ABS(R(I)))
   80     CONTINUE
C                                          CHECK IF CONVERGED
          IF (RMAX.LE.1.D-10*R0MAX) THEN
             PRINT *, ' Number of iterations = ',ITER
             RETURN
          ENDIF
   90 CONTINUE
C                                          DCG DOES NOT CONVERGE
      PRINT 100
  100 FORMAT('***** DCG DOES NOT CONVERGE *****')
      RETURN
      END
```

Figure 1.10.1

When DCG was used to solve the sparse, positive definite, system 1.9.10 (the diagonal for this system is a constant, so $D(I) = 1$ was used), it required 25 iterations with $M = 10$, 56 iterations with $M = 20$, and 115 with $M = 40$. So, like successive overrelaxation with optimal ω, the conjugate-gradient method appears to require only $O(M)$ iterations (this can also be verified theoretically for this problem), and since the work per iteration is proportional to $NZ < 7N = 7(M-1)^3$, the total work is $O(M^4)$ or $O(N^{1.333})$. And the conjugate-gradient algorithm does not require us to find the optimal value for a key parameter.

1.11 Problems

1. The Gaussian elimination algorithm could be modified so that each pivot a_{ii} is used to eliminate the elements below *and* above it, in column i (Gauss–Jordan). Then at the end of the forward-elimination phase we have reduced A to diagonal rather than triangular form. Calculate the approximate number of multiplications that this algorithm requires to solve $A\boldsymbol{x} = \boldsymbol{b}$ and compare with the $\frac{1}{3}N^3$ used by DLINEQ (Figure

1.2.1). Why is Gauss–Jordan a particularly bad idea for banded systems?

2. Write a routine to calculate the inverse of an arbitrary N by N matrix A. Use the fact that the kth column of A^{-1} is the solution to $A\boldsymbol{x}_k = \boldsymbol{e}_k$, where $\boldsymbol{e}_k$ is the kth column of the identity matrix. Call DLINEQ to solve $A\boldsymbol{x}_1 = \boldsymbol{e}_1$, and DRESLV (Figure 1.3.1) to solve $A\boldsymbol{x}_k = \boldsymbol{e}_k$ for $k = 2, \ldots, N$. By carefully examining subroutines DLINEQ and DRESLV, determine the (approximate) total number of multiplications used to calculate A^{-1} in this manner, and compare this with the number $\frac{1}{3}N^3$ required to calculate the LU decomposition of A. Test your routine by solving the system below in the form $\boldsymbol{x} = A^{-1}\boldsymbol{b}$ (exact solution is $\boldsymbol{x} = (1, 1, 1, 1, 1)$):

$$\begin{bmatrix} 16 & 8 & 4 & 2 & 1 \\ 8 & 16 & 4 & 2 & 1 \\ 4 & 4 & 16 & 2 & 1 \\ 2 & 2 & 2 & 16 & 1 \\ 1 & 1 & 1 & 1 & 16 \end{bmatrix} \begin{bmatrix} x1 \\ x2 \\ x3 \\ x4 \\ x5 \end{bmatrix} = \begin{bmatrix} 31 \\ 31 \\ 27 \\ 23 \\ 20 \end{bmatrix}.$$

3. Verify that $L = M_{12}^{-1} \ldots M_{N-1,N}^{-1}$ has the form shown in 1.4.8, where the matrices M_{ij} are as shown in 1.4.1. (Hint: start with $L = I$ and start premultiplying by the elementary matrices, and note that when you premultiply by M_{ij}^{-1}, which adds l_{ji} times row i to row j, row i of L is still equal to row i of I.) Show by an example, however, that $L^{-1} = M_{N-1,N} \ldots M_{12}$ does *not* have the (perhaps expected) form

$$\begin{bmatrix} 1 & & & & \\ -l_{21} & 1 & & & \\ -l_{31} & -l_{32} & 1 & & \\ \vdots & \vdots & \vdots & \ddots & \\ -l_{N1} & -l_{N2} & -l_{N3} & \ldots & 1 \end{bmatrix}.$$

4. a. Show that, if Gaussian elimination is done on a symmetric matrix without pivoting, after the first m columns have been zeroed, the $N - m$ by $N - m$ submatrix in the lower right-hand corner is still symmetric. (Hint: When a multiple $-a_{j1}/a_{11}$ of row 1 is added to row j, to zero a_{j1}, element a_{jk} is replaced by $a_{jk} - (a_{j1}/a_{11})a_{1k}$. Show that the new a_{jk} and the new a_{kj} are equal, and use induction to extend the argument.)

   b. Show that, if A is symmetric, the LU decomposition of A computed with no pivoting has $U = DL^{\mathrm{T}}$, where D is the diagonal part of U, and so $A = LDL^{\mathrm{T}}$. (Hint: When $a_{ji} (j > i)$ is eliminated, the negative of the multiplier used to zero it is saved as $l_{ji} = a_{ji}/a_{ii}$, where the elements of A mentioned here represent the values immediately before a_{ji} is zeroed and not the original values. By part (a), $a_{ji} = a_{ij}$; so $l_{ji} = a_{ij}/a_{ii} = u_{ij}/u_{ii}$.)

5. Modify subroutine DLINEQ so that it solves a symmetric system without pivoting, taking full advantage of symmetry. According to Problem 4a, the subdiagonal element $a_{ji} (j > i)$ will remain equal to the superdiagonal element a_{ij} until it is eliminated; so DLINEQ does not need ever to modify or access the subdiagonal elements. The total work can be cut in half if only the superdiagonal elements are manipulated. Test your program on the system of Problem 2. The coefficient matrix is positive-definite, so it is appropriate to solve this system without pivoting.

6. a. Using hand calculations, without pivoting, find the LU decomposition of

$$B = \begin{bmatrix} -4 & -4 & 4 \\ 3 & -2 & 2 \\ 1 & 0 & 1 \end{bmatrix}.$$

   b. Notice that the diagonal element was already the largest potential pivot (in absolute value) in each case, so you actually used partial pivoting in part (a). Now suppose the original matrix had been

$$A = \begin{bmatrix} 3 & -2 & 2 \\ 1 & 0 & 1 \\ -4 & -4 & 4 \end{bmatrix}.$$

   Find a permutation matrix P such that $PA = B(= LU)$. Thus if you had started with A and done partial pivoting, you would have ended up computing the LU decomposition of PA. Note that $A = P^{-1}LU$, where P^{-1} is also a permutation matrix. What is P^{-1} for this example?

   c. Now use the LU decomposition of PA to solve $A\boldsymbol{x} = \boldsymbol{b}$, where $\boldsymbol{b} = (5, 1, 20)$; that is, first find $P\boldsymbol{b}$, then solve $L\boldsymbol{y} = P\boldsymbol{b}$ using forward substitution, then solve $U\boldsymbol{x} = \boldsymbol{y}$ using back substitution.

7. Consider the 1D boundary value problem $-U_{xx} + U = 2\sin(x)$ with boundary conditions $U(0) = U(2\pi) = 0$. This differential equation can be approximated using the finite difference equation:

$$\frac{-U_{i+1} + 2U_i - U_{i-1}}{h^2} + U_i = 2\sin(x_i)$$

for $i = 1, ..., N-1$, where $x_i = ih, h = 2\pi/N$, and U_i is an approximation to $U(x_i)$. The boundary conditions imply $U_0 = U_N = 0$.

a. The resulting linear system is tridiagonal, so it could be solved using DBAND, with $NLD = NUD = 1$, but write your own tridiagonal linear system solver to solve it. Your routine should hold the three diagonals in three vectors, where $a(i) = A_{i,i-1}, b(i) = A_{i,i}, c(i) = A_{i,i+1}$. You may assume no pivoting is necessary; in fact, this $N-1$ by $N-1$ system does not require pivoting. The forward elimination and backward substitution may have 10 or fewer lines of code! If N is large, your solution should be close to the true solution of the differential equation $U_i = \sin(x_i)$.

b. Now change the boundary conditions to "periodic" boundary conditions, $U(0) = U(2\pi), U_x(0) = U_x(2\pi)$. These conditions can be approximated by $U_0 = U_N$ and $(U_{N+1} - U_N)/h = (U_1 - U_0)/h$, or $U_{N+1} = U_1$. Thus, in the finite difference equation corresponding to $i = 1$, U_0 can be replaced by U_N, and in the equation corresponding to $i = N$, U_{N+1} can be replaced by U_1, and we then have N equations for the N unknowns $U_1, ..., U_N$. Now the system is almost still tridiagonal, except that $A_{1,N}$ and $A_{N,1}$ are also nonzero, so the bandwidth is the maximum possible. Nevertheless, modify your program from part (a) to solve this linear system without pivoting, using $O(N)$ work and $O(N)$ storage. (Hint: The only fill-in is in the last column and last row.) The differential equation solution is still $U(x) = \sin(x)$.

c. Show that the matrix A in part (a) is diagonal-dominant; thus the Jacobi and Gauss–Seidel iterative methods will converge when applied to these equations (A is also positive-definite, so SOR converges as well). Write a program to solve the linear system of part (a) using Gauss–Seidel. How does the computer time for this method compare with the direct solver?

Part (b) shows that direct (Gauss elimination-based) solvers can sometimes be used to efficiently solve linear systems that are sparse but not banded.

8. An alternative way to keep up with the row switches in DLINEQ is to simply set $IPERM(I) = L$ after loop 15, to remember which row was switched with row I, and then in DRESLV, instead of permuting the elements of C at the beginning, switch $C\_(I)$ and $C\_(L)$ ($L = IPERM(I)$) immediately before loop 15, replicating the switches done to $B\_$ in DLINEQ. We now must only switch the "nonzero" portions of rows I and L of A in DLINEQ, and not the multipliers saved in the previous columns, that is, loop 20 should only run from $K = I$ to N. We also need to remove the code that switches $IPERM(I)$ and $IPERM(L)$. The new approach has the advantage that if A is a band matrix and pivoting is done, you can save the multipliers without fill-in below the

band in the lower triangle. However, note that the lower triangle of A returned by the new DLINEQ is no longer equal to the lower triangle of the L matrix of the LU decomposition (of PA).

a. Make the changes to DLINEQ and DRESLV suggested above and solve $A\boldsymbol{x} = \boldsymbol{b}$ using the new DLINEQ, where A is a 6 by 6 tridiagonal matrix with 1 in each main diagonal position and 2 in each sub- and super-diagonal, and $\boldsymbol{b}$ is chosen so that the exact solution is $(1, 1, 1, 1, 1, 1)$. Then also solve $A\boldsymbol{x} = 2\boldsymbol{b}$ using the new DRESLV. Print out the matrix A after calling the modified DLINEQ, and note that there is no fill-in below the band. Repeat this problem using the original versions of DLINEQ and DRESLV, and you will see that the saved multipliers do cause fill-in below the band.

b. Make two copies of the band solver DBAND, call one DBLINEQ and the other DBRESLV; each should have an additional argument $IPERM$. Set $A(J, I - J) = LJI$ immediately before the end of loop 30 in DBLINEQ so that the multiplier used to zero each element is saved in the same location, as done in DLINEQ, and set $IPERM(I) = L$ after loop 15 to keep track of the row switches, as done in part (a). Now modify DBRESLV so that it solves another system with the same matrix, using A and $IPERM$ as output by DBLINEQ. DBRESLV should set $L = IPERM(I)$ before $B\_(L)$ and $B\_(I)$ are switched, and set $LJI = A(J, I - J)$ in loop 30 to retreive the multiplier. All code that changes elements of A should be eliminated from DBRESLV, and code that checks for zero pivots can also be removed.

   Test DBLINEQ and DBRESLV using the same band matrix as in part (a), and solve the same two systems.

   DBLINEQ will do the same number of multiplications as DBAND (about $N\ N_{LD}(N_{UD} + N_{LD})$), to solve the first system. Approximately how many multiplications will DBRESLV do to solve each additional system?

9. As mentioned in the text, the condition number of $C\boldsymbol{x} = \boldsymbol{c}$ (see 1.8.1) is very large (about 10^{10}) even though this system represents the intersection of two nearly perpendicular lines, and Gaussian elimination produces very little roundoff error when $C\boldsymbol{x} = \boldsymbol{c}$ is solved. However, show by an example that $\|\Delta\boldsymbol{x}\|/\|\boldsymbol{x}\|$ can indeed be very large when $\|\Delta\boldsymbol{c}\|/\|\boldsymbol{c}\|$ is very small, so that Theorem 1.8.1 is not misleading, in this respect. What is misleading is our measure $\|\Delta\boldsymbol{c}\|/\|\boldsymbol{c}\|$ of the relative error in $\boldsymbol{c}$. You may be able to make $\|\Delta\boldsymbol{x}\|/\|\boldsymbol{x}\|$ large without making

$\|\Delta \boldsymbol{c}\|/\|\boldsymbol{c}\|$ large, but not without making a large relative change in some of the components of $\boldsymbol{c}$.

10. Verify 1.8.2 and 1.8.3 for an arbitrary 2 by 2 matrix A. (Hint: For the second formula, start with the cross-product formula

$$\|\boldsymbol{u} \times \boldsymbol{v}\|_2 = \|\boldsymbol{u}\|_2 \|\boldsymbol{v}\|_2 |sin(\theta)|,$$

where $\boldsymbol{u} = (a_{11}, a_{12}, 0)$ and $\boldsymbol{v} = (a_{21}, a_{22}, 0)$.)

11. As mentioned in Section 1.8, it has been shown that, when $A\boldsymbol{x} = \boldsymbol{b}$ is solved using Gaussian elimination with partial pivoting, the resulting approximate solution $\boldsymbol{x}_1$ satisfies $(A + \Delta A)\boldsymbol{x}_1 = \boldsymbol{b}$ exactly, for some small matrix ΔA whose norm is of the same order of magnitude as $\epsilon\|A\|$ (ϵ is the machine relative precision). Use this information to show that, if the elements of $A, \boldsymbol{b}$, and the exact solution $\boldsymbol{x}$ are all of moderate size (norms are $O(1)$, as for the Hilbert matrix example), the relative residual

$$\frac{\|A\boldsymbol{x}_1 - \boldsymbol{b}\|}{\|\boldsymbol{b}\|}$$

will be small, even if A is very ill-conditioned and $\boldsymbol{x}_1$ is a poor approximation to $\boldsymbol{x}$. For the case $N = 2$, give a geometric explanation for why the residual may be very small even though the computed solution is far from the true solution.

12. Verify the assertions made in the proof of Theorem 1.9.4 that, if A is a positive-definite matrix and D is its diagonal part, then $\bar{\boldsymbol{z}}^{\mathrm{T}} A\boldsymbol{z}$ and $\bar{\boldsymbol{z}}^{\mathrm{T}} D\boldsymbol{z}$ are real and positive, for any nonzero (possibly complex) $\boldsymbol{z}$. (Hint: Show that D has positive elements, and use the fact that $A = S^{\mathrm{T}} ES$, where E is a diagonal matrix containing the positive eigenvalues of A.)

13. Show that the coefficient matrix of the linear system 1.9.10 is positive-definite. (Hint: Given that the eigenvectors are

$$U_{i,j,k} = \sin\left(\frac{l\pi i}{M}\right) \sin\left(\frac{m\pi j}{M}\right) \sin\left(\frac{n\pi k}{M}\right), \qquad l, m, n = 1, \ldots, M-1,$$

find all the eigenvalues, and show that they are positive. You will need to use the trigonometric identity

$$\sin(a + b) = \sin(a)\cos(b) + \cos(a)\sin(b).)$$

14. For the Jacobi method, applied to the linear system 1.9.10, $B = D = 6I$. If the eigenvalues of the coefficient matrix A are $\lambda_{l,m,n}$, what are the eigenvalues of $I - B^{-1}A = I - A/6$? Using the eigenvalues of A as calculated in Problem 13, show that the largest eigenvalue, in absolute value, of $I - B^{-1}A$ is $1 - O(1/M^2)$. Since the factor by which the error decreases each iteration is (approximately) the absolute value of the largest eigenvalue of $I - B^{-1}A$, show that the number of iterations required to reduce the error by a factor of ϵ is $O(M^2)$.

15. Prove the formulas given for J_i^n in Section 1.9, for the 2 by 2 and 3 by 3 cases, by writing $J_i^n = (\lambda_i I + N)^n$, where N is the nilpotent matrix with ones on the first superdiagonal and zeros elsewhere.

16. a. Prove that the Gauss–Seidel iteration 1.9.5 converges when A is diagonal-dominant. (Hint: Show that

$$|(e_i)_{n+1}| \leq r_i \|\boldsymbol{e}_{n+1}\|_\infty + s_i \|\boldsymbol{e}_n\|_\infty,$$

where $r_i = \sum_{j<i} |a_{ij}|/|a_{ii}|$ and $s_i = \sum_{j>i} |a_{ij}|/|a_{ii}|$. Then let i be the index that maximizes $|(e_i)_{n+1}|$.)

b. Show that the matrix

$$A = \begin{bmatrix} 1 & 1 & 1 \\ 1 & 2 & 1 \\ 1 & 1 & 3 \end{bmatrix}.$$

is positive definite (but not diagonal dominant), by computing its eigenvalues. But show that the Jacobi iteration does not converge when used to solve $A\boldsymbol{x} = \boldsymbol{b}$, by computing the eigenvalues of $I - D^{-1}A$. Thus the Jacobi method is **not** guaranteed to converge when A is positive definite.

17. Nick Trefethen of Oxford University published a "100-dollar, 100-digit Challenge" set of problems in the *SIAM News* [Trefethen 2002], which consisted of ten numerical analysis problems. The answer to each was a single real number; the challenge was to compute it to ten significant digits. One of the problems was as follows: "Let A be the 20,000 × 20,000 matrix whose entries are zero everywhere except for the primes 2, 3, 5, 7,..., 224737 along the main diagonal and the number 1 in all the positions a_{ij} with $|i-j| = 1, 2, 4, 8, ..., 16384$. What is the (1,1) entry of A^{-1}?"

Since there are only about 15 nonzeros in each row, this is a very sparse matrix, so we don't want to actually calculate the full matrix A^{-1}, we want to solve $A\boldsymbol{x} = \boldsymbol{b}$ for a certain $\boldsymbol{b}$ (what $\boldsymbol{b}$?). Since A is not a band matrix, we shouldn't use a band solver. The matrix is not diagonal-dominant, but it is heavy along the diagonal, so an iterative method (or sparse direct method) seems indicated. Try solving this system using the Gauss–Seidel iterative method.

2

Linear Least Squares Problems

2.1 Introduction

The program DLINEQ in Figure 1.2.1 will only allow us to solve linear systems $A\boldsymbol{x} = \boldsymbol{b}$ where A is square (and then only if A is nonsingular), and probably most applications have the same number of equations as unknowns; however, sometimes there are more equations than unknowns (A has more rows than columns), in which case there usually is no solution, and sometimes there are more unknowns than equations (A has more columns than rows), in which case there are usually many solutions. Even if we have the same number of equations as unknowns, $A\boldsymbol{x} = \boldsymbol{b}$ may have many solutions or none. If there are many solutions, we may be satisfied to find any one of them and, if there are no solutions, we may be happy to find a vector $\boldsymbol{x}$ which *nearly* solves the system $A\boldsymbol{x} = \boldsymbol{b}$, in some sense. A standard linear system solver such as DLINEQ will be of no help in either case.

In this chapter we want to develop algorithms for solving the linear least squares problem

$$\text{minimize } \|A\boldsymbol{x} - \boldsymbol{b}\|_2 \tag{2.1.1}$$

Our algorithms will always return a solution, one of the (possibly many) vectors $\boldsymbol{x}$ which solve $A\boldsymbol{x} = \boldsymbol{b}$ with as little residual as possible, measured in the L_2-norm. (Algorithms for minimizing the residual in the L_∞- and L_1-norms are described in Problems 3 and 4 of Chapter 4.)

As an example application, consider the problem of fitting a polynomial $p(x) = a_1 + a_2x + a_3x^2 + \ldots + a_Nx^{N-1}$ of degree $N-1$ through M data points $(x_i, y_i), i = 1, \ldots, M$. The coefficients $a_1, \ldots, a_N$ must satisfy the M linear equations

$$\begin{bmatrix} 1 & x_1 & x_1^2 & \dots & x_1^{N-1} \\ 1 & x_2 & x_2^2 & \dots & x_2^{N-1} \\ \vdots & \vdots & \vdots & & \vdots \\ 1 & x_M & x_M^2 & \dots & x_M^{N-1} \end{bmatrix} \begin{bmatrix} a_1 \\ a_2 \\ a_3 \\ \vdots \\ a_N \end{bmatrix} = \begin{bmatrix} y_1 \\ y_2 \\ \vdots \\ y_M \end{bmatrix}.$$

If there are the same number of polynomial coefficients (N) as data points (M), this is a square system, and it can be shown that (provided that the x_i are all distinct) the matrix is nonsingular and so DLINEQ can be used to solve it, although DLINEQ may experience numerical difficulties, since the linear system is ill-conditioned. However, often we may want to pass a low-order polynomial through many data points ($M \gg N$). This is usually impossible to do, but we can determine the polynomial that fits the M data points as closely as possible, in the least squares sense, by solving 2.1.1.

Another example application is the ranking of college football teams. If we assign a point rating p_k to each team, we might ideally like to have $p_k - p_i = s_{ki}$ if the point spread was s_{ki} when team k and team i played (s_{ki} is positive if team k won). Since the "transitive law of football scores" does not always hold (often team 1 will beat team 2 and team 2 will beat team 3, but then team 1 will lose to team 3), such an assignment is impossible. However, if we can settle for a point assignment that comes as close as possible (in the least squares sense) to satisfying the ideal equations $p_k - p_i = s_{ki}$, we can calculate ratings for each team by solving a linear least squares problem. (I actually did this for a number of years, and the results were quite reasonable.) Note that the solution to this least squares problem cannot be unique, as we can add any constant to all ratings without changing their relative positions.

The following theorem suggests that 2.1.1 is very easy to solve. We shall see subsequently that things are not quite as simple as this theorem might seem to imply.

Theorem 2.1.1.

(a) $\boldsymbol{x}$ *is a solution to the linear least squares problem*

$$\textit{minimize } \|A\boldsymbol{x} - \boldsymbol{b}\|_2$$

if and only if it satisfies $A^{\mathrm{T}}A\boldsymbol{x} = A^{\mathrm{T}}\boldsymbol{b}$.

(b) $A^{\mathrm{T}}A\boldsymbol{x} = A^{\mathrm{T}}\boldsymbol{b}$ *always has solutions, and it will have a unique solution if the columns of* A *are linearly independent.*

Proof:

(a) Suppose $A^{\mathrm{T}}A\boldsymbol{x} = A^{\mathrm{T}}\boldsymbol{b}$, where A is an M by N matrix. Then, if $\boldsymbol{e}$ is an arbitrary N-vector,

$$\begin{aligned}\|A(\boldsymbol{x}+\boldsymbol{e})-\boldsymbol{b}\|_2^2 &= (A\boldsymbol{x}-\boldsymbol{b}+A\boldsymbol{e})^{\mathrm{T}}(A\boldsymbol{x}-\boldsymbol{b}+A\boldsymbol{e})\\ &= (A\boldsymbol{x}-\boldsymbol{b})^{\mathrm{T}}(A\boldsymbol{x}-\boldsymbol{b})+2(A\boldsymbol{e})^{\mathrm{T}}(A\boldsymbol{x}-\boldsymbol{b})+(A\boldsymbol{e})^{\mathrm{T}}(A\boldsymbol{e})\\ &= \|A\boldsymbol{x}-\boldsymbol{b}\|_2^2+\|A\boldsymbol{e}\|_2^2+2\boldsymbol{e}^{\mathrm{T}}(A^{\mathrm{T}}A\boldsymbol{x}-A^{\mathrm{T}}\boldsymbol{b})\\ &= \|A\boldsymbol{x}-\boldsymbol{b}\|_2^2+\|A\boldsymbol{e}\|_2^2 \geq \|A\boldsymbol{x}-\boldsymbol{b}\|_2^2.\end{aligned}$$

Since $\boldsymbol{e}$ is arbitrary, this shows that $\boldsymbol{x}$ minimizes the 2-norm of $A\boldsymbol{y}-\boldsymbol{b}$ over all N-vectors $\boldsymbol{y}$. The converse is established in Problem 1.

(b) First, note that $A^{\mathrm{T}}A$ is a square N by N matrix. Now, if $\boldsymbol{v}$ is an arbitrary vector perpendicular to the range of $A^{\mathrm{T}}A$, then $0 = \boldsymbol{v}^{\mathrm{T}}(A^{\mathrm{T}}A\boldsymbol{v}) = (A\boldsymbol{v})^{\mathrm{T}}(A\boldsymbol{v}) = \|A\boldsymbol{v}\|_2^2$, and $A\boldsymbol{v} = \boldsymbol{0}$. If the columns of A are linearly independent, $A\boldsymbol{v} = \boldsymbol{0}$ implies that $\boldsymbol{v} = \boldsymbol{0}$, and so the only vector perpendicular to the range space of $A^{\mathrm{T}}A$ is the zero vector, which means that the range is all of R^N, and thus $A^{\mathrm{T}}A$ is a nonsingular (square) matrix. In this case, $A^{\mathrm{T}}A\boldsymbol{x} = A^{\mathrm{T}}\boldsymbol{b}$ has a unique solution.

On the other hand, if the columns of A are not linearly independent, we still have $A\boldsymbol{v} = \boldsymbol{0}$ for any vector $\boldsymbol{v}$ perpendicular to the range space of $A^{\mathrm{T}}A$, and thus $0 = (A\boldsymbol{v})^{\mathrm{T}}\boldsymbol{b} = \boldsymbol{v}^{\mathrm{T}}(A^{\mathrm{T}}\boldsymbol{b})$; however, this means that $A^{\mathrm{T}}\boldsymbol{b}$ is perpendicular to every vector $\boldsymbol{v}$ normal to the range space of $A^{\mathrm{T}}A$, and thus $A^{\mathrm{T}}\boldsymbol{b}$ lies in the range of $A^{\mathrm{T}}A$. Therefore, even if $A^{\mathrm{T}}A$ is a singular matrix, $A^{\mathrm{T}}A\boldsymbol{x} = A^{\mathrm{T}}\boldsymbol{b}$ still has solutions. ■

It appears from Theorem 2.1.1 that all one has to do to find the least squares solution to $A\boldsymbol{x} = \boldsymbol{b}$ is simply to multiply both sides by A^{T} and to solve the resulting square N by N system, called the "normal equations". Indeed, this approach is probably satisfactory in the majority of cases. However, there are two potential problems.

First, if the columns of A are not linearly independent, the matrix $A^{\mathrm{T}}A$ is singular and, even though $A^{\mathrm{T}}A\boldsymbol{x} = A^{\mathrm{T}}\boldsymbol{b}$ will still have solutions, a standard linear equation solver such as DLINEQ will detect the singularity and stop without returning any solution. Second, even if $A^{\mathrm{T}}A$ is nonsingular, the normal equations will be more poorly conditioned than the original system $A\boldsymbol{x} = \boldsymbol{b}$. To appreciate this, assume first that A is square ($M = N$) and nonsingular, and let us compare the condition numbers of A and $A^{\mathrm{T}}A$. Recall from Section 1.8 that the condition number is defined by $c(B) = \|B\|\|B^{-1}\|$. If we use 2-norms in this definition,

$$\begin{aligned}c(A) &= \|A\|_2\|A^{-1}\|_2,\\ c(A^{\mathrm{T}}A) &= \|A^{\mathrm{T}}A\|_2\|(A^{\mathrm{T}}A)^{-1}\|_2.\end{aligned}$$

Recall also from Section 0.3 that the 2-norm of a general square matrix A is the square root of the maximum eigenvalue of the positive-semidefinite matrix $A^{\mathrm{T}}A$, and the 2-norm of a square symmetric matrix (eg, A^TA or its inverse) is the absolute value of its dominant eigenvalue. Thus

$$c(A) = [\lambda_{\max}(A^{\mathrm{T}}A)\lambda_{\max}((A^{\mathrm{T}}A)^{-1})]^{1/2} \tag{2.1.2}$$

$$c(A^{\mathrm{T}}A) = \lambda_{\max}(A^{\mathrm{T}}A)\lambda_{\max}((A^{\mathrm{T}}A)^{-1}). \tag{2.1.3}$$

In 2.1.2 we have used the fact that $A^{-\mathrm{T}}A^{-1}$ and $(A^{\mathrm{T}}A)^{-1}$ have the same eigenvalues (they are similar); so by comparing 2.1.2 and 2.1.3 we see that

$$c(A^{\mathrm{T}}A) = [c(A)]^2. \tag{2.1.4}$$

The condition number of a matrix is always greater than or equal to one, and so 2.1.4 shows that the condition of $A^{\mathrm{T}}A$ is always worse than that of A. If $c(A) = 10^m$, $c(A^{\mathrm{T}}A) = 10^{2m}$, and according to the discussion in Section 1.8 we expect to lose about m significant decimal digits when $A\boldsymbol{x} = \boldsymbol{b}$ is solved, and around $2m$ digits when $A^{\mathrm{T}}A\boldsymbol{x} = A^{\mathrm{T}}\boldsymbol{b}$ is solved.

Although 2.1.4 was derived assuming A to be square, if $M > N$ but A^TA is still nonsingular, 2.1.2 can still be used to define the condition number of A, so 2.1.4 still holds, and solving the normal equations still amplifies roundoff error unnecessarily, compared with the methods of the next two sections, which attack the problem $A\boldsymbol{x} \approx \boldsymbol{b}$ directly.

2.2 Orthogonal Reduction

The basic strategy behind nearly all the algorithms outlined in this book is this: We find a transformation that preserves the property that we are interested in, and then we transform the matrix or system into a form in which that property is obvious. If we could transform A into "row echelon" form, the linear least squares solution of $A\boldsymbol{x} = \boldsymbol{b}$ would then be obvious. Row echelon form means the first nonzero entry in each row (called a "nonzero pivot") is farther to the right than the first nonzero entry in the previous row. An example of a linear system whose matrix is in row echelon form is shown in Figure 2.2.1.

Note the characteristic "stair-step" nonzero structure. The least squares problem is to minimize $\boldsymbol{r}_1^2 + \ldots + \boldsymbol{r}_M^2$, where $\boldsymbol{r} \equiv A\boldsymbol{x} - \boldsymbol{b}$ is the residual vector. In the above example, the last eight components of the residual vector are

$$\boldsymbol{r}_i = 0x_1 + \ldots + 0x_{12} - b_i \qquad (i = 7, \ldots, 14).$$

Now these last components are going to be $(-b_7, \ldots, -b_{14})$ no matter how we choose the unknowns; we have no control over these components. On the other hand, we can choose $x_1, \ldots, x_{12}$ so that the first six equations are satisfied exactly, and so $r_1 = \ldots = r_6 = 0$. Clearly, then, the residual is

$$
\begin{bmatrix}
0 & P & X & X & X & X & X & X & X & X & X & X \\
0 & 0 & P & X & X & X & X & X & X & X & X & X \\
0 & 0 & 0 & P & X & X & X & X & X & X & X & X \\
0 & 0 & 0 & 0 & 0 & P & X & X & X & X & X & X \\
0 & 0 & 0 & 0 & 0 & 0 & P & X & X & X & X & X \\
0 & 0 & 0 & 0 & 0 & 0 & 0 & 0 & 0 & 0 & P & X \\
\cdots & & & & & & & & & & & \cdots \\
0 & 0 & 0 & 0 & 0 & 0 & 0 & 0 & 0 & 0 & 0 & 0 \\
0 & 0 & 0 & 0 & 0 & 0 & 0 & 0 & 0 & 0 & 0 & 0 \\
0 & 0 & 0 & 0 & 0 & 0 & 0 & 0 & 0 & 0 & 0 & 0 \\
0 & 0 & 0 & 0 & 0 & 0 & 0 & 0 & 0 & 0 & 0 & 0 \\
0 & 0 & 0 & 0 & 0 & 0 & 0 & 0 & 0 & 0 & 0 & 0 \\
0 & 0 & 0 & 0 & 0 & 0 & 0 & 0 & 0 & 0 & 0 & 0 \\
0 & 0 & 0 & 0 & 0 & 0 & 0 & 0 & 0 & 0 & 0 & 0 \\
0 & 0 & 0 & 0 & 0 & 0 & 0 & 0 & 0 & 0 & 0 & 0
\end{bmatrix}
\begin{bmatrix} x_1 \\ x_2 \\ x_3 \\ x_4 \\ x_5 \\ x_6 \\ x_7 \\ x_8 \\ x_9 \\ x_{10} \\ x_{11} \\ x_{12} \end{bmatrix}
=
\begin{bmatrix} b_1 \\ b_2 \\ b_3 \\ b_4 \\ b_5 \\ b_6 \\ \cdot\cdot \\ b_7 \\ b_8 \\ b_9 \\ b_{10} \\ b_{11} \\ b_{12} \\ b_{13} \\ b_{14} \end{bmatrix}
$$

Figure 2.2.1 Row Echelon Form
P, nonzero pivot; X, not necessarily zero.

minimized (in the 2-norm or in any other norm) by choosing the unknowns so that the first six equations are satisfied exactly. In the present example, there are many ways to do this; we can assign arbitrary values to the "non-pivot" unknowns $x_1, x_5, x_8, x_9, x_{10}$, and x_{12} and solve for the "pivot" variables $x_{11}, x_7, x_6, x_4, x_3, x_2$ in reverse order. (A variable is a pivot variable if there is a pivot in the corresponding column.) Thus the least squares solution is not unique, for this example. A matrix in row echelon form is always upper triangular; if the diagonal elements are all nonzero (and assuming $M \geq N$), then all the variables are pivot variables and the least squares solution is unique. In this case, there are no variables that need to be assigned arbitrary values, and the least squares solution is found by ordinary back substitution.

Most introductory texts on matrix algebra explain how to reduce an arbitrary rectangular matrix to row echelon form using elementary row operations. (Their objective is to determine whether there are solutions or not, and to find the general solution if there are solutions.) We can therefore transform an arbitrary linear system $A\boldsymbol{x} = \boldsymbol{b}$ to its row echelon form by premultiplying both sides by the M_{ij} matrices introduced in Section 1.4. The only problem is that this type of transformation does not preserve the property we are interested in; there is no reason why the least squares solution of the row echelon system (which is obvious) should be the same as the least squares solution of the original system $A\boldsymbol{x} = \boldsymbol{b}$.

However, if we can reduce $A\boldsymbol{x} = \boldsymbol{b}$ to row echelon form by multiplying both sides by *orthogonal* matrices, the least squares solution of the reduced problem *will* be the same as the least squares solution of the original problem.

This is because, if Q is orthogonal ($Q^{\mathrm{T}}Q = I$), then

$$\begin{aligned}\|Q A\boldsymbol{x} - Q\boldsymbol{b}\|_2^2 &= (QA\boldsymbol{x} - Q\boldsymbol{b})^{\mathrm{T}}(QA\boldsymbol{x} - Q\boldsymbol{b}) = (A\boldsymbol{x} - \boldsymbol{b})^{\mathrm{T}}Q^{\mathrm{T}}Q(A\boldsymbol{x} - \boldsymbol{b}) \\ &= (A\boldsymbol{x} - \boldsymbol{b})^{\mathrm{T}}(A\boldsymbol{x} - \boldsymbol{b}) = \|A\boldsymbol{x} - \boldsymbol{b}\|_2^2\end{aligned}$$

and so a vector x that minimizes $\|QA\boldsymbol{x} - Q\boldsymbol{b}\|_2$ also minimizes $\|A\boldsymbol{x} - \boldsymbol{b}\|_2$.

The simplest useful orthogonal matrices are the "Givens rotation" matrices Q_{ij}, which have the form (j is always taken to be greater than i)

$$Q_{ij} = \begin{array}{c} \\ \\ \\ \text{row } i \\ \\ \\ \text{row } j \\ \\ \\ \\ \end{array} \left[\begin{array}{ccccccccc} 1 & & & & & & & & \\ & 1 & & & & & & & \\ & & 1 & & & & & & \\ & & & c & & & -s & & \\ & & & & 1 & & & & \\ & & & & & 1 & & & \\ & & & s & & & c & & \\ & & & & & & & 1 & \\ & & & & & & & & 1 \\ & & & & & & & & & 1 \end{array}\right] \quad (c^2 + s^2 = 1)$$

(column i is the column containing the upper c and s; column j is the column containing $-s$ and the lower c.)

Since we shall always require that $c^2 + s^2 = 1$, c and s represent the cosine and sine of some angle θ, but we shall not have any use for the angle θ itself. It is easy to verify that $Q_{ij}^{\mathrm{T}}Q_{ij} = I$, and so Q_{ij} and Q_{ij}^{T} are both orthogonal.

Let us see what happens to the matrix A when it is premultiplied by a (transposed) Givens matrix $B = Q_{ij}^{\mathrm{T}}A$:

$$\begin{bmatrix} \vdots & & \vdots & & \vdots \\ b_{i1} & \dots & b_{ik} & \dots & b_{iN} \\ \vdots & & \vdots & & \vdots \\ b_{j1} & \dots & b_{jk} & \dots & b_{jN} \\ \vdots & & \vdots & & \vdots \end{bmatrix} = \begin{bmatrix} 1 & & & & & & & & & \\ & 1 & & & & & & & & \\ & & 1 & & & & & & & \\ & & & c & & & s & & & \\ & & & & 1 & & & & & \\ & & & & & 1 & & & & \\ & & & -s & & & c & & & \\ & & & & & & & 1 & & \\ & & & & & & & & 1 & \\ & & & & & & & & & 1 \end{bmatrix} \begin{bmatrix} \vdots & & \vdots & & \vdots \\ a_{i1} & \dots & a_{ik} & \dots & a_{iN} \\ \vdots & & \vdots & & \vdots \\ a_{j1} & \dots & a_{jk} & \dots & a_{jN} \\ \vdots & & \vdots & & \vdots \end{bmatrix}$$

We see that only rows i and j change and that

$$\begin{aligned} b_{ik} &= ca_{ik} + sa_{jk}, \\ b_{jk} &= -sa_{ik} + ca_{jk}, \end{aligned} \tag{2.2.1}$$

for any $k = 1, \dots, N$.

Premultiplication by M_{ij} has the effect of adding a multiple of row i to row j; premultiplication by Q_{ij}^{T} has the effect of replacing row i by a linear combination of the old rows i and j, and replacing row j by another linear combination of old rows i and j. If we choose (for any l)

$$\begin{aligned} c &= \frac{a_{il}}{(a_{il}^2 + a_{jl}^2)^{1/2}}, \\ s &= \frac{a_{jl}}{(a_{il}^2 + a_{jl}^2)^{1/2}}, \end{aligned} \tag{2.2.2}$$

we can make b_{jl} equal to zero (note that $c^2 + s^2 = 1$, as required). Thus we can use "pivots" (a_{il} here) to knock out lower elements (a_{jl}) in the pivot column, in much the same way as before; only now the pivot element (and entire pivot row) changes also. Note that, if the pivot a_{il} is zero, that is not a problem. The denominators in 2.2.2 will be zero only if *both* the pivot and the element to be eliminated are zero but, if the element to be zeroed is already zero, no transformation is needed. In fact, if the pivot a_{il} is zero and the element a_{jl} to be eliminated is not, note that c will be 0 and s will be ± 1; so the effect of the Givens rotation will be essentially a row switch!

So we can reduce an arbitrary rectangular matrix to row echelon form using orthogonal transformations, in much the same way as is commonly done using elementary row operations (i.e., premultiplying with M_{ij} matrices). First we use a_{11} to knock out all elements below it. Then we check to see whether a_{11} is zero (it will be zero only if all elements in the first column were zero to start with—see Problem 7); if it is zero, we move on to the next entry (a_{12}) in the first row and use it to knock out everything below it. If a_{12} is zero after the second column has been zeroed, we move on to a_{13}, and so on, until we find a nonzero pivot (identified by P in Figure 2.2.1) in the first row. Then we move down to the second row and continue this process in that row. Note that, since we eliminate elements column by column, starting with the first column, by the time we use a_{il} as a pivot, all elements in rows i to M of the previous columns 1 to $l-1$ are already zero. Thus, while we are using a_{il} to knock out lower elements a_{jl}, we are just adding multiples of zero to zero in columns 1 to $l-1$ (we can skip these calculations, in fact); so the elements in the previous columns which were zeroed earlier remain zero, just as in the Gaussian elimination algorithm. Naturally, when we premultiply A by Q_{ij}^{T}, we also premultiply the right-hand side $\boldsymbol{b}$ by this matrix.

Continuing this process (which is familiar to anyone who has reduced a matrix to row echelon form using Gaussian elimination), we eventually reach a form like that shown in Figure 2.2.1, from which the least squares solution can easily be extracted.

Figure 2.2.2 displays a FORTRAN program that solves the linear least squares problem 2.1.1 using the algorithm described above. Using Givens rotations (equations 2.2.1 and 2.2.2), it reduces an arbitrary M by N linear system $A\boldsymbol{x} = \boldsymbol{b}$ to its row echelon form. Then it computes the least squares solution to the reduced system, by ignoring the last equations, corresponding to zero rows, and solving the other equations by back substitution, with nonpivot variables arbitrarily assigned a value of zero. Thus, whether there are more equations than unknowns, or more unknowns than equations, and whether our linear system has no solution or many solutions (or a unique solution), DLLSQR will return a solution that will be the best (in the least squares sense), or one of the best.

```
      SUBROUTINE DLLSQR(A,M,N,X,B)
      IMPLICIT DOUBLE PRECISION (A-H,O-Z)
C                                   DECLARATIONS FOR ARGUMENTS
      DOUBLE PRECISION A(M,N),X(N),B(M)
      INTEGER M,N
C                                   DECLARATIONS FOR LOCAL VARIABLES
      DOUBLE PRECISION B_(M)
      INTEGER PIVOT(M)
C
C  SUBROUTINE DLLSQR SOLVES THE LINEAR LEAST SQUARES PROBLEM
C
C         MINIMIZE  2-NORM OF (A*X-B)
C
C
C  ARGUMENTS
C
C             ON INPUT                          ON OUTPUT
C             --------                          ---------
C
C    A      - THE M BY N MATRIX.                DESTROYED.
C
C    M      - THE NUMBER OF ROWS IN A.
C
C    N      - THE NUMBER OF COLUMNS IN A.
C
C    X      -                                   AN N-VECTOR CONTAINING
C                                               THE LEAST SQUARES
C                                               SOLUTION.
C
C    B      - THE RIGHT HAND SIDE M-VECTOR.
C
C-----------------------------------------------------------------------
C                                   EPS = MACHINE FLOATING POINT RELATIVE
C                                         PRECISION
C *****************************
      DATA EPS/2.D-16/
C *****************************
C                                   AMAX = MAXIMUM ELEMENT OF A
      AMAX = 0.0
      DO 10 I=1,M
C                                   COPY B TO B_, SO B WILL NOT BE ALTERED
         B_(I) = B(I)
         DO 5 J=1,N
```

```
            AMAX = MAX(AMAX,ABS(A(I,J)))
    5    CONTINUE
   10 CONTINUE
      ERRLIM = 1000*EPS*AMAX
C                                REDUCTION TO ROW ECHELON FORM
      CALL REDQ(A,M,N,B_,PIVOT,NPIVOT,ERRLIM)
C                                CAUTION USER IF SOLUTION NOT UNIQUE.
      IF (NPIVOT.NE.N) THEN
         PRINT 15
   15    FORMAT (' NOTE: SOLUTION IS NOT UNIQUE ')
      ENDIF
C                                ASSIGN VALUE OF ZERO TO NON-PIVOT
C                                VARIABLES.
      DO 20 K=1,N
         X(K) = 0.0
   20 CONTINUE
C                                SOLVE FOR PIVOT VARIABLES USING BACK
C                                SUBSTITUTION.
      DO 30 I=NPIVOT,1,-1
         L = PIVOT(I)
         SUM = 0.0
         DO 25 K=L+1,N
            SUM = SUM + A(I,K)*X(K)
   25    CONTINUE
         X(L) = (B_(I)-SUM)/A(I,L)
   30 CONTINUE
      RETURN
      END

      SUBROUTINE REDQ(A,M,N,B,PIVOT,NPIVOT,ERRLIM)
      IMPLICIT DOUBLE PRECISION (A-H,O-Z)
C                                DECLARATIONS FOR ARGUMENTS
      DOUBLE PRECISION A(M,N),B(M),ERRLIM
      INTEGER PIVOT(M),M,N,NPIVOT
C                                USE GIVENS ROTATIONS TO REDUCE A
C                                TO ROW ECHELON FORM
      I = 1
      DO 15 L=1,N
C                                USE PIVOT A(I,L) TO KNOCK OUT ELEMENTS
C                                I+1 TO M IN COLUMN L.
         DO 10 J=I+1,M
            IF (A(J,L).EQ.0.0) GO TO 10
            DEN = SQRT(A(I,L)**2+A(J,L)**2)
            C = A(I,L)/DEN
            S = A(J,L)/DEN
C                                PREMULTIPLY A BY Qij**T
            DO 5 K=L,N
               BIK = C*A(I,K) + S*A(J,K)
```

```
                 BJK =-S*A(I,K) + C*A(J,K)
                 A(I,K) = BIK
                 A(J,K) = BJK
    5          CONTINUE
C                                  PREMULTIPLY B BY Qij**T
               BI = C*B(I) + S*B(J)
               BJ =-S*B(I) + C*B(J)
               B(I) = BI
               B(J) = BJ
   10       CONTINUE
C                                  PIVOT A(I,L) IS NONZERO AFTER PROCESSING
C                                  COLUMN L--MOVE DOWN TO NEXT ROW, I+1
            IF (ABS(A(I,L)).LE.ERRLIM) A(I,L) = 0.0
            IF (A(I,L).NE.0.0) THEN
               NPIVOT = I
               PIVOT(NPIVOT) = L
               I = I+1
               IF (I.GT.M) RETURN
            ENDIF
   15 CONTINUE
      RETURN
      END
```

Figure 2.2.2

In fact, the only advantage that DLINEQ has over DLLSQR for square systems is that it is four times faster, because DLINEQ requires $N-i$ multiplications to knock out a_{ji} (loop 25 of DLINEQ in Figure 1.2.1), while DLLSQR has to do $4(N-i+1)$ multiplications (loop 5 of subroutine REDQ in Figure 2.2.2) to knock out a_{ji} (for nonsingular square systems, $i=l$). Since DLINEQ does $\frac{1}{3}N^3$ multiplications, DLLSQR requires $\frac{4}{3}N^3$ multiplications, for a nonsingular square system. DLLSQR also has to compute $O(N^2)$ square roots, while DLINEQ has no square roots to calculate.

If A happens to be a band matrix, orthogonal reduction has the same effect on the band structure as Gaussian elimination with partial pivoting (see Section 1.5), namely, the elements below the N_{LD} nonzero subdiagonals remain zero, but the number of nonzero superdiagonals increases from N_{UD} to $N_{UD}+N_{LD}$. (We are assuming here that the diagonal elements in the final upper triangular matrix are all nonzero.)

We have seen how to reduce A to its row echelon form, which we shall call R (for "right" triangular, which is the same as upper triangular), by premultiplying it by a series of orthogonal Givens matrices. In other words,

$$Q_L^{\mathrm{T}} Q_{L-1}^{\mathrm{T}} \cdots Q_2^{\mathrm{T}} Q_1^{\mathrm{T}} A = R,$$

where each Q_k represents one of the Givens Q_{ij} matrices. Once we have done this, we have found what is called the QR decomposition of A, since

$A = QR$, where $Q = Q_1Q_2 \dots Q_{L-1}Q_L$. Since the product of orthogonal matrices is orthogonal ($(Q_1Q_2)^{\mathrm{T}}(Q_1Q_2) = Q_2^{\mathrm{T}}Q_1^{\mathrm{T}}Q_1Q_2 = I$, if $Q_1^{\mathrm{T}}Q_1 = I$ and $Q_2^{\mathrm{T}}Q_2 = I$) Q is orthogonal, and R is upper triangular (or right triangular).

The QR decomposition plays a role similar to that played by the LU decomposition. One does not need to form the LU decomposition explicitly in order to solve a square nonsingular system $A\boldsymbol{x} = \boldsymbol{b}$, but it is obtained at no extra cost when one such linear system is solved and, if another system $A\boldsymbol{x} = \boldsymbol{c}$ with the same coefficient matrix is encountered subsequently, it can be solved quickly by solving the triangular systems $L\boldsymbol{y} = \boldsymbol{c}$ and $U\boldsymbol{x} = \boldsymbol{y}$. In a similar manner, we do not need to form the QR decomposition of a rectangular matrix A explicitly in order to solve the problem (minimize $\|A\boldsymbol{x} - \boldsymbol{b}\|_2$) but, once we have it, another problem (minimize $\|A\boldsymbol{x} - \boldsymbol{c}\|_2$) can be solved quickly by first solving $Q\boldsymbol{y} = \boldsymbol{c}$ ($\boldsymbol{y} = Q^{\mathrm{T}}\boldsymbol{c}$) and then finding the least squares solution to $R\boldsymbol{x} = \boldsymbol{y}$ (found by back substitution, since R is in row echelon form).

The computer program DLLSQR does not explicitly form the QR decomposition of A, but it could easily be modified to do so (Problem 8). We could start with $Q = I$, and every time that A is multiplied by a Q_{ij}^{T}, postmultiply Q by Q_{ij}.

2.3 Reduction Using Householder Transformations

A rectangular matrix can alternatively be reduced to row echelon form using the orthogonal "Householder" matrices, which have the form $H = I - 2\boldsymbol{\omega}\boldsymbol{\omega}^{\mathrm{T}}$, where $\boldsymbol{\omega}$ is a *unit* M-vector. First note that H is symmetric, since $H^{\mathrm{T}} = (I - 2\boldsymbol{\omega}\boldsymbol{\omega}^{\mathrm{T}})^{\mathrm{T}} = I - 2\boldsymbol{\omega}\boldsymbol{\omega}^{\mathrm{T}} = H$. Then, to verify that the M by M matrix H is orthogonal, for any unit vector $\boldsymbol{\omega}$, note that

$$\begin{aligned} H^{\mathrm{T}}H = HH &= (I - 2\boldsymbol{\omega}\boldsymbol{\omega}^{\mathrm{T}})(I - 2\boldsymbol{\omega}\boldsymbol{\omega}^{\mathrm{T}}) = I - 4\boldsymbol{\omega}\boldsymbol{\omega}^{\mathrm{T}} + 4\boldsymbol{\omega}(\boldsymbol{\omega}^{\mathrm{T}}\boldsymbol{\omega})\boldsymbol{\omega}^{\mathrm{T}} \\ &= I - 4\boldsymbol{\omega}\boldsymbol{\omega}^{\mathrm{T}} + 4\boldsymbol{\omega}(1)\boldsymbol{\omega}^{\mathrm{T}} = I. \end{aligned}$$

The M-vector $\boldsymbol{\omega}$ will be chosen so that $H_i \equiv I - 2\boldsymbol{\omega}\boldsymbol{\omega}^{\mathrm{T}}$ will zero *all* elements below a given pivot a_{il}, when it is used to premultiply an M by N matrix A. To this end, we first stipulate that the first $i - 1$ components of $\boldsymbol{\omega}$ must be zero, so that H_i has the nonzero structure

$$H_i = \begin{array}{r} \\ \\ \\ \\ \\ \text{row } i \\ \\ \\ \text{row } M \end{array} \left[\begin{array}{ccccccccc} 1 &&&&&&&& \\ & 1 &&&&&&& \\ && 1 &&&&&& \\ &&& 1 &&&&& \\ &&&& 1 &&&& \\ &&&&& X & X & X & X \\ &&&&& X & X & X & X \\ &&&&& X & X & X & X \\ &&&&& X & X & X & X \end{array} \right] .$$

(with "column i" labelling the first column of X entries and "column M" the last)

Clearly, premultiplication by H_i will change only rows i to M, no matter how we choose components i to M of $\boldsymbol{\omega}$. However, we want to choose $\boldsymbol{\omega}$ so that, when A is premultiplied by H_i, components $i+1$ to M of column l (i.e., everything below the pivot a_{il}) all become zero. Now, if $\boldsymbol{a}_l$ denotes column l of A, then

$$H_i \boldsymbol{a}_l = (I - 2\boldsymbol{\omega}\boldsymbol{\omega}^{\mathrm{T}})\boldsymbol{a}_l = \boldsymbol{a}_l - 2\alpha\boldsymbol{\omega},$$

where $\alpha \equiv \boldsymbol{\omega}^{\mathrm{T}}\boldsymbol{a}_l$. This means that we must set components $i+1$ to M of $\boldsymbol{\omega}$ equal to $l/2\alpha$ times the corresponding components of $\boldsymbol{a}_l$. If the components of $\boldsymbol{\omega}$ are denoted by ω_j and the components of $\boldsymbol{a}_l$ by a_j, we have from the definition of α (recall that the first $i-1$ components of $\boldsymbol{\omega}$ are zero):

$$\omega_i a_i + \omega_{i+1} a_{i+1} + \ldots + \omega_M a_M = \alpha, \tag{2.3.1}$$

but, since $\boldsymbol{\omega}$ must be a unit vector, we also have

$$\omega_i^2 + \omega_{i+1}^2 + \ldots + \omega_M^2 = 1. \tag{2.3.2}$$

Let us define s and β by the equations

$$\begin{aligned} s &= a_i^2 + \ldots + a_M^2, \\ \omega_i &= \frac{a_i + \beta}{2\alpha}. \end{aligned} \tag{2.3.3a}$$

Recall also that

$$\begin{aligned} \omega_j &= 0, \qquad j = 1, \ldots, i-1, \\ \omega_j &= \frac{a_j}{2\alpha}, \qquad j = i+1, \ldots, M. \end{aligned} \tag{2.3.3b}$$

Using 2.3.3a and 2.3.3b to replace the components of $\boldsymbol{\omega}$ in equations 2.3.1 and 2.3.2 gives

$$\begin{aligned} a_i\beta + s &= 2\alpha^2, \\ 2a_i\beta + \beta^2 + s &= 4\alpha^2. \end{aligned} \tag{2.3.4}$$

Now the two equations 2.3.4 can be solved for the two unknowns α and β, and the result is

$$
\begin{aligned}
\beta &= \pm(s)^{1/2}, \\
2\alpha &= [2\beta(a_i+\beta)]^{1/2}.
\end{aligned}
\tag{2.3.5}
$$

If a_{i+1} to a_M are almost zero already, $s \approx a_i^2$ and β will be nearly $\pm|a_i|$. We shall take β to have the same sign as a_i, otherwise $a_i + \beta \approx 0, \alpha \approx 0$, and all components of $\boldsymbol{\omega}$ will be approximately of the (numerically dangerous) indeterminate form 0/0. Notice that $\alpha = 0$ only if $s = 0$, that is, if $a_i = a_{i+1} = ... = a_M = 0$, in which case no transformation is necessary.

Now we can use $H_i = I - 2\boldsymbol{\omega}\boldsymbol{\omega}^{\mathrm{T}}$ to eliminate everything directly below the pivot a_{il}. Before premultiplying by H_i, A will look something like this (cf. Figure 2.2.1):

$$
\begin{array}{r}
\\ \\ \\ \text{row } i \\ \\ \\ \\ \\ \\ \\ \\ \\ \\ \text{row } M
\end{array}
\begin{array}{c}
\begin{array}{cccccccccccc}
 & & & & & \text{column} & & & & & & \text{column} \\
 & & & & & l & & & & & & N
\end{array} \\
\left[\begin{array}{cccccccccccc}
0 & P & a & a & a & a & a & a & a & a & a & a \\
0 & 0 & P & a & a & a & a & a & a & a & a & a \\
0 & 0 & 0 & P & a & a & a & a & a & a & a & a \\
0 & 0 & 0 & 0 & 0 & x & x & x & x & x & x & x \\
0 & 0 & 0 & 0 & 0 & Z & x & x & x & x & x & x \\
0 & 0 & 0 & 0 & 0 & Z & x & x & x & x & x & x \\
0 & 0 & 0 & 0 & 0 & Z & x & x & x & x & x & x \\
0 & 0 & 0 & 0 & 0 & Z & x & x & x & x & x & x \\
0 & 0 & 0 & 0 & 0 & Z & x & x & x & x & x & x \\
0 & 0 & 0 & 0 & 0 & Z & x & x & x & x & x & x \\
0 & 0 & 0 & 0 & 0 & Z & x & x & x & x & x & x \\
0 & 0 & 0 & 0 & 0 & Z & x & x & x & x & x & x \\
0 & 0 & 0 & 0 & 0 & Z & x & x & x & x & x & x \\
0 & 0 & 0 & 0 & 0 & Z & x & x & x & x & x & x
\end{array}\right]
\end{array}.
$$

After premultiplication, the elements in column l marked Z will be zero, by design, and those marked x may be changed. The previous columns $\boldsymbol{a}_j, j = 1, \ldots, l-1$, will be unchanged because $\boldsymbol{\omega}^{\mathrm{T}}\boldsymbol{a}_j = 0$ (components 1 to $i-1$ of $\boldsymbol{\omega}$ are zero, while components i to M of $\boldsymbol{a}_j$ are zero); so $H_i\boldsymbol{a}_j = (I - 2\boldsymbol{\omega}\boldsymbol{\omega}^{\mathrm{T}})\boldsymbol{a}_j = \boldsymbol{a}_j$. Thus the zeros introduced by the previous transformations will not be destroyed by H_i.

To use Householder transformations in place of Givens rotations in the computer program in Figure 2.2.2, replace subroutine REDQ by the subroutines REDH and CALW in Figure 2.3.1.

```
      SUBROUTINE REDH(A,M,N,B,PIVOT,NPIVOT,ERRLIM)
      IMPLICIT DOUBLE PRECISION (A-H,O-Z)
C                                 DECLARATIONS FOR ARGUMENTS
      DOUBLE PRECISION A(M,N),B(M),ERRLIM
      INTEGER PIVOT(M),M,N,NPIVOT
C                                 DECLARATIONS FOR LOCAL VARIABLES
      DOUBLE PRECISION W(M)
```

```
C                                  USE HOUSEHOLDER TRANSFORMATIONS TO
C                                  REDUCE A TO ROW ECHELON FORM
      I = 1
      DO 30 L=1,N
C                                  USE PIVOT A(I,L) TO KNOCK OUT ELEMENTS
C                                  I+1 TO M IN COLUMN L.
         IF (I+1.LE.M) THEN
C                                  CHOOSE UNIT M-VECTOR W (WHOSE FIRST
C                                  I-1 COMPONENTS ARE ZERO) SUCH THAT WHEN
C                                  COLUMN L IS PREMULTIPLIED BY
C                                  H = I - 2W*W**T, COMPONENTS I+1 THROUGH
C                                  M ARE ZEROED.
            CALL CALW(A(1,L),M,W,I)
C                                  PREMULTIPLY A BY H = I - 2W*W**T
            DO 15 K=L,N
               WTA = 0.0
               DO 5 J=I,M
                  WTA = WTA + W(J)*A(J,K)
    5          CONTINUE
               TWOWTA = 2*WTA
               DO 10 J=I,M
                  A(J,K) = A(J,K) - TWOWTA*W(J)
   10          CONTINUE
   15       CONTINUE
C                                  PREMULTIPLY B BY H = I - 2W*W**T
            WTA = 0.0
            DO 20 J=I,M
               WTA = WTA + W(J)*B(J)
   20       CONTINUE
            TWOWTA = 2*WTA
            DO 25 J=I,M
               B(J) = B(J) - TWOWTA*W(J)
   25       CONTINUE
         ENDIF
C                                  PIVOT A(I,L) IS NONZERO AFTER PROCESSING
C                                  COLUMN L--MOVE DOWN TO NEXT ROW, I+1
         IF (ABS(A(I,L)).LE.ERRLIM) A(I,L) = 0.0
         IF (A(I,L).NE.0.0) THEN
            NPIVOT = I
            PIVOT(NPIVOT) = L
            I = I+1
            IF (I.GT.M) RETURN
         ENDIF
   30 CONTINUE
      RETURN
      END

      SUBROUTINE CALW(A,M,W,I)
```

```
      IMPLICIT DOUBLE PRECISION (A-H,O-Z)
      DOUBLE PRECISION A(M),W(M)
C                                 SUBROUTINE CALW CALCULATES A UNIT
C                                 M-VECTOR W (WHOSE FIRST I-1 COMPONENTS
C                                 ARE ZERO) SUCH THAT PREMULTIPLYING THE
C                                 VECTOR A BY H = I - 2W*W**T ZEROES
C                                 COMPONENTS I+1 THROUGH M.
      S = 0.0
      DO 5 J=I,M
         S = S + A(J)**2
         W(J) = A(J)
    5 CONTINUE
      IF (A(I) .GE. 0.0) THEN
         BETA = SQRT(S)
      ELSE
         BETA = -SQRT(S)
      ENDIF
      W(I) = A(I) + BETA
      TWOALP = SQRT(2*BETA*W(I))
C                                 TWOALP=0 ONLY IF A(I),...,A(M) ARE ALL
C                                 ZERO.  IN THIS CASE, RETURN WITH W=0
      IF (TWOALP.EQ.0.0) RETURN
C                                 NORMALIZE W
      DO 10 J=I,M
         W(J) = W(J)/TWOALP
   10 CONTINUE
      RETURN
      END
```

Figure 2.3.1

If we actually had to form the M by M matrix H_i, and to multiply it by the M by N matrix A in the usual way, we would have to do $O(M^2N)$ work per column, making this algorithm useless, but we do not; the premultiplication by H_i is done in the following manner:

$$H_iA = (I - 2\boldsymbol{\omega}\boldsymbol{\omega}^{\mathrm{T}})A = A - 2\boldsymbol{\omega}(\boldsymbol{\omega}^{\mathrm{T}}A).$$

The formation of $\boldsymbol{\omega}$ (in subroutine CALW) requires only $O(M)$ work. Then only columns l to N of A are processed, since the previous columns have already been reduced and will not be altered by this premultiplication. Premultiplication of each column $\boldsymbol{a}_k (k = l, \ldots, N)$ by H_i involves first calculating the scalar product $\omega ta = \boldsymbol{\omega}^{\mathrm{T}}\boldsymbol{a}_k$ (loop 5) and then subtracting $2\omega ta$ times $\boldsymbol{\omega}$ from $\boldsymbol{a}_k$ (loop 10). Each of these calculations requires $M - (i-1)$ multiplications, since the first $i-1$ components of $\boldsymbol{\omega}$ are zero. So about $2(M-i)$ multiplications per column are required and, since there are about $N-l$ columns to process (actually $N-l+2$ columns counting $\boldsymbol{b}$), the total work to multiply H_iA is about $2(N-l)(M-i)$ multiplications. Let us assume for simplicity that the columns of A are linearly independent (hence the least squares solution is unique, and $M \geq N$); so no zero pivots will be encountered, and we can take $l = i$. Then, since transformations involving $H_1, \ldots, H_N$ are required, the total number of multiplications required to orthogonally reduce an M by N matrix A to upper triangular form using Householder transformations is about (see 0.1.1)

$$\begin{aligned}\sum_{i=1}^{N}[2(N-i)(M-i)] &= \sum_{i=1}^{N}[2NM - 2(N+M)i + 2i^2] \\ &\approx 2N^2M - (N+M)N^2 + \frac{2}{3}N^3 = N^2(M - \frac{1}{3}N).\end{aligned}$$

The work done during back substitution is negligible, as it is for Gaussian elimination.

While each Householder transformation processes an entire column, a Givens rotation eliminates only one element at a time. In column l ($l = 1, \ldots, N$), there are about $M - i$ elements to zero, and it requires about $4(N - l)$ multiplications to knock out each (see loop 5 of REDQ in Figure 2.2.2). Assuming as above that $i = l$, the total number of multiplications required to reduce orthogonally a rectangular matrix to upper triangular form using Givens transformations is about

$$\sum_{i=1}^{N}[4(N-i)(M-i)] \approx 2N^2(M - \frac{1}{3}N).$$

So, while both methods require $O(N^2M)$ work, the Householder method is twice as fast, for large problems. Note that the Householder method requires only $O(N)$ square roots while, if we use Givens rotations, we must compute $O(N^2)$ square roots. Also note that, for square nonsingular systems ($M = N$), the Householder reduction is only twice as slow as Gaussian elimination.

As discussed in Section 2.1, we can also solve a linear least squares problem by forming and solving the normal equations $A^{\mathrm{T}}A\boldsymbol{x} = A^{\mathrm{T}}\boldsymbol{b}$. If A is M by N, forming the matrix $A^{\mathrm{T}}A$ requires N^2M multiplications; then solving the N by N linear system involves $\frac{1}{3}N^3$ multiplications, using Gaussian elimination. However, if we take advantage of symmetry, both the formation and the solution work (see Problem 5 of Chapter 1) can be cut in half, and the total number of multiplications is $\frac{1}{2}N^2(M + \frac{1}{3}N)$. If M is large compared with N, the normal equations approach is about twice as fast as Householder reduction. Thus the main argument for orthogonal reduction is not speed, but accuracy.

2.4 Least Squares Approximation with Cubic Splines

In Section 1.7 we saw that a "natural" cubic spline is uniquely determined by its values at the knots $x_1, \ldots, x_N$. Thus, if we define $\Phi_j(x)$ to be the natural cubic spline with $\Phi_j(x_j) = 1$ and $\Phi_j(x_k) = 0$ for $k \neq j$, it is clear that any natural cubic spline with the same knots can be expressed as a linear combination of the functions $\Phi_j(x)$:

$$s(x) = \sum_{j=1}^{N} y_j \Phi_j(x). \tag{2.4.1}$$

If we want to calculate a natural cubic spline $s(x)$ that passes through the M data points $(xd_i, yd_i), i = 1, \ldots, M$, we have to solve the M by N linear system

$$\begin{bmatrix} \Phi_1(xd_1) & \Phi_2(xd_1) & \ldots & \Phi_N(xd_1) \\ \Phi_1(xd_2) & \Phi_2(xd_2) & \ldots & \Phi_N(xd_2) \\ \vdots & \vdots & & \vdots \\ \Phi_1(xd_M) & \Phi_2(xd_M) & \ldots & \Phi_N(xd_M) \end{bmatrix} \begin{bmatrix} y_1 \\ y_2 \\ \vdots \\ y_N \end{bmatrix} = \begin{bmatrix} yd_1 \\ yd_2 \\ \vdots \\ yd_M \end{bmatrix}. \tag{2.4.2}$$

In Section 1.7 the knots x_i were chosen to coincide with the data points xd_i; in this case there is a unique solution to the linear system 2.4.2, and thus a unique natural cubic spline interpolant. (In fact, the coefficient matrix reduces to the identity, and $y_i = yd_i$.)

Now interpolation may be appropriate if the data points are known to be accurate but, if it is known that there is some experimental scatter in the data, forcing $s(x)$ to pass exactly through each point will produce a curve that fluctuates in an unreasonable manner. In this case, we should probably choose N (the number of knots) to be smaller than M (the number of data points). With fewer degrees of freedom, the spline cannot fluctuate as rapidly, and it will generally not be possible to force it to pass exactly through all the data points. In other words, 2.4.2 will generally have no solution. However, if we solve 2.4.2 in the least squares sense, we shall get the natural cubic spline, among all those with the given knots, which most nearly interpolates to the data.

Figure 2.4.1 gives a FORTRAN program that calculates the natural cubic spline with knots $x_1, \ldots, x_N$, which is the least squares approximation to the data points $(xd_i, yd_i), i = 1, \ldots, M$. For each j, it calls the subroutine DSPLN in Figure 1.7.2 to calculate the natural cubic spline $\Phi_j(x)$, and to evaluate Φ_j at the points $xd_1, \ldots, xd_M$. Each call to DSPLN thus gives us one column of the coefficient matrix of the linear system 2.4.2. Then DLLSQR (Figure 2.2.2) is called to solve 2.4.2 in the least squares sense, computing values for the coefficients y_j of the $\Phi_j(x)$. Now it is easy to see from 2.4.1 that y_j is just the value of the spline $s(x)$ at x_j, so DSPLN is called again to compute the coefficients of the natural cubic spline that takes the values y_j at the knots, and to evaluate this spline at the user-supplied output points.

DLSQSP (Figure 2.4.1) was used to calculate the least squares natural cubic spline approximation, with N equally spaced knots, to a set of $M = 20$ data points. The splines produced by DLSQSP with $N = 5$ and $N = 15$ are shown in Figures 2.4.2 and 2.4.3. As N increases, the number of degrees of freedom increases, and the spline becomes more wiggly and approximates the data points more closely. Which approximation is more reasonable depends on how much accuracy can be assumed for our data.

When N is larger than M, the system 2.4.2 has many solutions, and DLLSQR returns one of these many solutions. So the spline calculated by DLSQSP interpolates the data points exactly; however, it oscillates wildly between data points. No doubt there are other solutions (cf. Problem 9) that interpolate the data with less oscillation.

```
      SUBROUTINE DLSQSP(X,N,XD,YD,M,XOUT,YOUT,NOUT)
      IMPLICIT DOUBLE PRECISION (A-H,O-Z)
C                                 DECLARATIONS FOR ARGUMENTS
      DOUBLE PRECISION X(N),XD(M),YD(M),XOUT(NOUT),YOUT(NOUT)
      INTEGER N,M,NOUT
C                                 DECLARATIONS FOR LOCAL VARIABLES
      DOUBLE PRECISION Y(N),A(M,N)
C
C  SUBROUTINE DLSQSP CALCULATES A NATURAL CUBIC SPLINE WITH KNOTS AT
C    X(1),...,X(N) WHICH IS THE LEAST SQUARES FIT TO THE DATA POINTS
C    (XD(I),YD(I)), I=1,...,M, AND EVALUATES THIS SPLINE AT THE OUTPUT
C    POINTS XOUT(1),...,XOUT(NOUT).
C
C  ARGUMENTS
C
C             ON INPUT                          ON OUTPUT
C             --------                          ---------
C
C    X      - A VECTOR OF LENGTH N CONTAINING
C             THE SPLINE KNOTS.
C
C    N      - THE NUMBER OF KNOTS.
C             (N.GE.3).
C
C    XD     - A VECTOR OF LENGTH M CONTAINING
C             THE X-COORDINATES OF THE DATA
C             POINTS.
C
C    YD     - A VECTOR OF LENGTH M CONTAINING
C             THE Y-COORDINATES OF THE DATA
C             POINTS.
```

```
C
C     M      - THE NUMBER OF DATA POINTS.
C
C     XOUT   - A VECTOR OF LENGTH NOUT CONTAINING
C              THE X-COORDINATES AT WHICH THE
C              CUBIC SPLINE IS EVALUATED.  THE
C              ELEMENTS OF XOUT MUST BE IN
C              ASCENDING ORDER.
C
C     YOUT   -                                   A VECTOR OF LENGTH NOUT.
C                                                YOUT(I) CONTAINS THE
C                                                VALUE OF THE SPLINE
C                                                AT XOUT(I).
C
C     NOUT   - THE NUMBER OF OUTPUT POINTS.
C
C-----------------------------------------------------------------------
      ZERO = 0.0D0
      DO 5 J=1,N
         Y(J) = 0.0
    5 CONTINUE
      DO 10 J=1,N
         Y(J) = 1.0
C                              CALCULATE PHI(J,X), NATURAL CUBIC SPLINE
C                              WHICH IS EQUAL TO ONE AT KNOT X(J) AND
C                              ZERO AT OTHER KNOTS.  THEN SET
C                                  A(I,J) = PHI(J,XD(I)), I=1,...,M
         CALL DSPLN(X,Y,N,ZERO,ZERO,XD,A(1,J),M)
         Y(J) = 0.0
   10 CONTINUE
C                              CALL DLLSQR TO MINIMIZE NORM OF A*Y-YD
      CALL DLLSQR(A,M,N,Y,YD)
C                              LEAST SQUARES SPLINE IS
C                               Y(1)*PHI(1,X) + ... + Y(N)*PHI(N,X).
C                              EVALUATE SPLINE AT XOUT(1),...,XOUT(NOUT)
      CALL DSPLN(X,Y,N,ZERO,ZERO,XOUT,YOUT,NOUT)
      RETURN
      END
```

Figure 2.4.1

2.5 Problems

1. The least squares problem

$$\text{minimize } \|A\boldsymbol{x} - \boldsymbol{b}\|_2$$

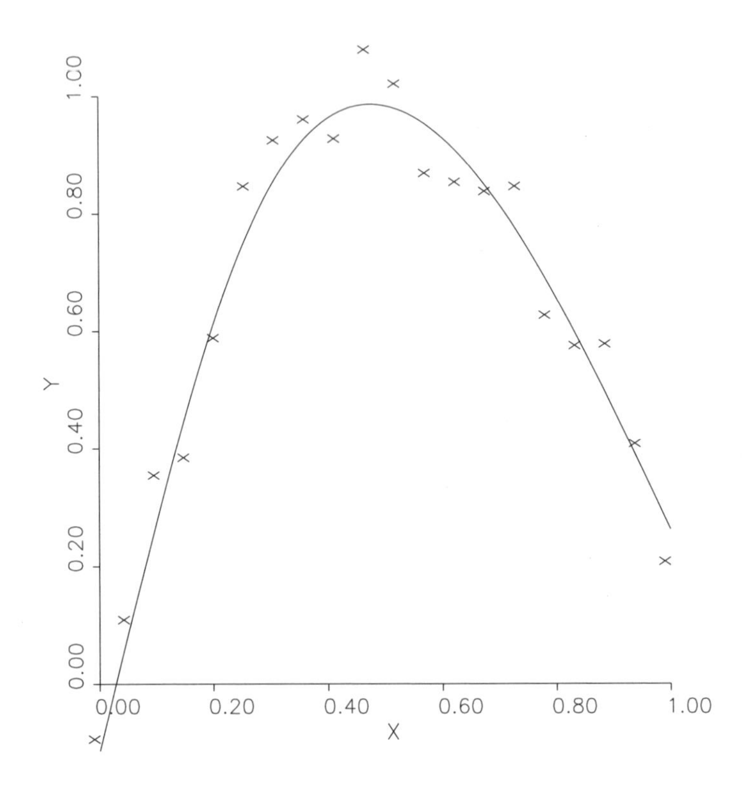

Figure 2.4.2
Least Squares Spline Approximation, with Five Knots

can be written in component form as

$$\text{minimize } f(x_1, \ldots, x_N) \equiv \sum_{i=1}^{M} (a_{i1}x_1 + \ldots + a_{iN}x_N - b_i)^2.$$

Using the fact that the partial derivatives of f must all be zero at a minimum, show that any solution $\boldsymbol{x}$ to the least squares problem satisfies $A^{\mathrm{T}}A\boldsymbol{x} = A^{\mathrm{T}}\boldsymbol{b}$.

2. If $N > M$, will the least squares problem ever have a unique solution? Why or why not?

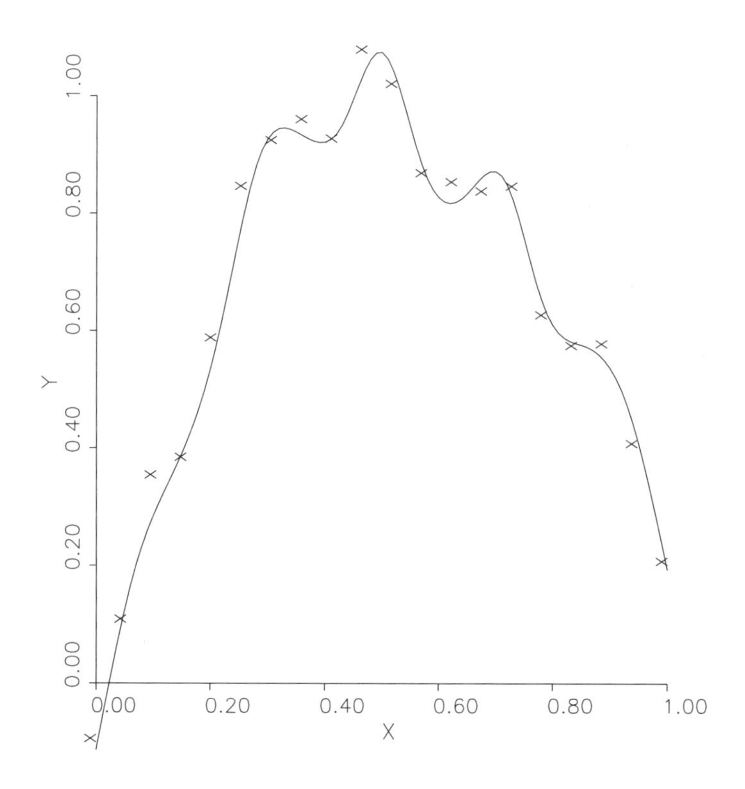

Figure 2.4.3
Least Squares Spline Approximation, with 15 Knots

3. Use DLLSQR (Figure 2.2.2) to solve all seven linear equations 1.1.1 and 1.1.2 for the five unknown currents.

4. Here are the scores from games between Texas A&M, Oklahoma, Texas, and UTEP:

   Texas A&M 14, Oklahoma 7
   Texas A&M 35, Texas 3
   UTEP 7, Texas A&M 7
   Texas 10, Oklahoma 7
   UTEP 19, Oklahoma 6
   UTEP 7, Texas 0.

   Calculate ratings for Texas A&M, Oklahoma, Texas, and UTEP based on the discussion in Section 2.1. Use DLLSQR to solve the linear least squares system that results. The solution is not unique: How does DLLSQR pick a particular solution? (In my original football poll I added the constraint that the average rating must be 25, to get a unique solution. In this way, one could compare the best teams from different years. For the first few weeks of the season, however, even this did not produce a unique solution. Think of the teams as vertices in a graph, with an edge between two vertices if the teams have played. Then until the graph is connected, a unique solution will not exist, even with the extra constraint. Explain in non-mathematical terms why this would be expected.)

5. Use DLLSQR to compute the polynomial of degree n which most closely fits the data points $(1,1)$, $(2,3)$, $(3,2)$, $(4,3)$, and $(5,4)$, for $n = 1, 3$, and 5. Plot your least squares straight line ($n = 1$) and compare it with the best L_∞ straight line, drawn in Figure 4.5.2. Note that, for $n = 5$, there are many polynomials that pass through the given five points; DLLSQR will return one of them.

6. Consider the least squares problem min $\|A\boldsymbol{x} - \boldsymbol{b}\|_2$, where

$$A = \begin{bmatrix} 3 & 4 \\ -4 & 3 \\ 0 & 12 \end{bmatrix}, \qquad \boldsymbol{b} = \begin{bmatrix} 1 \\ 1 \\ -1 \end{bmatrix}.$$

   a. Use two Givens rotations, with hand calculations, to solve this problem, and also display the QR decomposition of A.

   b. Solve the problem using the normal equations.

7. Show that the L_2-norm of each column of A does not change when A is premultiplied by an orthogonal matrix Q. (Postmultiplication by an orthogonal matrix leaves the norms of the *rows* unchanged.) Thus, during an orthogonal reduction process, when a pivot element is used to "knock out" the elements directly below it, since the elements above the pivot do not change, the pivot element grows in magnitude at the expense of the eliminated elements. Therefore a pivot can be zero *after* it has been used to knock out the elements below it only if the pivot and all elements below were zero to begin with.

8. a. Modify DLLSQR to return the QR factorization of A, by initializing $Q = I$ and postmultiplying by the Q_{ij} matrices. Test your QR factorization on the system in Problem 5, with $n = 3$. Verify that Q is orthogonal, and that Q times R (final A) does restore the original matrix.

b. Your program for part (a) will do $O(M^2N)$ arithmetic operations; if $M >> N$ this is unacceptable. It is possible to get the QR decomposition essentially for free while solving the first least squares problem, if you don't insist on actually forming the M by M matrix Q, but just save enough information to be able to compute $Q^T\mathbf{c}$ (so that $R\mathbf{x} = Q^T\mathbf{c}$ can then be solved in the least squares sense) when you need to solve a second least squares problem, min $\|A\mathbf{x} - \mathbf{c}\|_2$. Following the pattern used by DLINEQ/DRESLV, modify DLLSQR (in REDQ) to save the rotation angle (ATAN2(S,C)) in A(J,L), although it is understood that A(J,L) is really zero. Then create another routine, DLLSQ2, which takes A as output by DLLSQR and solves another least squares problem efficiently. DLLSQ2 should be identical to DLLSQR, except the loop that premultiplies A by Q_{ij}^T should be removed, and set S=SIN(A(J,L)), C= COS(A(J,L)) (also, set ERRLIM=0). With this approach, you can compute $Q^T\mathbf{c}$ in $O(MN)$ operations, rather than $O(M^2)$, also an important improvement when $M >> N$. Test your programs by calling your modified DLLSQR to solve the system in Problem 5 with $n = 3$, then call DLLSQ2 to solve it again.

9. A linear system $A\mathbf{x} = \mathbf{b}$ that has no solution is called an overdetermined system, while one that has many solutions is called underdetermined. The "underdetermined least squares problem" is to find, of all solutions to an underdetermined system $A\mathbf{x} = \mathbf{b}$, the one that minimizes $\|\mathbf{x}\|_2$. Show that, if $\mathbf{z}$ is any solution to the square system $AA^{\mathrm{T}}\mathbf{z} = \mathbf{b}$ (there will always be at least one solution), and $\mathbf{x} = A^{\mathrm{T}}\mathbf{z}$, then $\mathbf{x}$ is the solution to the underdetermined least squares problem. (Hint: Obviously $\mathbf{x}$ is a solution to $A\mathbf{x} = \mathbf{b}$. To show that it is the minimum norm solution, let $\mathbf{y}$ be any other solution to $A\mathbf{y} = \mathbf{b}$ and let $\mathbf{e} \equiv \mathbf{y} - \mathbf{x}$. Then show that $\|\mathbf{x} + \mathbf{e}\|_2^2 = \|\mathbf{x}\|_2^2 + \|\mathbf{e}\|_2^2$ following the pattern given in the proof of Theorem 2.1.1a.) (See Section 3.7 for how to find the minimum norm solution of a general least squares problem, where *min* $\|A\mathbf{x} - \mathbf{b}\|_2$ is not necessarily 0.)

10. Using the approach outlined in Problem 9, find the polynomial of degree $n = 5$ that passes through the five points given in Problem 5, which minimizes the sum of squares of the coefficients. While this polynomial may be better in some sense than the polynomial of degree 5 found by DLLSQR, neither is the polynomial interpolant of minimum curvature.

3

The Eigenvalue Problem

3.1 Introduction

The eigenvalues of a square N by N matrix A are those scalar values λ for which

$$A\boldsymbol{x} = \lambda\boldsymbol{x} \tag{3.1.1}$$

has one or more nonzero solutions $\boldsymbol{x}$. The corresponding solutions are called the eigenvectors. (We reject $\boldsymbol{x} = \mathbf{0}$ as an eigenvector because otherwise all scalars λ could claim to be eigenvalues.) Since 3.1.1 is equivalent to $(A - \lambda I)\boldsymbol{x} = \mathbf{0}$, we see that the eigenvalues are those scalars for which $\det(A - \lambda I) = 0$.

In theory, we can use $\det(A - \lambda I) = 0$ to calculate the eigenvalues of A. When the determinant is expanded, $\det(A - \lambda I)$ becomes a polynomial of degree N in λ; so calculation of the eigenvalues of A is reduced to finding the roots of a polynomial of degree N. Although this is of theoretical interest, because now we see that A will always have exactly N eigenvalues (counting multiple roots multiply) and that the eigenvalues (and therefore also the eigenvectors) of a real matrix may be complex, it is not practical to calculate the eigenvalues in this way, unless N is very small.

It has been proved that it is not possible to calculate the roots of an arbitrary Nth-degree polynomial exactly in a finite number of steps, for $N > 4$. Now for any given polynomial $p_N(\lambda) = \lambda^N + \alpha_1\lambda^{N-1} + \ldots + \alpha_{N-1}\lambda + \alpha_N$, we can construct a matrix A such that $\det(A - \lambda I) = \pm p_N(\lambda)$. In fact, such a matrix is given by

$$\begin{bmatrix} -\alpha_1 & -\alpha_2 & \dots & -\alpha_{N-1} & -\alpha_N \\ 1 & 0 & \dots & 0 & 0 \\ 0 & 1 & \dots & 0 & 0 \\ \vdots & \vdots & & \vdots & \vdots \\ 0 & 0 & \dots & 1 & 0 \end{bmatrix}.$$

Therefore we cannot hope to devise an algorithm that finds the eigenvalues of an arbitrary matrix A exactly in a finite number of steps; otherwise we could use this algorithm as a means to find the roots of a general polynomial, in violation of the known theory for polynomials.

This result sets the eigenvalue problem apart from the other problems in a text on computational linear algebra. Problems that are really "linear", such as linear systems, linear least squares problems, and linear programming (LP) problems, should be solvable exactly in a finite number of calculations, but the "linear" eigenvalue problem is not really linear; all methods for finding matrix eigenvalues are iterative. However, this distinction has little computational significance because, whether a computation is iterative or terminating in theory, on a computer the end result is the same: After a finite number of operations we have an answer that is correct to a finite number of digits.

Most methods for finding all the eigenvalues and/or eigenvectors of a matrix are based on the fact that the transformation $A \to Q^{-1}AQ$ does not alter the eigenvalues of A. Since $Q^{-1}AQ\boldsymbol{x} = \lambda\boldsymbol{x}$ implies $A(Q\boldsymbol{x}) = \lambda(Q\boldsymbol{x})$, we see that, if λ is an eigenvalue of $Q^{-1}AQ$, with eigenvector $\boldsymbol{x}$, then λ is also an eigenvalue of A, with eigenvector $Q\boldsymbol{x}$. Thus "transformation" methods attempt to find matrices Q such that $Q^{-1}AQ$ has a form that makes eigenvalue extraction trivial (the eigenvalues can be read off the diagonals of triangular and diagonal matrices) or at least easier (the eigenvalues of tridiagonal and Hessenberg matrices require a little extra work to calculate).

Many matrix eigenvalue problems arise from the discretization of ordinary or partial differential equation eigenvalue problems. For example, calculating the allowable energy states of an electron in an atom requires solving the Schrödinger partial differential equation, in which the energy levels are the eigenvalues and the corresponding electron probability distributions are the eigenfunctions. The approximate solution of this problem always requires the solution of a related matrix eigenvalue problem.

Let us consider another example. The vibrational frequencies of an elastic membrane (e.g., a drumhead) stretched over a square frame are the eigenvalues λ of the partial differential equation

$$\begin{aligned} -U_{xx} - U_{yy} &= \lambda U \qquad \text{in } (0,1)\times(0,1), \\ U &= 0 \qquad \text{on the boundary.} \end{aligned} \tag{3.1.2}$$

We can approximate the differential equation and boundary condition using finite differences as follows:

$$-\frac{U_{i+1,j}-2U_{i,j}+U_{i-1,j}}{h^2}-\frac{U_{i,j+1}-2U_{i,j}+U_{i,j-1}}{h^2}=\lambda U_{i,j},$$
$$U_{i,0}=U_{i,M}=U_{0,j}=U_{M,j}=0,$$

where $h = 1/M$, and $U_{i,j}$ approximates $U(ih, jh)$. This is a matrix eigenvalue problem, and when $M = 5$ it can be written in the form $A\boldsymbol{u} = \lambda\boldsymbol{u}$, where $\boldsymbol{u} = [U_{11}, U_{12}, U_{13}, U_{14}, U_{21}, U_{22}, U_{23}, U_{24}, U_{31}, U_{32}, U_{33}, U_{34}, U_{41}, U_{42}, U_{43}, U_{44}]$ and A is shown in Figure 3.1.1.

Many, perhaps most, eigenvalue problems that occur in applications involve symmetric matrices, like the one in Figure 3.1.1. As shown in Section 0.2, all the eigenvalues (and thus all eigenvectors) of a symmetric matrix are real. Symmetric matrix eigenvalue problems are so much easier to solve—in part because we can avoid complex arithmetic, but also for other reasons—and they arise so often in applications that they are generally solved using algorithms especially designed for symmetric matrices, such as the method presented in the next section.

$$\begin{bmatrix}
100 & -25 & 0 & 0 & -25 & 0 & 0 & 0 & 0 & 0 & 0 & 0 & 0 & 0 & 0 & 0\\
-25 & 100 & -25 & 0 & 0 & -25 & 0 & 0 & 0 & 0 & 0 & 0 & 0 & 0 & 0 & 0\\
0 & -25 & 100 & -25 & 0 & 0 & -25 & 0 & 0 & 0 & 0 & 0 & 0 & 0 & 0 & 0\\
0 & 0 & -25 & 100 & 0 & 0 & 0 & -25 & 0 & 0 & 0 & 0 & 0 & 0 & 0 & 0\\
-25 & 0 & 0 & 0 & 100 & -25 & 0 & 0 & -25 & 0 & 0 & 0 & 0 & 0 & 0 & 0\\
0 & -25 & 0 & 0 & -25 & 100 & -25 & 0 & 0 & -25 & 0 & 0 & 0 & 0 & 0 & 0\\
0 & 0 & -25 & 0 & 0 & -25 & 100 & -25 & 0 & 0 & -25 & 0 & 0 & 0 & 0 & 0\\
0 & 0 & 0 & -25 & 0 & 0 & -25 & 100 & 0 & 0 & 0 & -25 & 0 & 0 & 0 & 0\\
0 & 0 & 0 & 0 & -25 & 0 & 0 & 0 & 100 & -25 & 0 & 0 & -25 & 0 & 0 & 0\\
0 & 0 & 0 & 0 & 0 & -25 & 0 & 0 & -25 & 100 & -25 & 0 & 0 & -25 & 0 & 0\\
0 & 0 & 0 & 0 & 0 & 0 & -25 & 0 & 0 & -25 & 100 & -25 & 0 & 0 & -25 & 0\\
0 & 0 & 0 & 0 & 0 & 0 & 0 & -25 & 0 & 0 & -25 & 100 & 0 & 0 & 0 & -25\\
0 & 0 & 0 & 0 & 0 & 0 & 0 & 0 & -25 & 0 & 0 & 0 & 100 & -25 & 0 & 0\\
0 & 0 & 0 & 0 & 0 & 0 & 0 & 0 & 0 & -25 & 0 & 0 & -25 & 100 & -25 & 0\\
0 & 0 & 0 & 0 & 0 & 0 & 0 & 0 & 0 & 0 & -25 & 0 & 0 & -25 & 100 & -25\\
0 & 0 & 0 & 0 & 0 & 0 & 0 & 0 & 0 & 0 & 0 & -25 & 0 & 0 & -25 & 100
\end{bmatrix}$$

Figure 3.1.1

3.2 The Jacobi Method for Symmetric Matrices

The Jacobi method [Forsythe and Henrici 1960] for the symmetric eigenvalue problem is no longer considered state-of-the-art; there are other methods that are somewhat faster. However, it has the advantage that it is simple to program and to analyze, and it is no less stable or robust than the more sophisticated methods; it is guaranteed to find all the eigenvalues and eigenvectors of a symmetric matrix in a reasonable amount of time.

The Jacobi method constructs a sequence of similarity transformations $A_{n+1} = Q_n^{-1} A_n Q_n (A_0 = A)$, where the Q-matrices are Givens rotation matrices (see Section 2.2), of the form

$$
Q_{ij} = \begin{array}{c} \\ \\ \\ \text{row } i \\ \\ \\ \text{row } j \\ \\ \\ \\ \end{array}
\overset{\begin{array}{cc}\text{column} & \text{column} \\ i & j\end{array}}{\begin{bmatrix}
1 & & & & & & & & & \\
& 1 & & & & & & & & \\
& & 1 & & & & & & & \\
& & & c & & & -s & & & \\
& & & & 1 & & & & & \\
& & & & & 1 & & & & \\
& & & s & & & c & & & \\
& & & & & & & 1 & & \\
& & & & & & & & 1 & \\
& & & & & & & & & 1
\end{bmatrix}}
\quad (c^2 + s^2 = 1) \qquad (3.2.1)
$$

As verified earlier, Q_{ij} is orthogonal; that is, $Q_{ij}^{-1} = Q_{ij}^{\mathrm{T}}$, so the Jacobi iteration can be written $A_{n+1} = Q_{ij}^{\mathrm{T}} A_n Q_{ij}$. Thus this transformation preserves symmetry since, if $A_n^{\mathrm{T}} = A_n$, we have

$$
A_{n+1}^{\mathrm{T}} = (Q_{ij}^{\mathrm{T}} A_n Q_{ij})^{\mathrm{T}} = Q_{ij}^{\mathrm{T}} A_n^{\mathrm{T}} Q_{ij} = Q_{ij}^{\mathrm{T}} A_n Q_{ij} = A_{n+1}.
$$

Premultiplying a matrix A by Q_{ij}^{T} has the effect of replacing rows i and j by linear combinations of the original rows i and j, and we saw in Section 2.2 that c and s can be chosen so that a zero will be introduced into the (j,i)th position of A. However, to preserve the eigenvalues of A, we are forced to postmultiply by Q_{ij}, which has the effect of changing columns i and j, and the postmultiplication will normally cause the element just zeroed (a_{ji}) to become nonzero. This is not a problem if we plan ahead; we simply have to choose c and s so that the zero is not created until the end—after premultiplication *and* postmultiplication.

One Jacobi transformation can be represented by

$$
\begin{bmatrix}
& \vdots & & \vdots & \\
\cdots & b_{ii} & \cdots & b_{ij} & \cdots \\
& \vdots & & \vdots & \\
\cdots & b_{ji} & \cdots & b_{jj} & \cdots \\
& \vdots & & \vdots &
\end{bmatrix}
= B \equiv Q_{ij}^{\mathrm{T}} A Q_{ij} \qquad (3.2.2)
$$

$$
= \begin{bmatrix}
1 & & & & & & & \\
& 1 & & & & & & \\
& & c & & & s & & \\
& & & 1 & & & & \\
& & & & 1 & & & \\
& & & & & 1 & & \\
& & -s & & & c & & \\
& & & & & & 1 & \\
& & & & & & & 1
\end{bmatrix}
\begin{bmatrix}
& \vdots & & \vdots & \\
\cdots & a_{ii} & \cdots & a_{ij} & \cdots \\
& \vdots & & \vdots & \\
\cdots & a_{ji} & \cdots & a_{jj} & \cdots \\
& \vdots & & \vdots &
\end{bmatrix}
\begin{bmatrix}
1 & & & & & & & \\
& 1 & & & & & & \\
& & c & & & -s & & \\
& & & 1 & & & & \\
& & & & 1 & & & \\
& & & & & 1 & & \\
& & s & & & c & & \\
& & & & & & 1 & \\
& & & & & & & 1
\end{bmatrix}.
$$

Although all elements in rows i and j and columns i and j may be modified by the transformation, we shall primarily be interested in the new values b_{ii}, b_{jj}, and (especially) $b_{ji} = b_{ij}$:

$$\begin{aligned} b_{ii} &= c^2 a_{ii} + s^2 a_{jj} + 2sca_{ji}, \\ b_{jj} &= s^2 a_{ii} + c^2 a_{jj} - 2sca_{ji}, \\ b_{ji} &= b_{ij} = cs(a_{jj} - a_{ii}) + (c^2 - s^2)a_{ji}. \end{aligned} \tag{3.2.3}$$

If we want $b_{ji} = b_{ij} = 0$, we need to choose c and s so that

$$\frac{c^2 - s^2}{cs} = \frac{a_{ii} - a_{jj}}{a_{ji}} \equiv 2\beta \tag{3.2.4}$$

(we can assume that $a_{ji} \neq 0$; otherwise no transformation is necessary). Substituting $c = (1 - s^2)^{1/2}$ into 3.2.4 gives

$$s^4 - s^2 + \frac{1}{4 + 4\beta^2} = 0$$

or, using the quadratic equation,

$$s^2 = \frac{1}{2} \pm \frac{\frac{1}{2}\beta}{(1 + \beta^2)^{1/2}}.$$

Thus the following values for s and c will satisfy 3.2.4 and $s^2 + c^2 = 1$:

$$\begin{aligned} s &= \left(\frac{1}{2} - \frac{\frac{1}{2}\beta}{(1 + \beta^2)^{1/2}}\right)^{1/2}, \\ c &= \left(\frac{1}{2} + \frac{\frac{1}{2}\beta}{(1 + \beta^2)^{1/2}}\right)^{1/2}. \end{aligned} \tag{3.2.5}$$

Thus by proper choice of s and c we can introduce a zero into any specified off-diagonal position, while preserving the eigenvalues (and the symmetry) of A. Now it would be nice if we could zero all the off-diagonal elements of A in succession, and after $N(N-1)/2$ transformations have a diagonal matrix that is similar to A; then we could read the eigenvalues off the diagonal.

But there must be a catch because, as mentioned in the last section, it is not possible to devise an algorithm that finds the eigenvalues of an arbitrary matrix (even an arbitrary symmetric matrix) exactly, in a finite number of steps. The catch is that, when we zero a new element of A, a previously zeroed element may become nonzero. Every time we knock out one off-diagonal element, others pop back up; so it might seem that our algorithm is useless. Fortunately, as we shall now show, although our transformed matrices never become quite diagonal, they do make steady progress toward that goal.

Note that proving that the Jacobi method will always converge is essentially a constructive proof of a fundamental theorem of linear algebra, that any symmetric matrix is orthogonally similar to a diagonal matrix.

Theorem 3.2.1. *When the symmetric matrix A is transformed into $B = Q_{ij}^{\mathrm{T}} A Q_{ij}$, with Q_{ij} chosen so that $b_{ji} = 0$, the sum of the squares of the diagonal elements increases by $2a_{ji}^2$, while the sum of squares of the off-diagonal elements decreases by the same amount.*

Proof: Using equations 3.2.3 and the fact that $c^2 + s^2 = 1$, it is a straightforward but messy exercise in algebra (Problem 1) to show that

$$2b_{ji}^2 + b_{ii}^2 + b_{jj}^2 = 2a_{ji}^2 + a_{ii}^2 + a_{jj}^2.$$

Since $b_{ji} = 0$ by design,

$$b_{ii}^2 + b_{jj}^2 = 2a_{ji}^2 + a_{ii}^2 + a_{jj}^2.$$

Now, only rows i and j and columns i and j of A change during a Jacobi transformation; so a_{ii} and a_{jj} are the only diagonal elements to change. Thus

$$\sum_{k=1}^{N} b_{kk}^2 = 2a_{ji}^2 + \sum_{k=1}^{N} a_{kk}^2,$$

which proves the first part of the theorem.

Next we shall show that the transformation $Q_{ij}^{\mathrm{T}} A Q_{ij}$ does not change the sum of squares of all the elements of A. That will complete the proof since, if the sum of squares along the diagonal increases by $2a_{ji}^2$ and the total sum of squares is constant, the sum of squares of the off-diagonal elements must decrease by $2a_{ji}^2$.

If we denote the kth row of A by $\boldsymbol{a}_k^{\mathrm{T}}$ and the kth row of $P \equiv Q_{ij}^{\mathrm{T}} A$ by $\boldsymbol{p}_k^{\mathrm{T}}$, it is easy to see from 3.2.2 that only the ith and jth rows of A and P differ, and that

$$\begin{aligned} \boldsymbol{p}_i^{\mathrm{T}} &= c\boldsymbol{a}_i^{\mathrm{T}} + s\boldsymbol{a}_j^{\mathrm{T}}, \\ \boldsymbol{p}_j^{\mathrm{T}} &= -s\boldsymbol{a}_i^{\mathrm{T}} + c\boldsymbol{a}_j^{\mathrm{T}}, \end{aligned} \tag{3.2.6}$$

but then

$$\begin{aligned} \boldsymbol{p}_i^{\mathrm{T}}\boldsymbol{p}_i + \boldsymbol{p}_j^{\mathrm{T}}\boldsymbol{p}_j &= c^2\boldsymbol{a}_i^{\mathrm{T}}\boldsymbol{a}_i + 2cs\boldsymbol{a}_i^{\mathrm{T}}\boldsymbol{a}_j + s^2\boldsymbol{a}_j^{\mathrm{T}}\boldsymbol{a}_j \\ &\quad + s^2\boldsymbol{a}_i^{\mathrm{T}}\boldsymbol{a}_i - 2cs\boldsymbol{a}_i^{\mathrm{T}}\boldsymbol{a}_j + c^2\boldsymbol{a}_j^{\mathrm{T}}\boldsymbol{a}_j \\ &= \boldsymbol{a}_i^{\mathrm{T}}\boldsymbol{a}_i + \boldsymbol{a}_j^{\mathrm{T}}\boldsymbol{a}_j \end{aligned}$$

and, since the other rows of A and P are identical, we see that the premultiplication by Q_{ij}^{T} does not change the total sum of squares. In a similar fashion, the columns of P and $B = PQ_{ij} = Q_{ij}^{\mathrm{T}} A Q_{ij}$ are the same except for

$$\begin{aligned} \boldsymbol{b}_i &= c\boldsymbol{p}_i + s\boldsymbol{p}_j, \\ \boldsymbol{b}_j &= -s\boldsymbol{p}_i + c\boldsymbol{p}_j, \end{aligned} \tag{3.2.7}$$

where now the subscript indicates the column number. It can likewise be verified that this postmultiplication does not change the total sum of squares. ■

Now, to perform one Jacobi iteration, we first pick out a nonzero off-diagonal element a_{ji} ($j > i$) to knock out, then we use 3.2.4 and 3.2.5 to determine the proper values for c and s; finally we actually perform the transformation $Q_{ij}^{\mathrm{T}} A Q_{ij}$ using equations 3.2.6 and 3.2.7. The only thing missing is a strategy for determining which elements a_{ji} to zero, and in what order.

One obvious possibility is to zero all off-diagonal elements in a systematic order, for example, $\{(a_{ji}, j = i+1, \ldots, N), i = 1, \ldots, N-1\}$, and then to repeat this cycle until all off-diagonal elements are small. The only problem with this strategy is that we may be wasting time zeroing some elements that are already very small. Since the progress made toward diagonal form depends on the size of the element knocked out (see Theorem 3.2.1), our time is better spent zeroing the larger elements. Alternatively, we could search for the largest (in absolute value) off-diagonal element each time and zero it; however, just searching for the largest element requires $O(N^2)$ work (unless we are very clever), while the entire transformation $Q_{ij}^{\mathrm{T}} A Q_{ij}$ only requires $O(N)$ work!

We have chosen a third approach in our implementation: We check all the off-diagonal elements in a systematic order, zeroing only those whose squares exceed $\frac{1}{2}e_k/[N(N-1)]$, where e_k is given by

$$e_k = \sum_{i \neq j}\sum a_{ij}^2$$

after k iterations. In other words, we cyclically check all off-diagonal elements but skip over those whose squares are less than half the current average for all off-diagonal elements. (Note that there will always be elements greater than this threshold, because they cannot all have below-average squares.) To determine the rate of convergence of this algorithm, we need the following theorem.

Theorem 3.2.2. *If at each step of the Jacobi method the element to be zeroed satisfies $a_{ji}^2 \geq \frac{1}{2}e_k/[N(N-1)]$, then the convergence criterion*

$$\sum_{i \neq j}\sum a_{ij}^2 \leq \epsilon \sum_{i=1}^{N}\sum_{j=1}^{N} a_{ij}^2$$

will be satisfied after at most $L = N^2 \ln(1/\epsilon)$ iterations.

Proof: According to Theorem 3.2.1, $e_{k+1} = e_k - 2a_{ji}^2$, where a_{ji} is the element zeroed on the $(k+1)$st iteration. Thus

$$\begin{aligned} e_{k+1} &= e_k - 2a_{ji}^2 \le e_k - \frac{e_k}{N(N-1)} \\ &\le e_k\left(1 - \frac{1}{N^2}\right) \le e_k \exp\left(-\frac{1}{N^2}\right). \end{aligned}$$

The last inequality follows from the fact that $1-h \le \exp(-h)$ for all positive h (see Problem 2). Then, after L iterations,

$$\begin{aligned} e_L &\le \left[\exp\left(-\frac{1}{N^2}\right)\right]^L e_0 \le \exp\left[-\ln\left(\frac{1}{\epsilon}\right)\right] e_0 = \epsilon e_0 \\ &\le \epsilon \sum_{i=1}^{N}\sum_{j=1}^{N} (a_0)_{ij}^2 = \epsilon \sum_{i=1}^{N}\sum_{j=1}^{N} a_{ij}^2 \end{aligned}$$

■

According to Theorem 3.2.2, the number of iterations for convergence of the Jacobi method is $O(N^2)$ but, since the work per iteration is $O(N)$, the total work to find the eigenvalues using the Jacobi method is $O(N^3)$.

Our algorithm would not be practical if we had to calculate e_k, the sum of squares of the off-diagonal elements, directly each iteration, as it requires $O(N^2)$ work to sum the squares of the off-diagonal elements. Fortunately, we can calculate e_0 from the definition, and then use Theorem 3.2.1 to keep up with how e_k decreases each iteration (but see Problem 3 for a discussion of the risks involved in not calculating e_k directly). Now the Jacobi method can be summarized as follows:

$$\begin{aligned} A_0 &= A, \\ A_{n+1} &= Q_n^{\mathrm{T}} A_n Q_n, \end{aligned}$$

where each Q_n has the form 3.2.1 for some i, j. Once the iteration has converged, so that $A_L = D$, where D is nearly diagonal, we have

$$D = Q_L^{\mathrm{T}} Q_{L-1}^{\mathrm{T}} \dots Q_2^{\mathrm{T}} Q_1^{\mathrm{T}} A Q_1 Q_2 \dots Q_{L-1} Q_L.$$

If we define

$$Q \equiv Q_1 Q_2 \dots Q_{L-1} Q_L,$$

then

$$D = Q^{\mathrm{T}} A Q,$$

and, since Q is still orthogonal,

$$AQ = QD. \tag{3.2.8}$$

From 3.2.8, we see that not only are the eigenvalues of A contained in the diagonal of D, but also the columns of Q contain the corresponding eigenvectors. This is clear when we notice that 3.2.8 says that A times the kth column of Q is equal to d_{kk} times the kth column of Q. Hence, if we start with $X_0 = I$ and calculate $X_{n+1} = X_n Q_n$ each Jacobi iteration, the last X_L will be Q and will contain the orthogonal eigenvectors of A, already normalized.

Figure 3.2.1 displays a FORTRAN program that solves a symmetric eigenvalue problem using the Jacobi method described above. Although, as stated at the beginning of this section, it is no longer considered state-of-the-art, it is very easy to program and it is only marginally slower than the best codes for general symmetric eigenproblems (which also require $O(N^3)$ work). Since simplicity and not speed is DEGSYM's strongest point, we opted to sacrifice some further gains in speed to keep the program simple and did not even take advantage of symmetry in the calculations of loops 25 and 30.

DEGSYM was used to find the eigenvalues of the symmetric matrix displayed in Figure 3.1.1. The Jacobi method converged after 380 transformations, with diagonal elements as shown in Table 3.2.1. The exact eigenvalues of this matrix are known to be [Sewell 2005, Section 4.7]

$$\lambda = 100\left[\sin^2\left(\frac{k\pi}{10}\right) + \sin^2\left(\frac{l\pi}{10}\right)\right], \qquad k,l = 1,2,3,4,$$

in agreement with the results of Table 3.2.1 to the number of digits printed.

Table 3.2.1

Eigenvalues Calculated by DEGSYM

19.0983	75.0000	100.0000	130.9017
44.0983	75.0000	100.0000	155.9017
44.0983	100.0000	125.0000	155.9017
69.0983	100.0000	125.0000	180.9017

```
      SUBROUTINE DEGSYM(A,N,X)
      IMPLICIT DOUBLE PRECISION (A-H,O-Z)
C                                DECLARATIONS FOR ARGUMENTS
      DOUBLE PRECISION A(N,N),X(N,N)
      INTEGER N
C
C  SUBROUTINE DEGSYM SOLVES THE EIGENVALUE PROBLEM
C
C                A*X = LAMBDA*X
C
C    WHERE A IS A SYMMETRIC MATRIX.
C
C
C  ARGUMENTS
C
C               ON INPUT                           ON OUTPUT
```

```
C           --------                            ---------
C
C   A      - THE N BY N SYMMETRIC MATRIX.      A DIAGONAL MATRIX,
C                                              WITH THE EIGENVALUES
C                                              OF A ON THE DIAGONAL.
C
C   N      - THE SIZE OF MATRIX A.
C
C   X      -                                   AN N BY N MATRIX WHICH
C                                              CONTAINS THE EIGEN-
C                                              VECTORS OF A IN ITS
C                                              COLUMNS, IN THE SAME
C                                              ORDER AS THE EIGENVALUES
C                                              APPEAR ON THE DIAGONAL.
C
C-----------------------------------------------------------------------
C                             EPS = MACHINE FLOATING POINT RELATIVE
C                                   PRECISION
C ***************************
      DATA EPS/2.D-16/
C ***************************
C                             ANORM = SUM OF ALL SQUARES
C                             X INITIALIZED TO IDENTITY
      ANORM = 0.0
      DO 10 I=1,N
         DO 5 J=1,N
            ANORM = ANORM + A(I,J)**2
            X(I,J) = 0.0
    5    CONTINUE
         X(I,I) = 1.0
   10 CONTINUE
      ERRLIM = 1000*EPS*ANORM
C                             EK = SUM OF OFF-DIAGONAL SQUARES
      EK = 0.0
      DO 15 I=1,N
      DO 15 J=1,N
         IF (I .NE. J) EK = EK + A(I,J)**2
   15 CONTINUE
      IF (EK .LE. ERRLIM) RETURN
      THRESH = 0.5*EK/N/(N-1)
   20 CONTINUE
         DO 40 I=1,N-1
            DO 35 J=I+1,N
C                             IF A(J,I)**2 LESS THAN HALF THE
C                             AVERAGE FOR OFF-DIAGONALS, SKIP IT.
               IF (A(J,I)**2 .LE. THRESH) GO TO 35
C                             KNOCKING OUT A(J,I) WILL DECREASE OFF-
C                             DIAGONAL SUM OF SQUARES BY 2*A(J,I)**2.
```

```
               EK = EK - 2*A(J,I)**2
C                                   CALCULATE NEW THRESHOLD.
               THRESH = 0.5*EK/N/(N-1)
C                                   CALCULATE C,S
               BETA = (A(I,I)-A(J,J))/(2.*A(J,I))
               FRACT = 0.5*BETA/SQRT(1.0+BETA**2)
               S = SQRT(MAX(0.5-FRACT,0.D0))
               C = SQRT(MAX(0.5+FRACT,0.D0))
C                                   PREMULTIPLY A BY Qij**T
               DO 25 K=1,N
                  PIK =  C*A(I,K)+S*A(J,K)
                  PJK = -S*A(I,K)+C*A(J,K)
                  A(I,K) = PIK
                  A(J,K) = PJK
   25          CONTINUE
C                                   POSTMULTIPLY A AND X BY Qij
               DO 30 K=1,N
                  BKI =  C*A(K,I)+S*A(K,J)
                  BKJ = -S*A(K,I)+C*A(K,J)
                  A(K,I) = BKI
                  A(K,J) = BKJ
                  XKI =  C*X(K,I)+S*X(K,J)
                  XKJ = -S*X(K,I)+C*X(K,J)
                  X(K,I) = XKI
                  X(K,J) = XKJ
   30          CONTINUE
C                                   CHECK FOR CONVERGENCE
               IF (EK .LE. ERRLIM) RETURN
   35       CONTINUE
   40    CONTINUE
C                                   RETURN TO BEGINNING OF CYCLE
      GO TO 20
      END
```

Figure 3.2.1

3.3 The QR Method for General Real Matrices

In Section 3.2, we used the Givens rotation matrices Q_{ij} (see 3.2.1) to introduce zeros into positions a_{ji}. Unfortunately, the zeros introduced in this way pop back up on later iterations, although A_n does approach diagonal form in the limit. Thus, even when A is symmetric, the Jacobi method cannot produce a diagonal matrix in a finite number of steps. The Jacobi method can be modified to knock out a_{ji} even when $a_{ij} \neq a_{ji}$, and by iteratively knocking out subdiagonal elements it may transform a nonsymmetric matrix into upper triangular form (from which the eigenvalues can also be read off the diagonal), but certainly not in a finite number of steps, and usually not even in the limit. Indeed, if the nonsymmetric matrix has any complex eigenvalues, there is no hope of convergence to triangular form, for real matrix multiplications cannot produce a triangular matrix with complex diagonal entries.

If we are willing to be less ambitious, however, and instead of trying to achieve upper triangular form we aim for "upper Hessenberg" form, which has one extra diagonal below the upper triangle of nonzero elements, this we *can* achieve in a finite number of steps. Since we shall still use orthogonal matrices, which preserve symmetry, if the original matrix is symmetric, the final Hessenberg matrix will be also, which means it will be tridiagonal.

The eigenvalues of Hessenberg and tridiagonal matrices are not as easily extracted as those of upper triangular and diagonal matrices; so, once we reduce A to a similar Hessenberg (or tridiagonal, in the symmetric case) matrix, we still have work to do, but we shall see that these forms are much easier to deal with than the original full matrix.

We shall again use Givens rotation matrices to accomplish the reduction, but this time we shall use Q_{ij}, not to zero a_{ji}, but $a_{j,i-1}$. Since premultiplication by Q_{ij}^{T} modifies the ith and jth rows, and postmultiplication by Q_{ij} modifies the ith and jth columns, if we choose c and s so that the element in row j, column $i-1$ is zeroed by the premultiplication, it will not be modified by the postmultiplication. Note, however, that, since $j > i$, only the elements $a_{j,i-1} (j = i+1, \ldots, N)$ of column $i-1$ can be eliminated in this manner, and the elements $a_{i,i-1}$ on the first subdiagonal cannot be zeroed, but this is the price we have to pay to protect the zeros created now from destruction later.

To determine c and s, let us form the product $Q_{ij}^{\mathrm{T}} A Q_{ij}$:

$$\begin{bmatrix} & \vdots & & \vdots & \\ \ldots\ b_{i,i-1} & b_{ii} & \ldots & b_{ij} & \ldots \\ & \vdots & & \vdots & \\ \ldots\ b_{j,i-1} & b_{ji} & \ldots & b_{jj} & \ldots \\ & \vdots & & \vdots & \end{bmatrix} = B \equiv Q_{ij}^{\mathrm{T}} A Q_{ij} \tag{3.3.1}$$

$$= \begin{bmatrix} 1 &&&&&&&& \\ & 1 &&&&&&& \\ && c &&&& s && \\ &&& 1 &&&&& \\ &&&& 1 &&&& \\ &&&&& 1 &&& \\ && -s &&&& c && \\ &&&&&&& 1 & \\ &&&&&&&& 1 \end{bmatrix} \begin{bmatrix} & \vdots & & \vdots & \\ \ldots\ a_{i,i-1} & a_{ii} & \ldots & a_{ij} & \ldots \\ & \vdots & & \vdots & \\ \ldots\ a_{j,i-1} & a_{ji} & \ldots & a_{jj} & \ldots \\ & \vdots & & \vdots & \end{bmatrix} \begin{bmatrix} 1 &&&&&&&& \\ & 1 &&&&&&& \\ && c &&&& -s && \\ &&& 1 &&&&& \\ &&&& 1 &&&& \\ &&&&& 1 &&& \\ && s &&&& c && \\ &&&&&&& 1 & \\ &&&&&&&& 1 \end{bmatrix}.$$

Since the postmultiplication does not alter the element in row j, column $i-1$, we have

$$b_{j,i-1} = -s a_{i,i-1} + c a_{j,i-1},$$

and setting this to zero gives (using also $c^2 + s^2 = 1$)

$$\begin{aligned} c &= \frac{a_{i,i-1}}{(a_{i,i-1}^2 + a_{j,i-1}^2)^{1/2}}, \\ s &= \frac{a_{j,i-1}}{(a_{i,i-1}^2 + a_{j,i-1}^2)^{1/2}}. \end{aligned} \tag{3.3.2}$$

Since we shall only perform the transformation when $a_{j,i-1}$ is not already zero, the danger of a zero denominator is avoided.

Now, if we want the zeroed elements to remain zero, it is essential to knock them out in the proper order. When we reduced A to row echelon form in Section 2.2 by premultiplying with Q_{ij} matrices, it was important to zero the subdiagonal elements column by column, beginning with the first column. (The same is true of ordinary Gaussian elimination, where we premultiply A by M_{ij} matrices.) This ensures that, while we are knocking out subdiagonal elements in column i, we are only adding multiples of 0 to 0 back in columns 1 to $i-1$. In a similar manner, to make sure that later premultiplications do not destroy zeros created earlier, we shall zero subdiagonal elements of A column by column, beginning with the first column. That is, we shall eliminate elements in the order $\{(a_{j,i-1}, j = i+1, \ldots, N), i = 2, \ldots, N-1\}$. Furthermore, because we use Q_{ij}^{T} to knock out elements in the $(i-1)$st column, when we postmultiply by Q_{ij}, only columns i and j $(j > i)$ are altered, and the zeros in columns 1 to $i-1$ will not be destroyed. Thus neither the premultiplication nor postmultiplication will destroy any of the zeros created earlier (Figure 3.3.1).

In Section 3.2 we knocked out elements in the ith column using Q_{ij}, and that made it possible for us to eliminate *any* target subdiagonal element, whereas using Q_{ij} to knock out elements in the $(i-1)$st column means we have to leave the first diagonal below the main diagonal untouched. But in Section 3.2 we paid a price—the zeroed elements did not stay zero.

$$\begin{array}{c} \\ \\ i \\ j \\ \\ \\ \\ \end{array} \begin{array}{c} \begin{array}{ccccccc} & i & j & & & & \end{array} \\ \begin{bmatrix} a & x & x & a & a & a & a \\ x & x & x & x & x & x & x \\ Z & x & x & x & x & x & x \\ a & x & x & a & a & a & a \\ a & x & x & a & a & a & a \\ a & x & x & a & a & a & a \\ a & x & x & a & a & a & a \end{bmatrix} \end{array} \qquad \begin{array}{c} \\ \\ i \\ \\ j \\ \\ \\ \end{array} \begin{array}{c} \begin{array}{ccccccc} & i & & j & & & \end{array} \\ \begin{bmatrix} a & x & a & x & a & a & a \\ x & x & x & x & x & x & x \\ 0 & x & a & x & a & a & a \\ Z & x & x & x & x & x & x \\ a & x & a & x & a & a & a \\ a & x & a & x & a & a & a \\ a & x & a & x & a & a & a \end{bmatrix} \end{array} \qquad \begin{array}{c} \\ \\ \\ \\ \\ \\ i \\ j \end{array} \begin{array}{c} \begin{array}{ccccccc} & & & & & i & j \end{array} \\ \begin{bmatrix} a & a & a & a & a & x & x \\ a & a & a & a & a & x & x \\ 0 & a & a & a & a & x & x \\ 0 & 0 & a & a & a & x & x \\ 0 & 0 & 0 & a & a & x & x \\ 0 & 0 & 0 & 0 & x & x & x \\ 0 & 0 & 0 & 0 & Z & x & x \end{bmatrix} \end{array}$$

first iteration — second iteration — last iteration

Figure 3.3.1
Reduction to Upper Hessenberg Form
Z, zeroed by premultiplication (and not changed by postmultiplication)
x, may change during this iteration.

For a given i and j, and once we have chosen c and s (using 3.3.2), the transformation $B = Q_{ij}^{\mathrm{T}} A Q_{ij}$, or $P = Q_{ij}^{\mathrm{T}} A, B = P Q_{ij}$, is done exactly as before, using equations 3.2.6 and 3.2.7. Each transformation, as before, requires $O(N)$ operations and, as there are $(N-1)(N-2)/2$ elements to zero, the total work done reducing A to Hessenberg form is $O(N^3)$.

Now we have reduced our arbitrary real matrix to an upper Hessenberg (possibly tridiagonal) matrix with the same eigenvalues. There are various methods available to find the eigenvalues of this Hessenberg matrix, including the inverse power method discussed in Section 3.5. Most of these methods could be applied to the original full matrix but can be applied much more efficiently to a Hessenberg matrix. For example, the inverse power method requires solving a linear system each iteration, and this is much less expensive if the coefficient matrix is upper Hessenberg.

A popular and powerful method for finding the eigenvalues of a general real matrix (preferably after it has been reduced to upper Hessenberg form) is the QR method [Francis 1961]. To do one iteration of the QR method, we first reduce A to upper triangular form using orthogonal reduction; that is, we premultiply A by an orthogonal matrix, $Q^{\mathrm{T}}(=Q^{-1})$, designed so that $Q^{-1}A = R$ is upper triangular. Then we postmultiply by the inverse matrix, Q. Although the postmultiplication destroys the upper triangular structure created by the premultiplication, under certain conditions the QR method nevertheless makes gradual progress toward triangular form, as the following theorem states.

Theorem 3.3.1. *If the real N by N matrix A has N eigenvalues satisfying*

$$|\lambda_1| > |\lambda_2| > \ldots > |\lambda_N| > 0,$$

and the QR iteration is defined by

$$\begin{aligned} A_0 &= A, \\ A_n &= Q_{n-1}^{-1} A_{n-1} Q_{n-1}, \end{aligned}$$

where Q_{n-1} is orthogonal, and $R_{n-1} \equiv Q_{n-1}^{-1} A_{n-1}$ is upper triangular, then, for large n, A_n approaches upper triangular form.

Proof: First note that

$$A_n = Q_{n-1}^{-1} A_{n-1} Q_{n-1} = \ldots = Q_{n-1}^{-1} \ldots Q_0^{-1} A Q_0 \ldots Q_{n-1}, \tag{3.3.3}$$

and so A_n is similar to A. Also note that

$$Q_0 \ldots Q_{n-1} A_n = A Q_0 \ldots Q_{n-1}. \tag{3.3.4}$$

Then using $A_n = Q_n R_n$ and 3.3.4 repeatedly, we get

$$\begin{aligned} Q_0 \ldots Q_{n-1} R_{n-1} \ldots R_0 &= Q_0 \ldots Q_{n-2} A_{n-1} R_{n-2} \ldots R_0 \\ &= A Q_0 \ldots Q_{n-2} R_{n-2} \ldots R_0 \\ &= A Q_0 \ldots Q_{n-3} A_{n-2} R_{n-3} \ldots R_0 \qquad (3.3.5) \\ &= A^2 Q_0 \ldots Q_{n-3} R_{n-3} \ldots R_0 = \ldots = A^n. \end{aligned}$$

Meanwhile, since A has N distinct eigenvalues, by a well-known theorem of linear algebra, A is diagonalizable, $A = ZDZ^{-1}$, where we have arranged the eigenvalues $\lambda_i = d_{ii}$ in descending order in the diagonal matrix D and have packed the eigenvectors into the columns of Z in the corresponding order.

Now let us form the LU decomposition of $Z^{-1} = LU$, where L has ones along the diagonal (see Section 1.4). We have assumed that no pivoting is required during the formation of this LU decomposition (so $P = I$ in 1.4.7) and, since Z^{-1} is nonsingular, an exactly zero pivot is unlikely to arise. In any case, J.H. Wilkinson [1965], in his classic text *The Algebraic Eigenvalue Problem*, shows how to continue the proof even if pivoting is required.

Now from 3.3.5 we have

$$Q_0 \dots Q_{n-1} R_{n-1} \dots R_0 = A^n = ZD^nZ^{-1} = ZD^nLU.$$

Since $\lambda = 0$ was assumed not to be an eigenvalue of A, A^n is nonsingular, and therefore so are all the R_i. So,

$$Z^{-1}Q_0 \dots Q_{n-1} = D^nLUR_0^{-1} \dots R_{n-1}^{-1} \equiv H_n, \tag{3.3.6}$$

where 3.3.6 defines the matrix H_n.

From 3.3.3 then,

$$A_n = (ZH_n)^{-1}A(ZH_n) = H_n^{-1}(Z^{-1}AZ)H_n = H_n^{-1}DH_n. \tag{3.3.7}$$

Now let us look more closely at the matrix H_n. From 3.3.6,

$$H_n = (D^nLD^{-n})D^nUR_0^{-1} \dots R_{n-1}^{-1} \equiv (D^nLD^{-n})K_n \tag{3.3.8}$$

If the lower triangular matrix L has elements l_{ij} ($l_{ii} = 1$), the matrix in parentheses is

$$D^nLD^{-n} = \begin{bmatrix} 1 & 0 & 0 & 0 & \dots \\ l_{21}\left(\frac{d_{22}}{d_{11}}\right)^n & 1 & 0 & 0 & \dots \\ l_{31}\left(\frac{d_{33}}{d_{11}}\right)^n & l_{32}\left(\frac{d_{33}}{d_{22}}\right)^n & 1 & 0 & \dots \\ l_{41}\left(\frac{d_{44}}{d_{11}}\right)^n & l_{42}\left(\frac{d_{44}}{d_{22}}\right)^n & l_{43}\left(\frac{d_{44}}{d_{33}}\right)^n & 1 & \dots \\ \vdots & \vdots & \vdots & \vdots & \end{bmatrix}.$$

Since the eigenvalues $\lambda_i = d_{ii}$ are numbered in order of descending absolute value, the subdiagonal terms all tend to zero, and D^nLD^{-n} converges to the identity matrix, as $n \to \infty$. Thus for large n, $H_n \approx K_n$, where K_n is defined by 3.3.8. Since the inverse of an upper triangular matrix is upper triangular, and the product of upper triangular matrices is upper triangular (Theorem 0.1.2), we see that K_n is upper triangular.

Further, since multiplication by an orthogonal matrix does not change the total sum of squares of the elements of a matrix (cf. Problem 7 of Chapter

2), 3.3.3 and 3.3.6 tell us that each of A_n, H_n and H_n^{-1} has a constant sum of squares, as n increases. From this, and from 3.3.8, we conclude that if $H_n^{-1} = La_n + Ua_n$ and $H_n = Lb_n + Ub_n$ are decomposed into their subdiagonal parts (L) and diagonal and superdiagonal parts (U), La_n and Lb_n go to zero while Ua_n and Ub_n remain bounded, as $n \to \infty$. Then, from 3.3.7, $A_n = (La_n + Ua_n)D(Lb_n + Ub_n) = La_nDLb_n + Ua_nDLb_n + La_nDUb_n + Ua_nDUb_n$, and the first three terms go to zero, so the subdiagonal elements of A_n all go to zero. ■

Since A is real, complex eigenvalues will appear in conjugate pairs, with equal absolute values. Thus the assumptions in Theorem 3.3.1 are violated whenever A has complex eigenvalues, but then we should never have expected that using only real matrix multiplications we could converge to a triangular matrix if some of the eigenvalues are complex. Fortunately, however, the usefulness of the QR iteration is not limited to problems with real or distinct eigenvalues, as the following theorem shows.

Theorem 3.3.2. *If A is real, diagonalizable, and nonsingular, and if its eigenvalues occur in groups of at most two with equal absolute values, the QR iterate A_n approaches "quasitriangular" form, for large n. A quasitriangular matrix is upper Hessenberg with no two consecutive nonzero elements on the first subdiagonal.*

Proof: Under the assumptions on the eigenvalues d_{ii} of A, we can see from its expansion above that D^nLD^{-n} is quasitriangular for large n, since the only subdiagonal elements that do not converge to zero are on the first subdiagonal, and they are isolated from each other. Note that we are not saying that D^nLD^{-n} (or A_n) converges to *anything*; in fact the nonconvergent subdiagonal elements may well oscillate forever, but the other subdiagonal elements do converge to zero. It is left as a problem (Problem 5) to show that the nonzero structure of a quasitriangular matrix is preserved when it is multiplied by an upper triangular matrix or by another quasitriangular matrix of the same nonzero structure, and when it is inverted. Once these intuitively reasonable results are established, it follows from 3.3.8 that H_n is quasitriangular, and from 3.3.7 that A_n is quasitriangular, for large n. ■

Although the QR method convergence proofs given above are rather complex, it is possible to give an intuitive explanation for why this method works. Each transformation $A_n = Q_{n-1}^{-1}A_{n-1}Q_{n-1}$ is designed so that the premultiplication produces an upper triangular matrix (R_{n-1}). The ensuing postmultiplication destroys this triangular structure; however, while the premultiplication is systematically, purposefully shoveling nonzero "material" over into the upper triangle, the postmultiplication is scattering material about aimlessly. It is not hard to believe that the directed efforts of the premultiplication gradually win out over the undirected efforts of the postmultiplication, and the matrix gradually moves, *as far as it can*, toward upper triangular form.

An example of a quasitriangular matrix is given in Figure 3.3.2. We shall see that the real and complex eigenvalues can be calculated trivially once we reduce A to quasitriangular form.

$$\begin{bmatrix} X & X & X & X & X & X & X & X \\ 0 & X & X & X & X & X & X & X \\ 0 & S & X & X & X & X & X & X \\ 0 & 0 & 0 & X & X & X & X & X \\ 0 & 0 & 0 & S & X & X & X & X \\ 0 & 0 & 0 & 0 & 0 & X & X & X \\ 0 & 0 & 0 & 0 & 0 & 0 & X & X \\ 0 & 0 & 0 & 0 & 0 & 0 & 0 & X \end{bmatrix}$$

Figure 3.3.2
A Quasitriangular Matrix
S, nonzeros on first subdiagonal; X, other nonzero elements.

Our implementation of the QR method will use more Givens rotations, but now our strategy has changed. Previously, every time that we premultiplied our matrix by $Q_{ij}^{-1} = Q_{ij}^{\mathrm{T}}$ we immediately followed this with a postmultiplication by the corresponding inverse matrix Q_{ij}. Indeed, if we are to preserve the eigenvalues of A we must postmultiply by the inverse. However, with the QR method, we *temporarily* forget about the postmultiplications and simply premultiply by orthogonal Q_{ij}^{T} matrices until we have an upper triangular matrix. Only after we have reduced A to upper triangular form R do we go back and postmultiply by the corresponding Q_{ij} matrices. What difference does it make whether we do all the premultiplications first, or alternate between premultiplications and postmultiplications (as done for the Jacobi method or the reduction to Hessenberg form)? The only difference is that our choices for c and s will depend on the order in which we multiply these matrices.

The reduction to upper triangular form follows the pattern of the orthogonal reduction in Section 2.2, except that we begin with a Hessenberg matrix, which is already nearly triangular. In fact, we could apply the QR method directly to the original full matrix, but each reduction would require $O(N^3)$ operations. Since an upper Hessenberg matrix has at most $N-1$ nonzero subdiagonal elements, only $N-1$ premultiplications are required to reach triangular form, each of which requires $O(N)$ work to perform. There are an equal number of postmultiplications to be done afterward, and so the total work for one QR iteration is $O(N^2)$, if we start with a Hessenberg matrix. It is easy to verify (Figure 3.3.3) that a QR iteration does not destroy the Hessenberg structure. That is, although the $N-1$ postmultiplications destroy the upper triangular structure created by the premultiplications, the nonzeros do not spread beyond the first subdiagonal.

$$
\begin{array}{c}
\quad\quad i \quad i+1 \\
\begin{array}{c} \\ \\ i \\ i+1 \\ \\ \\ \\ \end{array}
\begin{bmatrix}
a & a & a & a & a & a & a \\
0 & a & a & a & a & a & a \\
0 & 0 & x & x & x & x & x \\
0 & 0 & Z & x & x & x & x \\
0 & 0 & 0 & a & a & a & a \\
0 & 0 & 0 & 0 & a & a & a \\
0 & 0 & 0 & 0 & 0 & a & a
\end{bmatrix}
\end{array}
\qquad
\begin{array}{c}
\quad\quad i \quad i+1 \\
\begin{array}{c} \\ \\ i \\ i+1 \\ \\ \\ \\ \end{array}
\begin{bmatrix}
a & a & x & x & a & a & a \\
a & a & x & x & a & a & a \\
0 & a & x & x & a & a & a \\
0 & 0 & N & x & a & a & a \\
0 & 0 & 0 & 0 & a & a & a \\
0 & 0 & 0 & 0 & 0 & a & a \\
0 & 0 & 0 & 0 & 0 & 0 & a
\end{bmatrix}
\end{array}
$$

ith premultiplication step $\qquad$ ith postmultiplication step

Figure 3.3.3
QR Iteration on a Hessenberg Matrix
Z, becomes zero; N, becomes nonzero; x, may change.

In the symmetric case, where we start with a tridiagonal matrix, the premultiplications eliminate the one subdiagonal of nonzeros but add one extra superdiagonal on top of the original structure. The postmultiplications return this triangular matrix to a symmetric tridiagonal form; this is as expected, since the QR iteration $A_n = Q_{n-1}^{\mathrm{T}} A_{n-1} Q_{n-1}$ preserves symmetry and each iterate is Hessenberg, and therefore tridiagonal. In the symmetric case, then, a QR iteration can be performed in only $O(N)$ operations.

Now, once we have a quasitriangular matrix, such as that shown in Figure 3.3.2, we can calculate the eigenvalues directly from the determinant equation

$$
\begin{aligned}
0 &= \det(A - \lambda I) \\
&= \det \begin{bmatrix}
a_{11}-\lambda & X & X & X & X & X & X & X \\
0 & a_{22}-\lambda & a_{23} & X & X & X & X & X \\
0 & a_{32} & a_{33}-\lambda & X & X & X & X & X \\
0 & 0 & 0 & a_{44}-\lambda & a_{45} & X & X & X \\
0 & 0 & 0 & a_{54} & a_{55}-\lambda & X & X & X \\
0 & 0 & 0 & 0 & 0 & a_{66}-\lambda & X & X \\
0 & 0 & 0 & 0 & 0 & 0 & a_{77}-\lambda & X \\
0 & 0 & 0 & 0 & 0 & 0 & 0 & a_{88}-\lambda
\end{bmatrix}
\end{aligned}
\tag{3.3.9}
$$

It can be shown (Problem 8) that the above determinant can be expanded as follows:

$$\det(A - \lambda I) = (a_{11} - \lambda) \det \begin{bmatrix} a_{22} - \lambda & a_{23} \\ a_{32} & a_{33} - \lambda \end{bmatrix}$$
$$\times \det \begin{bmatrix} a_{44} - \lambda & a_{45} \\ a_{54} & a_{55} - \lambda \end{bmatrix} (a_{66} - \lambda)(a_{77} - \lambda)(a_{88} - \lambda). \tag{3.3.10}$$

Thus we see that a_{11}, a_{66}, a_{77}, and a_{88} are real eigenvalues of A, and the four eigenvalues of the two 2 by 2 blocks complete the list of eigenvalues.

Figure 3.3.4 displays a FORTRAN program that finds the eigenvalues of a general real matrix, using the algorithms described above. First the matrix is reduced to upper Hessenberg form (in subroutine HESSQ), and then the QR method is iterated to convergence (in subroutine QR), that is, until the first subdiagonal contains only zeros and *isolated* nonzeros. Then the eigenvalues are computed directly from the quasitriangular matrix and returned in a complex vector. The eigenvectors are not calculated by DEGNON but, once we have an eigenvalue λ, the corresponding eigenvector(s) are easily calculated, by solving $(A - \lambda I)\boldsymbol{x} = \mathbf{0}$. In Section 3.5 a subroutine (DPOWER in Figure 3.5.2) will be given which can be used to find these eigenvectors, even when λ is complex.

```
      SUBROUTINE DEGNON(A,N,EIG)
      IMPLICIT DOUBLE PRECISION (A-H,O-Z)
C                                    DECLARATIONS FOR ARGUMENTS
      COMPLEX*16 EIG(N)
      DOUBLE PRECISION A(N,N)
      INTEGER N
C
C  SUBROUTINE DEGNON SOLVES THE EIGENVALUE PROBLEM
C
C                    A*X = LAMBDA*X
C
C    WHERE A IS A GENERAL REAL MATRIX.
C
C
C  ARGUMENTS
C
C                ON INPUT                        ON OUTPUT
C                --------                        ---------
C
C    A       - THE N BY N MATRIX.                DESTROYED.
C
C    N       - THE SIZE OF MATRIX A.
C
C    EIG     -                                   A COMPLEX N-VECTOR
C                                                CONTAINING THE EIGEN-
C                                                VALUES OF A.
```

```
C
C-----------------------------------------------------------------------
C                               EPS = MACHINE FLOATING POINT RELATIVE
C                                     PRECISION
C *****************************
      DATA EPS/2.D-16/
C *****************************
C                               AMAX = MAXIMUM ELEMENT OF A
      AMAX = 0.0
      DO 5 I=1,N
      DO 5 J=1,N
    5 AMAX = MAX(AMAX,ABS(A(I,J)))
      ERRLIM = SQRT(EPS)*AMAX
C                               REDUCTION TO HESSENBERG FORM
      CALL HESSQ(A,N)
C                               REDUCTION TO QUASI-TRIANGULAR FORM
      CALL QR(A,N,ERRLIM)
C                               EXTRACT EIGENVALUES OF QUASI-TRIANGULAR
C                               MATRIX
      I = 1
      DO WHILE (I.LE.N-1)
         IF (A(I+1,I).EQ.0.0) THEN
C                               1 BY 1 BLOCK ON DIAGONAL
            EIG(I) = A(I,I)
            I = I+1
         ELSE
C                               2 BY 2 BLOCK ON DIAGONAL
            DISC = (A(I,I)-A(I+1,I+1))**2 + 4.0*A(I,I+1)*A(I+1,I)
            TERM =  0.5*(A(I,I)+A(I+1,I+1))
            IF (DISC.GE.0.0) THEN
               EIG(I)  = TERM + 0.5*SQRT(DISC)
               EIG(I+1)= TERM - 0.5*SQRT(DISC)
            ELSE
               EIG(I)  = TERM + 0.5*SQRT(-DISC)*CMPLX(0.0,1.0)
               EIG(I+1)= TERM - 0.5*SQRT(-DISC)*CMPLX(0.0,1.0)
            ENDIF
            I = I+2
         ENDIF
      END DO
      IF (I.EQ.N) EIG(N) = A(N,N)
      RETURN
      END


      SUBROUTINE HESSQ(A,N)
      IMPLICIT DOUBLE PRECISION (A-H,O-Z)
C                               DECLARATIONS FOR ARGUMENTS
      DOUBLE PRECISION A(N,N)
```

```
      INTEGER N
      IF (N.LE.2) RETURN
C                                 USE GIVENS ROTATIONS TO REDUCE A
C                                 TO UPPER HESSENBERG FORM
      DO 20 I=2,N-1
         DO 15 J=I+1,N
            IF (A(J,I-1).EQ.0.0) GO TO 15
            DEN = SQRT(A(I,I-1)**2+A(J,I-1)**2)
            C = A(I,I-1)/DEN
            S = A(J,I-1)/DEN
C                                 PREMULTIPLY BY Qij**T
            DO 5 K=I-1,N
               PIK = C*A(I,K) + S*A(J,K)
               PJK =-S*A(I,K) + C*A(J,K)
               A(I,K) = PIK
               A(J,K) = PJK
    5       CONTINUE
C                                 POSTMULTIPLY BY Qij
            DO 10 K=1,N
               BKI = C*A(K,I) + S*A(K,J)
               BKJ =-S*A(K,I) + C*A(K,J)
               A(K,I) = BKI
               A(K,J) = BKJ
   10       CONTINUE
   15    CONTINUE
   20 CONTINUE
      RETURN
      END


      SUBROUTINE QR(A,N,ERRLIM)
      IMPLICIT DOUBLE PRECISION (A-H,O-Z)
C                                 DECLARATIONS FOR ARGUMENTS
      DOUBLE PRECISION A(N,N),ERRLIM
      INTEGER N
C                                 DECLARATIONS FOR LOCAL VARIABLES
      DOUBLE PRECISION SAVE(2,N)
      IF (N.LE.2) RETURN
C                                 USE QR ITERATION TO REDUCE HESSENBERG
C                                 MATRIX A TO QUASI-TRIANGULAR FORM
      NITER = 1000*N
      DO 35 ITER=1,NITER
C                                 REDUCE A TO UPPER TRIANGULAR FORM USING
C                                 ORTHOGONAL REDUCTION (PREMULTIPLY BY
C                                 Qij**T MATRICES)
         DO 10 I=1,N-1
            IF (A(I+1,I).EQ.0.0) THEN
               C = 1.0
```

```
            S = 0.0
         ELSE
            DEN = SQRT(A(I,I)**2 + A(I+1,I)**2)
            C = A(I,I)/DEN
            S = A(I+1,I)/DEN
         ENDIF
C                             USE SAVE TO SAVE C,S FOR POST-
C                             MULTIPLICATION PHASE
         SAVE(1,I) = C
         SAVE(2,I) = S
         IF (S.EQ.0.0) GO TO 10
C                             IF MATRIX SYMMETRIC, LIMITS ON K
C                             CAN BE:  K = I , MIN(I+2,N)
         DO 5 K=I,N
            PIK = C*A(I,K) + S*A(I+1,K)
            PJK =-S*A(I,K) + C*A(I+1,K)
            A(I,K)   = PIK
            A(I+1,K) = PJK
    5    CONTINUE
   10 CONTINUE
C                             NOW POSTMULTIPLY BY Qij MATRICES
      DO 20 I=1,N-1
         C = SAVE(1,I)
         S = SAVE(2,I)
         IF (S.EQ.0.0) GO TO 20
C                             IF MATRIX SYMMETRIC, LIMITS ON K
C                             CAN BE:  K = MAX(1,I-1) , I+1
         DO 15 K=1,I+1
            BKI = C*A(K,I) + S*A(K,I+1)
            BKJ =-S*A(K,I) + C*A(K,I+1)
            A(K,I)   = BKI
            A(K,I+1) = BKJ
   15    CONTINUE
   20 CONTINUE
C                             SET NEARLY ZERO SUBDIAGONALS TO ZERO,
C                             TO AVOID UNDERFLOW.
      DO 25 I=1,N-1
         IF (ABS(A(I+1,I)).LT.ERRLIM) A(I+1,I) = 0.0
   25 CONTINUE
C                             CHECK FOR CONVERGENCE TO "QUASI-
C                             TRIANGULAR" FORM.
      ICONV = 1
      DO 30 I=2,N-1
         IF (A(I,I-1).NE.0.0 .AND. A(I+1,I).NE.0.0) ICONV = 0
   30 CONTINUE
      IF (ICONV.EQ.1) RETURN
   35 CONTINUE
C                             HAS NOT CONVERGED IN NITER ITERATIONS
```

```
      PRINT 40
   40 FORMAT (' ***** QR ITERATION DOES NOT CONVERGE *****')
      RETURN
      END
```

Figure 3.3.4

Unless there are more than two eigenvalues of the same modulus, the QR iteration will (eventually) reduce the Hessenberg matrix to quasitriangular form, and DEGNON is able to extract automatically the eigenvalues of the 1 by 1 and 2 by 2 irreducible diagonal blocks in the final matrix. If there are groups of $k > 2$ eigenvalues with the same modulus, the final matrix may (or may not!) contain irreducible k by k diagonal blocks, which do not reveal their eigenvalues as easily as the 1 by 1 and 2 by 2 blocks of a quasitriangular matrix. Nevertheless, the QR iteration will almost always introduce some new zeros into the first subdiagonal of the Hessenberg matrix, and DEGNON will still calculate the eigenvalues of the 1 by 1 and 2 by 2 blocks automatically; the eigenvalues of the larger blocks can be calculated "by hand", if they are not too large.

Although DEGNON will nearly always eventually find all the eigenvalues of A, convergence can sometimes be quite slow. The rate of convergence to zero of the subdiagonal element $a_{i+1,i}$ depends on the magnitude of the ratio $|\lambda_{i+1}|/|\lambda_i|$. Note, however, that DEGNON stops when A reaches quasitriangular form; so, if there are two eigenvalues of nearly equal (or equal) absolute values, this will not hurt anything. Only if there are three clustered closely together will we have slow convergence, since only then will two consecutive subdiagonal elements converge to zero slowly.

More sophisticated codes employ a "shifted QR" method, which may speed convergence substantially. The shifted QR method is based on the transformation

$$\begin{aligned} A_n &= Q_{n-1}^{-1}(A_{n-1} - \sigma_{n-1}I)Q_{n-1} + \sigma_{n-1}I \\ &= Q_{n-1}^{-1}A_{n-1}Q_{n-1}, \end{aligned} \tag{3.3.11}$$

where $R_{n-1} \equiv Q_{n-1}^{-1}(A_{n-1} - \sigma_{n-1}I)$ is upper triangular.

The philosophy behind the shifts is easily seen by looking at the expansion for D^nLD^{-n} given earlier and noting that, when we do several iterations of the shifted QR method with a constant shift σ, we are really just applying the ordinary QR method to the matrix $B = A - \sigma I$ (and restoring the σI at the end). This shifts the eigenvalues from d_{ii} to $d_{ii} - \sigma$, and, by choosing σ close to an eigenvalue d_{kk}, we can cause the subdiagonal elements of D^nLD^{-n} which contain $d_{kk} - \sigma$ in the numerator to converge rapidly to zero. By cleverly varying σ_n, then, we can speed selected subdiagonal elements on their way to zero (see Problem 9).

The QR method with shifts is considered by many to be the best method for the calculation of all the eigenvalues of a general real matrix. It is implemented by the most popular eigenvalue codes, such as those in the IMSL Library and EISPACK [Smith et al. 1976].

DEGNON was used to find the eigenvalues of the *symmetric* matrix in Figure 3.1.1, 101 QR iterations were required, and the eigenvalues returned agreed with those calculated by DEGSYM, shown in Table 3.2.1, to the number of digits shown there. Note that the QR iteration converged despite the fact that this matrix has four double eigenvalues, and one real eigenvalue of multiplicity 4.

We also used DEGNON to find the eigenvalues of the nonsymmetric matrix shown below:

$$\begin{bmatrix} -5 & 7 & 3 & 4 & -8 \\ 5 & 8 & 3 & 6 & 8 \\ 3 & -7 & 9 & -4 & 5 \\ -3 & 0 & 4 & 5 & 3 \\ 7 & 4 & 5 & 9 & 5 \end{bmatrix}. \tag{3.3.12}$$

The eigenvalues output by DEGNON were

$$\begin{aligned} &13.1406621 + 4.9368807\, i \\ &13.1406621 - 4.9368807\, i \\ &-4.5805667 + 6.9420509\, i \\ &-4.5805667 - 6.9420509\, i \\ &4.8798093 \end{aligned}$$

In Section 3.5 we shall see how eigenvectors of this matrix can be calculated, using inverse iteration.

3.4 Alternative Methods for General Matrices

The transformation of a general real matrix to Hessenberg form can also be accomplished using the symmetric orthogonal "Householder matrices" that were introduced in Section 2.3. Recall that the Householder matrix H_i $(= I - 2\boldsymbol{\omega}\boldsymbol{\omega}^{\mathrm{T}})$ has the form

$$H_i = \begin{array}{r} \\ \\ \\ \\ \\ \text{row } i \\ \\ \\ \text{row } N \end{array} \begin{bmatrix} 1 & & & & & & & & \\ & 1 & & & & & & & \\ & & 1 & & & & & & \\ & & & 1 & & & & & \\ & & & & 1 & & & & \\ & & & & & X & X & X & X \\ & & & & & X & X & X & X \\ & & & & & X & X & X & X \\ & & & & & X & X & X & X \end{bmatrix}.$$

(column i is the first column of the X block; column N is the last column.)

Since $H_i^{-1} = H_i^{\mathrm{T}}$ (because H_i is orthogonal) and $H_i^{\mathrm{T}} = H_i$ (because it is symmetric), a Householder transformation has the form $A_{n+1} = H_i A_n H_i$. Note that premultiplication by H_i has the effect of changing only rows i to N, while postmultiplication by H_i changes columns i to N. In Section 2.3 the nonzero portion of H_i was chosen so that premultiplication by H_i would zero the elements in rows $i+1$ to M of the pivot column i (we assume that the columns of A are independent, so that $l = i$). However, to preserve the eigenvalues of A, we now have to postmultiply by H_i, and the postmultiplication, since it changes columns i to N, will destroy the zeros in column i created by the premultiplication. Recall that we faced a similar problem using the Givens Q_{ij} matrices and, to ensure that the postmultiplication did not destroy the zeros created by the premultiplication, we used Q_{ij} to create (during premultiplication) zeros back in the $(i-1)$st column, where they are safe from the postmultiplication, which changes only columns i and j $(j > i)$. In a similar manner, we shall have to use H_i to zero the elements in rows $i+1$ to N, not of column i but of the previous column $i-1$, where they will be safe during the postmultiplication phase, when columns i to N are altered. This means, again, that the elements $a_{i,i-1}$ on the first subdiagonal cannot be zeroed (Figure 3.4.1).

$$
\begin{array}{r}
\\ \\ \\ \\ \\ \text{row } i \\ \\ \\ \text{row } N
\end{array}
\begin{array}{c}
\begin{array}{ccccccccc}
 & & & & & \text{column} & & & \text{column} \\
 & & & & & i & & & N
\end{array} \\
\left[\begin{array}{ccccccccc}
a & a & a & a & a & x & x & x & x \\
a & a & a & a & a & x & x & x & x \\
0 & a & a & a & a & x & x & x & x \\
0 & 0 & a & a & a & x & x & x & x \\
0 & 0 & 0 & a & a & x & x & x & x \\
0 & 0 & 0 & 0 & x & x & x & x & x \\
0 & 0 & 0 & 0 & Z & x & x & x & x \\
0 & 0 & 0 & 0 & Z & x & x & x & x \\
0 & 0 & 0 & 0 & Z & x & x & x & x
\end{array}\right]
\end{array}
$$

Figure 3.4.1
One Householder Iteration, Using H_i
Z, zeroed by premultiplication (and not changed by postmultiplication)
x, may change during this iteration.

Since the H_i matrices are orthogonal, symmetry is preserved and, if we start with a symmetric matrix, we shall end up with a tridiagonal matrix, as before.

To use Householder transformations in place of Givens rotations in DEGNON, we replace subroutine HESSQ in Figure 3.3.4 by the subroutine HESSH shown in Figure 3.4.2. Subroutine CALW in Figure 2.3.1 must also be loaded; we use this subroutine to determine the unit vector $\boldsymbol{\omega}$ such that premultiplication by $H_i = I - 2\boldsymbol{\omega}\boldsymbol{\omega}^{\mathrm{T}}$ will zero components $i+1$ to N of column $i-1$.

The premultiplication is done (as in Section 2.3) in the following manner:

$$H_i A = (I - 2\boldsymbol{\omega}\boldsymbol{\omega}^{\mathrm{T}})A = A - 2\boldsymbol{\omega}(\boldsymbol{\omega}^{\mathrm{T}} A).$$

The formation of $\boldsymbol{\omega}$ (in subroutine CALW) requires only $O(N)$ work. Then only columns $i-1$ to N of A are processed, since the previous columns have already been reduced to their Hessenberg form and will not be altered by the premultiplication (see Figure 3.4.1). Premultiplication of each column $\boldsymbol{a}_k$ $(k = i-1, \ldots, N)$ by H_i involves first calculating the scalar product $\omega ta = \boldsymbol{\omega}^{\mathrm{T}}\boldsymbol{a}_k$ (loop 5) and then subtracting $2\omega ta$ times $\boldsymbol{\omega}$ from $\boldsymbol{a}_k$ (loop 10). Each of these calculations requires $N-(i-1)$ multiplications, since the first $i-1$ components of $\boldsymbol{\omega}$ are zero. Thus about $2(N-i)$ multiplications per column are required and, since there are about $N-i$ columns to process, the total work during the premultiplication stage is about $2(N-i)^2$ multiplications.

```
      SUBROUTINE HESSH(A,N)
      IMPLICIT DOUBLE PRECISION (A-H,O-Z)
C                                DECLARATIONS FOR ARGUMENTS
      DOUBLE PRECISION A(N,N)
      INTEGER N
C                                DECLARATIONS FOR LOCAL VARIABLES
      DOUBLE PRECISION W(N)
      IF (N.LE.2) RETURN
C                                USE HOUSEHOLDER TRANSFORMATIONS TO
C                                REDUCE A TO UPPER HESSENBERG FORM
      DO 35 I=2,N-1
C                                CHOOSE UNIT N-VECTOR W (WHOSE FIRST
C                                I-1 COMPONENTS ARE ZERO) SUCH THAT WHEN
C                                COLUMN I-1 IS PREMULTIPLIED BY
C                                H = I - 2W*W**T, COMPONENTS I+1 THROUGH
C                                N ARE ZEROED.
         CALL CALW (A(1,I-1),N,W,I)
C                                PREMULTIPLY A BY H = I - 2W*W**T
         DO 15 K=I-1,N
            WTA = 0.0
            DO 5 J=I,N
               WTA = WTA + W(J)*A(J,K)
    5       CONTINUE
            TWOWTA = 2*WTA
            DO 10 J=I,N
               A(J,K) = A(J,K) - TWOWTA*W(J)
   10       CONTINUE
   15    CONTINUE
C                                POSTMULTIPLY A BY H = I - 2W*W**T
         DO 30 K=1,N
            ATW = 0.0
            DO 20 J=I,N
               ATW = ATW + A(K,J)*W(J)
```

```
20          CONTINUE
            TWOATW = 2*ATW
            DO 25 J=I,N
               A(K,J) = A(K,J) - TWOATW*W(J)
25          CONTINUE
30       CONTINUE
35    CONTINUE
      RETURN
      END
```

Figure 3.4.2

The postmultiplication is done similarly:

$$AH_i = A(I - 2\boldsymbol{\omega}\boldsymbol{\omega}^{\mathrm{T}}) = A - 2(A\boldsymbol{\omega})\boldsymbol{\omega}^{\mathrm{T}}.$$

Now all N rows have to be processed. Postmultiplication of each row $\boldsymbol{a}_k^{\mathrm{T}}$ ($k = 1, \ldots, N$) by H_i involves first calculating the scalar product $at\omega = \boldsymbol{a}_k^{\mathrm{T}}\boldsymbol{\omega}$ (loop 20), and subtracting $2at\omega$ times $\boldsymbol{\omega}^{\mathrm{T}}$ from $\boldsymbol{a}_k^{\mathrm{T}}$ (loop 25). Each of these calculations requires $N-(i-1)$ multiplications and, since there are N rows to process, the total work during the postmultiplication by H_i is about $2N(N-i)$. Since transformations involving $H_2, \ldots, H_{N-1}$ have to be performed, the total number of multiplications done in the reduction to Hessenberg form is about (see 0.1.1)

$$\sum_{i=2}^{N-1} [2(N-i)^2 + 2N(N-i)] = \sum_{i=2}^{N-1} (4N^2 - 6Ni + 2i^2) \approx \frac{5}{3}N^3.$$

While each Householder transformation reduces an entire column to its Hessenberg form, a Givens rotation eliminates only one element at a time. For each i ($i = 2, \ldots, N-1$), there are $N - i$ elements to zero, and it requires $4[(N - (i - 2)] + 4N$ multiplications to knock out each (see loops 5 and 10 of subroutine HESSQ in Figure 3.3.4). Thus the total number of multiplications required to reduce a general matrix to Hessenberg form using Givens transformations is about

$$\sum_{i=2}^{N-1} (N-i)[4(N-i) + 4N] \approx \frac{10}{3}N^3.$$

So, while both methods require $O(N^3)$ work, the Householder method is twice as fast, for large problems. However, for most problems, the time spent finding the eigenvalues of the Hessenberg matrix (using, e.g., the QR method) will be substantially greater than the computer time spent reducing the original matrix to Hessenberg form; so replacing HESSQ by HESSH will speed up DEGNON by a factor of much less than 2. There is nothing to be gained by using Householder transformations to QR factor the upper Hessenberg matrix, since only one element per column is zeroed.

The elementary matrices M_{ij} of Section 1.4 can also be used to transform a general real matrix to Hessenberg form. Recall that M_{ij} and M_{ij}^{-1} have the forms

$$
M_{ij} = \begin{array}{r} \\ \\ \text{row } i \\ \\ \\ \text{row } j \\ \\ \\ \end{array}
\begin{bmatrix}
1 &&&&&&& \\
& 1 &&&&&& \\
&& 1 &&&&& \\
&&& 1 &&&& \\
&&&& 1 &&& \\
&& r &&& 1 && \\
&&&&&& 1 & \\
&&&&&&& 1
\end{bmatrix}
\begin{array}{l} \text{(column } i \text{, column } j\text{)} \end{array},
\qquad
M_{ij}^{-1} = \begin{array}{r} \\ \\ \text{row } i \\ \\ \\ \text{row } j \\ \\ \\ \end{array}
\begin{bmatrix}
1 &&&&&&& \\
& 1 &&&&&& \\
&& 1 &&&&& \\
&&& 1 &&&& \\
&&&& 1 &&& \\
&& -r &&& 1 && \\
&&&&&& 1 & \\
&&&&&&& 1
\end{bmatrix}
$$

Each transformation has the form $A_{n+1} = M_{ij}^{-1} A_n M_{ij}$. Note that premultiplication by M_{ij}^{-1} has the effect of subtracting r times row i from row j, while postmultiplication by M_{ij} has the effect of adding r times column j to column i. In Section 1.4 we were able to reduce A to upper triangular form by premultiplying with M_{ij} matrices, but now the postmultiplication threat forces us to follow the tactics outlined in Section 3.3, and to use M_{ij} to zero $a_{j,i-1}$ rather than a_{ji}. If we choose $r = a_{j,i-1}/a_{i,i-1}$, we shall introduce a zero into row j, column $i-1$ during the premultiplication stage and, since the postmultiplication alters only column i, the zero introduced into column $i-1$ will survive the postmultiplication (Figure 3.4.3).

$$
\begin{bmatrix}
1 &&&&&&&& \\
& 1 &&&&&&& \\
&& 1 &&&&&& \\
&&& 1 &&&&& \\
&&&& 1 &&&& \\
&&&&& 1 &&& \\
&& -r &&&& 1 && \\
&&&&&&& 1 & \\
&&&&&&&& 1
\end{bmatrix}
\begin{bmatrix}
& \vdots & & \vdots & \\
\cdots\ a_{i,i-1} & a_{ii} & \cdots & a_{ij} & \cdots \\
& \vdots & & \vdots & \\
\cdots\ a_{j,i-1} & a_{ji} & \cdots & a_{jj} & \cdots \\
& \vdots & & \vdots &
\end{bmatrix}
\begin{bmatrix}
1 &&&&&&&& \\
& 1 &&&&&&& \\
&& 1 &&&&&& \\
&&& 1 &&&&& \\
&&&& 1 &&&& \\
&&&&& 1 &&& \\
&& r &&&& 1 && \\
&&&&&&& 1 & \\
&&&&&&&& 1
\end{bmatrix}
$$

Figure 3.4.3

We can now proceed to reduce A to upper Hessenberg form, eliminating subdiagonal elements column by column, beginning with the first column, just as we did using Givens transformations. However, when we calculated the elements c and s of Q_{ij} using equations 3.3.2, there was no danger of a zero denominator (unless the element to be eliminated was already zero, in which case we simply skip the transformation); now we have a problem if $a_{i,i-1}$ is zero.

During Gaussian elimination we use the diagonal elements a_{ii} as pivots, to "knock out" the elements below and, if one of the pivots is zero, we have to switch rows to move a nonzero element to the pivot position. Here we are essentially using the elements $a_{i,i-1}$ on the first *subdiagonal* as pivots, to knock out the elements below the first subdiagonal. Now, if $a_{i,i-1} = 0$, then $r = \infty$, and we must switch row i with a row $l > i$, to bring a nonzero element into the pivot position. However, switching rows i and l is equivalent to premultiplying A by a permutation matrix P_{il} (P_{il} is just the identity matrix with rows i and l switched), and so we must also postmultiply by $P_{il}^{-1} = P_{il}$, which has the effect of switching columns i and l. Fortunately, switching columns i and l ($l > i$) does not change any elements in columns 1 to $i-1$.

As we learned in our study of Gaussian elimination, it is a good idea not to wait until a pivot becomes exactly zero to switch rows. It is better to always bring the potential pivot (in this case the potential pivots are $a_{i,i-1}, \ldots, a_{N,i-1}$) that has the largest absolute value up to the pivot position before knocking out the elements below the pivot. If all the potential pivots are zero, this is not a problem; go on to the next column!

In summary, the reduction of a general full matrix to upper Hessenberg form using M_{ij} and P_{il} matrices is very much similar to Gaussian elimination (which reduces A to upper triangular form) with partial pivoting, except for two things. First, we use the elements on the first subdiagonal, rather than those on the main diagonal, to knock out the lower elements. Second, every time that we take a multiple $-r$ of row i and add it to row j, we must then add r times column j to column i and, every time that we switch rows i and l, we must also switch columns i and l.

If subroutine HESSQ in Figure 3.3.4 is replaced by subroutine HESSM in Figure 3.4.4, the reduction to Hessenberg form will be done by M_{ij} and P_{il} matrices. By comparing loops 5 and 10 of subroutine HESSQ with loops 20 and 25 of HESSM, we see that the transformation $A_{n+1} = M_{ij}^{-1} A_n M_{ij}$ requires only a quarter as many multiplications as a Givens transformation, and the same number of transformations are required. Thus this approach is four times as fast as using Givens rotations, and twice as fast as using Householder transformations. On the other hand, the M_{ij} transformations do not preserve symmetry; so, even if we begin with a symmetric matrix, the result will not be a tridiagonal matrix.

There are also alternatives to the QR method for extracting the eigenvalues of an upper Hessenberg (or tridiagonal) matrix. The LR method [Rutishauser 1958] is closely related to the QR method, but it is based not on the QR decomposition, but on the LU (or "LR") decomposition (see 1.4.7). Recall that one iteration of the QR method involves premultiplication by several elementary orthogonal matrices (usually Givens Q_{ij} matrices) to reduce A to upper triangular form, followed by several postmultiplications by the inverses (transposes) of the same elementary orthogonal matrices. Similarly, to do one iteration of the LR method, we reduce A to upper triangular form using Gaussian elimination, that is, we premultiply A by several elementary M_{ij} matrices, and *then* we postmultiply by the inverses of these same matrices. Although the postmultiplications destroy the upper triangular structure created by the premultiplications, the LR iteration, like the QR iteration, usually makes gradual progress toward triangular, or quasitriangular, form.

```
      SUBROUTINE HESSM(A,N)
      IMPLICIT DOUBLE PRECISION (A-H,O-Z)
C                                 DECLARATIONS FOR ARGUMENTS
      DOUBLE PRECISION A(N,N)
      INTEGER N
      IF (N.LE.2) RETURN
C                                 USE Mij TRANSFORMATIONS TO REDUCE A
C                                 TO UPPER HESSENBERG FORM
      DO 35 I=2,N-1
C                                 SEARCH FROM A(I,I-1) ON DOWN FOR
C                                 LARGEST POTENTIAL PIVOT, A(L,I-1)
         BIG = ABS(A(I,I-1))
         L = I
         DO 5 J=I+1,N
            IF (ABS(A(J,I-1)).GT.BIG) THEN
               BIG = ABS(A(J,I-1))
               L = J
            ENDIF
    5    CONTINUE
C                                 IF ALL SUBDIAGONAL ELEMENTS IN COLUMN
C                                 I-1 ALREADY ZERO, GO ON TO NEXT COLUMN
         IF (BIG.EQ.0.0) GO TO 35
C                                 PREMULTIPLY BY Pil
C                                 (SWITCH ROWS I AND L)
         DO 10 K=I-1,N
            TEMP = A(L,K)
            A(L,K) = A(I,K)
            A(I,K) = TEMP
   10    CONTINUE
C                                 POSTMULTIPLY BY Pil**(-1) = Pil
C                                 (SWITCH COLUMNS I AND L)
         DO 15 K=1,N
            TEMP = A(K,L)
            A(K,L) = A(K,I)
            A(K,I) = TEMP
   15    CONTINUE
         DO 30 J=I+1,N
            R = A(J,I-1)/A(I,I-1)
            IF (R.EQ.0.0) GO TO 30
C                                 PREMULTIPLY BY Mij**(-1)
C                                 (SUBTRACT R TIMES ROW I FROM ROW J)
            DO 20 K=I-1,N
               A(J,K) = A(J,K) - R*A(I,K)
   20       CONTINUE
C                                 POSTMULTIPLY BY Mij
C                                 (ADD R TIMES COLUMN J TO COLUMN I)
            DO 25 K=1,N
               A(K,I) = A(K,I) + R*A(K,J)
```

```
25          CONTINUE
30       CONTINUE
35 CONTINUE
   RETURN
   END
```

Figure 3.4.4

In fact, the assumption in the convergence Theorem 3.3.1 that Q_n is orthogonal was only needed to ensure that H_n and H_n^{-1} remain bounded as they approach upper triangular form; formulas 3.3.7-8 still hold if Q_n is any nonsingular matrix which makes $Q_n^{-1}A_n$ upper triangular. Thus it is reasonable to expect convergence of the LR iteration for many problems.

If A is upper Hessenberg before the premultiplications reduce it to upper triangular form, the postmultiplications will return A to its Hessenberg form (cf. Figure 3.3.3). In fact, if A is tridiagonal, it is easy to see that it will remain tridiagonal throughout the LR iteration, provided pivoting is not allowed; the QR iteration, by contrast, only preserves *symmetric* tridiagonal structures.

Figure 3.4.5 shows a subroutine LR that can be used to replace subroutine QR in Figure 3.3.4. This subroutine uses the LR iteration to reduce an upper Hessenberg matrix to triangular, or quasitriangular, form. Note that, during the premultiplication phase, we only switch rows i and $i+1$ (and then switch columns i and $i+1$ later, during the postmultiplication phase) when a (nearly) zero pivot is encountered. Partial pivoting is not done, because we have observed that (surprisingly) this often makes the LR iteration *less* stable numerically (see Problem 13).

Thus the calculation of the eigenvalues of a general real matrix can be done in two phases: First the matrix is reduced to upper Hessenberg form, in a finite number ($O(N^3)$) of computations, and then an iterative method is used to compute the eigenvalues of the Hessenberg matrix. We see now that for each phase there are available methods based on orthogonal reduction, and methods based on Gaussian elimination. The Gaussian elimination-based methods are faster; nevertheless, the orthogonal reduction methods are much more widely used, even for nonsymmetric problems, where preservation of symmetry cannot be cited as a reason for preferring orthogonal transformations.

The reason for this is that orthogonal transformations are more stable with respect to roundoff error. We spent some time in Chapter 1 worrying about roundoff error, because that is the only error that we have when $A\boldsymbol{x} = \boldsymbol{b}$ is solved using direct methods. In this chapter we have ignored the problem of roundoff error because we have other errors to worry about when eigenvalue problems are solved. The reader is referred to the book by Wilkinson [1965] for a thorough treatment of this topic, and for a rigorous demonstration of the superiority of orthogonal transformations. The heart of the matter is that an orthogonal transformation does not change the Euclidean norm (2-norm) of a matrix, for, if $B = Q^T AQ$, where $Q^T Q = I$, then $B^T B = Q^T A^T AQ$ and by Theorem 0.3.1, since $B^T B$ and $A^T A$ are similar, A and B have the same 2-norm. On the other hand, multiplication by nonorthogonal matrices can cause large elements to arise (see Problem 11).

```
      SUBROUTINE LR(A,N,ERRLIM)
      IMPLICIT DOUBLE PRECISION (A-H,O-Z)
C                                DECLARATIONS FOR ARGUMENTS
      DOUBLE PRECISION A(N,N),ERRLIM
      INTEGER N
C                                DECLARATIONS FOR LOCAL VARIABLES
      DOUBLE PRECISION SAVE(N)
      INTEGER IPERM(N)
      LOGICAL PIVOT
      IF (N.LE.2) RETURN
C                                USE LR ITERATION TO REDUCE HESSENBERG
C                                MATRIX A TO QUASI-TRIANGULAR FORM
C
C                                PIVOT = .TRUE. IF PIVOTING ALLOWED
      PIVOT = .TRUE.
      NITER = 1000*N
      DO 45 ITER=1,NITER
C                                REDUCE A TO UPPER TRIANGULAR FORM USING
C                                GAUSSIAN ELIMINATION (PREMULTIPLY BY
C                                Mij**(-1) MATRICES)
         DO 15 I=1,N-1
            IF (ABS(A(I,I)).LT.ERRLIM .AND. PIVOT) THEN
C                                SWITCH ROWS I AND I+1 IF NECESSARY
C                                (PREMULTIPLY BY Pi,i+1)
               IPERM(I) = I+1
               DO 5 K=I,N
                  TEMP = A(I+1,K)
                  A(I+1,K) = A(I,K)
                  A(I,K) = TEMP
    5          CONTINUE
            ELSE
               IPERM(I) = I
            ENDIF
            IF (ABS(A(I+1,I)).LT.ERRLIM) THEN
               R = 0
            ELSE
               IF (ABS(A(I,I)).LT.ERRLIM) GO TO 50
               R = A(I+1,I)/A(I,I)
            ENDIF
C                                USE SAVE TO SAVE R FOR POST-
C                                MULTIPLICATION PHASE
            SAVE(I) = R
            IF (R.EQ.0.0) GO TO 15
```

```
C                                   IF MATRIX TRIDIAGONAL, AND PIVOTING NOT
C                                   DONE, LIMITS ON K CAN BE:  K = I , I+1
            DO 10 K=I,N
               A(I+1,K) = A(I+1,K) - R*A(I,K)
   10       CONTINUE
   15    CONTINUE
C                                   NOW POSTMULTIPLY BY Mij MATRICES
         DO 30 I=1,N-1
            IF (IPERM(I).NE.I) THEN
C                                   SWITCH COLUMNS I AND I+1 IF NECESSARY
C                                   (POSTMULTIPLY BY Pi,i+1**(-1) = Pi,i+1)
               DO 20 K=1,I+1
                  TEMP = A(K,I+1)
                  A(K,I+1) = A(K,I)
                  A(K,I) = TEMP
   20          CONTINUE
            ENDIF
            R = SAVE(I)
            IF (R.EQ.0.0) GO TO 30
C                                   IF MATRIX TRIDIAGONAL, AND PIVOTING NOT
C                                   DONE, LIMITS ON K CAN BE:  K = I , I+1
            DO 25 K=1,I+1
               A(K,I) = A(K,I) + R*A(K,I+1)
   25       CONTINUE
   30    CONTINUE
C                                   SET NEARLY ZERO SUBDIAGONALS TO ZERO,
C                                   TO AVOID UNDERFLOW.
         DO 35 I=1,N-1
            IF (ABS(A(I+1,I)).LT.ERRLIM) A(I+1,I) = 0.0
   35    CONTINUE
C                                   CHECK FOR CONVERGENCE TO "QUASI-
C                                   TRIANGULAR" FORM.
         ICONV = 1
         DO 40 I=2,N-1
            IF (A(I,I-1).NE.0.0 .AND. A(I+1,I).NE.0.0) ICONV = 0
   40    CONTINUE
         IF (ICONV.EQ.1) RETURN
   45 CONTINUE
C                                   HAS NOT CONVERGED IN NITER ITERATIONS
   50 PRINT 55
   55 FORMAT (' ***** LR ITERATION DOES NOT CONVERGE *****')
      RETURN
      END
```

Figure 3.4.5

There are other alternatives as well, for finding the eigenvalues of an upper Hessenberg matrix. One is the inverse power method (Section 3.5), which finds only one eigenvalue at a time. Another obvious alternative is to use a good nonlinear equation solver to find the roots of $f(\lambda) = \det(A - \lambda I)$ and, for each given λ, to evaluate this determinant directly. If we are prepared to do complex arithmetic, there is no reason why we cannot find the complex eigenvalues of A in this manner.

The fastest way to evaluate a determinant is to reduce the matrix to upper triangular form, using ordinary Gaussian elimination. As long as we only add multiples of one row to another, the determinant does not change and, since the determinant of an upper triangular matrix is the product of its diagonal entries, this product gives the determinant of the original matrix. (If we have to switch rows, each row switch reverses the sign of the determinant.) Now $O(N^3)$ operations are required to reduce a full matrix to upper triangular form, and thus to calculate its determinant, but it takes only $O(N^2)$ operations to reduce an upper Hessenberg matrix to triangular form, and $O(N)$ to reduce a tridiagonal matrix. Thus, once a matrix has been reduced to Hessenberg (or tridiagonal) form, the function $f(\lambda) = \det(A - \lambda I)$ can be evaluated much more economically, and this approach may be feasible if an efficient and robust root finder is available.

Figure 3.6.1 provides a subroutine DEGENP which actually calculates the coefficients of the characteristic polynomial $f(\lambda) = \det(A - \lambda B)$, where A is upper Hessenberg and B is upper triangular, in $O(N^3)$ operations ($O(N^2)$ if A is tridiagonal and B is diagonal). Thus this routine could be called with $B = I$, and a good polynomial root finder could then be used to compute the eigenvalues, but only for small or moderate size N, since finding roots of a high degree polynomial is an unstable process numerically.

3.5 The Power and Inverse Power Methods

Often in applications we are only interested in one or a few of the eigenvalues of A—usually the largest or smallest. For example, only the smallest few eigenvalues of the matrix displayed in Figure 3.1.1 are reasonable approximations to the eigenvalues of the partial differential equation eigenproblem 3.1.2; so only the first few eigenvalues hold any interest for us. (This is not surprising, since the PDE eigenproblem has an infinite number of eigenvalues $\lambda_{kl} = (k^2 + l^2)\pi^2, k, l = 1, 2, \ldots$, while the matrix eigenvalue problem has only 16.) In such cases, the power or inverse power methods, which quickly find one eigenvalue at a time, may be more appropriate than the transformation methods in Sections 3.2–3.4.

The power method is marvelously simple: We just pick a starting vector $\boldsymbol{v}_0$ and start multiplying it by the matrix A. The following theorem states that, unless we are very unlucky in our choice of starting vector, the power method will find the eigenvalue of A of largest modulus, provided only that there is an eigenvalue of largest modulus.

Theorem 3.5.1. *Suppose the matrix A has one eigenvalue λ_1, which is greater in absolute value than all other eigenvalues, and let $\boldsymbol{v}_{n+1} = A\boldsymbol{v}_n$. Then, if the nonzero vectors $\boldsymbol{v}_0$ and $\boldsymbol{u}$ are chosen randomly, $\boldsymbol{u}^{\mathrm{T}}\boldsymbol{v}_{n+1}/\boldsymbol{u}^{\mathrm{T}}\boldsymbol{v}_n$ will converge to λ_1, with probability one.*

Proof: Let us first suppose A is diagonalizable, which will be true, for example, if A is symmetric or has distinct eigenvalues. The proof for this special case is much easier and more instructive, and it will motivate the proof for the general case.

We assume, then, that $A = SDS^{-1}$, where D is a diagonal matrix containing the eigenvalues of A in order of descending modulus, and so $A^n = SD^nS^{-1}$. If the first L diagonal elements of D are equal to λ_1 (note that the assumptions of Theorem 3.5.1 do not prevent λ_1 from being a multiple eigenvalue; what must be avoided are unequal eigenvalues of equal absolute values), then the fastest-growing components of D^n are equal to λ_1^n. So we divide D^n by λ_1^n, take the limit as $n \to \infty$, and get

$$\lim_{n\to\infty}\left(\frac{D^n}{\lambda_1^n}\right) = \lim_{n\to\infty}\begin{bmatrix} \left(\frac{\lambda_1}{\lambda_1}\right)^n & & & & & & \\ & \ddots & & & & & \\ & & \left(\frac{\lambda_1}{\lambda_1}\right)^n & & & & \\ & & & \left(\frac{\lambda_2}{\lambda_1}\right)^n & & & \\ & & & & \ddots & & \\ & & & & & \left(\frac{\lambda_3}{\lambda_1}\right)^n & \\ & & & & & & \ddots \end{bmatrix}$$

$$= \begin{bmatrix} 1 & & & & & & \\ & \ddots & & & & & \\ & & 1 & & & & \\ & & & 0 & & & \\ & & & & \ddots & & \\ & & & & & 0 & \\ & & & & & & \ddots \end{bmatrix} \equiv E.$$

Then, since $A^n = SD^nS^{-1}$, we have

$$\lim_{n\to\infty}\left(\frac{A^n}{\lambda_1^n}\right) = SES^{-1} \equiv G.$$

Now G is not the zero matrix, because otherwise $E = S^{-1}GS$ would be zero also, and it is not. Therefore the null space of G is a subspace of dimension at most $N-1$, and hence, if $\boldsymbol{v}_0$ is chosen at random, the probability is one that $G\boldsymbol{v}_0 \neq \boldsymbol{0}$. Then, if $\boldsymbol{u}$ is chosen randomly, the probability is one that it is not perpendicular to $G\boldsymbol{v}_0$; so $\boldsymbol{u}^{\mathrm{T}}G\boldsymbol{v}_0 \neq 0$. Then

$$\begin{aligned}\lim_{n\to\infty}\left(\frac{\boldsymbol{u}^{\mathrm{T}}\boldsymbol{v}_{n+1}}{\boldsymbol{u}^{\mathrm{T}}\boldsymbol{v}_n}\right) &= \lim_{n\to\infty}\left(\frac{\boldsymbol{u}^{\mathrm{T}}(A^{n+1}\boldsymbol{v}_0)}{\boldsymbol{u}^{\mathrm{T}}(A^n\boldsymbol{v}_0)}\right) = \lim_{n\to\infty}\left(\frac{\boldsymbol{u}^{\mathrm{T}}(A^{n+1}\boldsymbol{v}_0/\lambda_1^{n+1})}{\boldsymbol{u}^{\mathrm{T}}(A^n\boldsymbol{v}_0/\lambda_1^n)}\lambda_1\right)\\ &= \frac{\boldsymbol{u}^{\mathrm{T}}G\boldsymbol{v}_0}{\boldsymbol{u}^{\mathrm{T}}G\boldsymbol{v}_0}\lambda_1 = \lambda_1.\end{aligned}$$

Now we have shown that $\boldsymbol{u}^{\mathrm{T}}\boldsymbol{v}_{n+1}/\boldsymbol{u}^{\mathrm{T}}\boldsymbol{v}_n$ will (almost always) converge to the eigenvalue λ_1, provided that A is diagonalizable.

If A is *not* diagonalizable, the proof follows the same pattern, but it is more involved, and begins with $A = SJS^{-1}$, where J is the Jordan canonical form of A. It is known that *any* square matrix is similar to a matrix J of the form

$$\begin{bmatrix} J_1 & & & \\ & J_2 & & \\ & & \ddots & \\ & & & J_m \end{bmatrix}, \tag{3.5.1a}$$

where each J_i is an α_i by α_i square block of the form

$$\begin{bmatrix} \lambda_i & 1 & & & \\ & \lambda_i & 1 & & \\ & & \ddots & & \\ & & & \lambda_i & 1 \\ & & & & \lambda_i \end{bmatrix}. \tag{3.5.1b}$$

The diagonal elements of J_i (which are eigenvalues of the upper triangular matrix J, and thus also of A) are all equal. The λ_i corresponding to different blocks may or may not be equal. Now we form (cf. Theorem 0.1.4)

$$A^n = SJ^nS^{-1} = S\begin{bmatrix} J_1^n & & & \\ & J_2^n & & \\ & & \ddots & \\ & & & J_m^n \end{bmatrix}S^{-1}$$

and look at an individual block J_i^n. If J_i is 2 by 2, J_i^n has the form (see Problem 15 of Chapter 1)

$$J_i^n = \begin{bmatrix} \lambda_i & 1 \\ 0 & \lambda_i \end{bmatrix}^n = \begin{bmatrix} \lambda_i^n & n\lambda_i^{n-1} \\ 0 & \lambda_i^n \end{bmatrix},$$

and if J_i is 3 by 3, J_i^n has the form

$$J_i^n = \begin{bmatrix} \lambda_i & 1 & 0 \\ 0 & \lambda_i & 1 \\ 0 & 0 & \lambda_i \end{bmatrix}^n = \begin{bmatrix} \lambda_i^n & n\lambda_i^{n-1} & \frac{1}{2}n(n-1)\lambda_i^{n-2} \\ 0 & \lambda_i^n & n\lambda_i^{n-1} \\ 0 & 0 & \lambda_i^n \end{bmatrix}.$$

The structure for larger blocks is essentially clear from these two examples. If J_i is α_i by α_i, the fastest-growing component of J_i^n, of order $O(n^{\alpha_i-1}\lambda_i^n)$, will be in its upper right-hand corner. If we designate by $J_1, \ldots, J_L$ the *largest* blocks (of size α by α) containing λ_1 on the diagonal, then the fastest-growing elements in all of J^n are in the upper right-hand corners of $J_1^n, \ldots, J_L^n$, and they are of order $O(n^{\alpha-1}\lambda_1^n)$. So we divide J^n by $n^{\alpha-1}\lambda_1^n$ and get

$$\lim_{n\to\infty}\left(\frac{J^n}{n^{\alpha-1}\lambda_1^n}\right) = \begin{bmatrix} E_1 & & & & & \\ & \ddots & & & & \\ & & E_L & & & \\ & & & 0 & & \\ & & & & \ddots & \\ & & & & & 0 \end{bmatrix} \equiv E, \tag{3.5.2}$$

where E has the same block structure as J (3.5.1a), and every E_i is a zero matrix except $E_1, \ldots, E_L$, which are nonzero only in their upper right-hand corners. Then, from 3.5.2, since $A^n = SJ^nS^{-1}$, we have

$$\lim_{n\to\infty}\left(\frac{A^n}{n^{\alpha-1}\lambda_1^n}\right) = SES^{-1} \equiv G.$$

Now, since E is a nonzero matrix, so is G, and therefore $\boldsymbol{u}^{\mathrm{T}}G\boldsymbol{v}_0$ is nonzero with probability one again. So, even if A is not diagonalizable,

$$\begin{aligned}
\lim_{n\to\infty}\left(\frac{\boldsymbol{u}^{\mathrm{T}}\boldsymbol{v}_{n+1}}{\boldsymbol{u}^{\mathrm{T}}\boldsymbol{v}_n}\right) &= \lim_{n\to\infty}\left(\frac{\boldsymbol{u}^{\mathrm{T}}(A^{n+1}\boldsymbol{v}_0)}{\boldsymbol{u}^{\mathrm{T}}(A^n\boldsymbol{v}_0)}\right) \\
&= \lim_{n\to\infty}\left[\frac{\boldsymbol{u}^{\mathrm{T}}(A^{n+1}\boldsymbol{v}_0/(n+1)^{\alpha-1}/\lambda_1^{n+1})}{\boldsymbol{u}^{\mathrm{T}}(A^n\boldsymbol{v}_0/n^{\alpha-1}/\lambda_1^n)}\left(\frac{n+1}{n}\right)^{\alpha-1}\lambda_1\right] \\
&= \frac{\boldsymbol{u}^{\mathrm{T}}G\boldsymbol{v}_0}{\boldsymbol{u}^{\mathrm{T}}G\boldsymbol{v}_0}\lambda_1 = \lambda_1.
\end{aligned}$$

Now we have shown that $\boldsymbol{u}^{\mathrm{T}}\boldsymbol{v}_{n+1}/\boldsymbol{u}^{\mathrm{T}}\boldsymbol{v}_n$ will (almost always) converge to the eigenvalue λ_1, even if A is not diagonalizable. ■

We shall normally choose $\boldsymbol{u} = \boldsymbol{v}_n$ (there is nothing in the proof of Theorem 3.5.1 that prevents us from letting $\boldsymbol{u}$ vary with n). In Problem 14 it is shown that, if A is symmetric, this is a particularly good choice for $\boldsymbol{u}$.

Since, for large n, $\boldsymbol{u}^{\mathrm{T}}A\boldsymbol{v}_n = \boldsymbol{u}^{\mathrm{T}}\boldsymbol{v}_{n+1} \approx \lambda_1\boldsymbol{u}^{\mathrm{T}}\boldsymbol{v}_n$, for (essentially) arbitrary $\boldsymbol{u}$, we conclude that $A\boldsymbol{v}_n \approx \lambda_1\boldsymbol{v}_n$. Thus, for large $n, \boldsymbol{v}_n$ is an approximate eigenvector corresponding to the eigenvalue λ_1.

It is clear from the proof of Theorem 3.5.1 that, if λ_1 occurs only in blocks of size one (e.g., if it is a simple root or if A is symmetric), the convergence of the inverse power method is exponential. In fact, the most slowly decaying component in the limit equation 3.5.2 goes to zero as $|\lambda_2/\lambda_1|^n$, where λ_2 is the eigenvalue of second-largest modulus. Since the power method may not converge at all when there are two eigenvalues of equal modulus, it seems reasonable that convergence would be slowed by another eigenvalue of nearly equal magnitude.

However, if, in the Jordan canonical form 3.5.1, λ_1 occurs in blocks of size $\alpha > 1$, the most stubborn elements go to zero much more slowly, like $1/n$.

Since the complex eigenvalues of a real matrix occur in conjugate pairs, with equal absolute values, the assumptions of Theorem 3.5.1 are violated if A is real and its largest eigenvalues are complex but, if $\boldsymbol{v}_0$ is real, it is hardly surprising that the power iteration $\boldsymbol{v}_{n+1} = A\boldsymbol{v}_n$ is unable to find complex roots. However, Theorem 3.5.2 suggests a way to use the power method not only to find complex roots but also to locate any eigenvalue of A.

Theorem 3.5.2. *Suppose the matrix A has one eigenvalue λ_p that is closer to p than all other eigenvalues, and let $(A-pI)\boldsymbol{v}_{n+1} = \boldsymbol{v}_n$. Then, if the nonzero vectors $\boldsymbol{v}_0$ and $\boldsymbol{u}$ are chosen randomly, $\boldsymbol{u}^{\mathrm{T}}\boldsymbol{v}_n/\boldsymbol{u}^{\mathrm{T}}\boldsymbol{v}_{n+1} + p$ will converge to λ_p, with probability one.*

Proof: Since $\boldsymbol{v}_{n+1} = (A - pI)^{-1}\boldsymbol{v}_n$, by Theorem 3.5.1, $\boldsymbol{u}^{\mathrm{T}}\boldsymbol{v}_{n+1}/\boldsymbol{u}^{\mathrm{T}}\boldsymbol{v}_n$ converges to the eigenvalue of $(A-pI)^{-1}$ of largest modulus. Now the eigenvalues of $(A - pI)^{-1}$ are $1/(\lambda_i - p)$, where the λ_i are eigenvalues of A, and so the largest eigenvalue of $(A-pI)^{-1}$ is $1/(\lambda_p - p)$. Thus $\boldsymbol{u}^{\mathrm{T}}\boldsymbol{v}_{n+1}/\boldsymbol{u}^{\mathrm{T}}\boldsymbol{v}_n$ converges to $1/(\lambda_p - p)$, and $\boldsymbol{u}^{\mathrm{T}}\boldsymbol{v}_n/\boldsymbol{u}^{\mathrm{T}}\boldsymbol{v}_{n+1} + p$ converges to λ_p. ■

If we again choose $\boldsymbol{u} = \boldsymbol{v}_n$, we get the "shifted" inverse power iteration, with its estimate λ of λ_p:

$$n = 0, 1, 2, \ldots \quad \begin{cases} (A - pI)\boldsymbol{v}_{n+1} = \boldsymbol{v}_n \\ \lambda = p + \dfrac{\boldsymbol{v}_n^{\mathrm{T}}\boldsymbol{v}_n}{\boldsymbol{v}_n^{\mathrm{T}}\boldsymbol{v}_{n+1}} \\ \boldsymbol{v}_{n+1} = \dfrac{\boldsymbol{v}_{n+1}}{\|\boldsymbol{v}_{n+1}\|_2} \end{cases} \tag{3.5.3}$$

Note that we renormalize $\boldsymbol{v}_{n+1}$ each step, to avoid underflow or overflow.

If we take $p = 0$, the inverse power method $A\boldsymbol{v}_{n+1} = \boldsymbol{v}_n$ allows us to find the eigenvalue of A which is smallest in modulus (closest to 0) and, if we choose p to be complex, we can even converge to complex eigenvalues, provided that we are willing to do some complex arithmetic. Using the shifted inverse power method we can find *any* eigenvalue of A, by choosing p to be closer to the desired eigenvalue than any other. Since we do not usually know exactly where the eigenvalues are, this process is somewhat like going fishing: We throw out a p and catch whatever "fish" is nearest to p.

In fact, it is clear that the closer we choose p to the eigenvalue we are fishing for, the faster the convergence. The largest eigenvalue (in modulus) of $(A - pI)^{-1}$ is $1/(\lambda_p - p)$, while the second largest is $1/(\lambda_q - p)$, where λ_q is the next-closest eigenvalue (of A) to p. The ratio of the second-largest to largest eigenvalue (of the iteration matrix $(A-pI)^{-1}$), which governs the rate of convergence, is then $|\lambda_p - p|/|\lambda_q - p|$, and moving p closer to λ_p will clearly decrease this ratio and speed convergence.

In fact, this suggests that the following iteration, which updates p and sets it equal to the current best approximation of λ_p, might locate λ_p even faster:

$$
n = 0, 1, 2, \ldots \quad
\begin{aligned}
(A - p_n I)\boldsymbol{v}_{n+1} &= \boldsymbol{v}_n \\
\lambda &= p_n + \frac{\boldsymbol{v}_n^{\mathrm{T}}\boldsymbol{v}_n}{\boldsymbol{v}_n^{\mathrm{T}}\boldsymbol{v}_{n+1}} \\
p_{n+1} &= \lambda \\
\boldsymbol{v}_{n+1} &= \frac{\boldsymbol{v}_{n+1}}{\|\boldsymbol{v}_{n+1}\|_2}
\end{aligned}
\tag{3.5.4}
$$

Indeed it does converge more rapidly, once p_n is close to λ_p. However, the constant-p algorithm 3.5.3 has two advantages that may offset this faster convergence, at least in part. First, we can LU factor $A - pI$ once, and then use it to solve each $(A - pI)\boldsymbol{v}_{n+1} = \boldsymbol{v}_n$ so that each iteration after the first, we only do $O(N^2)$ work. If we vary p, we have to refactor each iteration, and refactoring requires $O(N^3)$ operations, if A is a full matrix. Second, 3.5.3 has the advantage that it is guaranteed to converge to the eigenvalue closest to p, whereas 3.5.4 may converge to another (possibly unwanted) eigenvalue. Both of these considerations suggest that the best strategy might be to fix p_n for several iterations between each update.

Naturally, 3.5.4 can be applied more efficiently after the original matrix has been reduced through similarity transformations (Sections 3.3 and 3.4) to upper Hessenberg or tridiagonal form, as then only $O(N^2)$ or $O(N)$ work, respectively, is required to solve $(A - p_n I)\boldsymbol{v}_{n+1} = \boldsymbol{v}_n$.

In Figure 3.5.2 a subroutine DPOWER is given which implements the shifted inverse power method, with p_n updated every IUPDAT iterations, where IUPDAT is a user-supplied parameter. DPOWER allows p_n and $\boldsymbol{v}_n$ to be complex, so that complex eigenvalue–eigenvector pairs can be found. The system $(A-p_n I)\boldsymbol{v}_{n+1} = \boldsymbol{v}_n$ is solved using versions of DLINEQ and DRESLV (CLINEQ and CRESLV) which have been modified to solve complex linear systems. As shown in Figure 3.5.2, only the type statements of DLINEQ and DRESLV had to be modified. The user supplies p_0 (EIG) and $\boldsymbol{v}_0$ (V) although, if $\boldsymbol{v}_0 = \mathbf{0}$ on input, it will be replaced by a starting vector of random components. The iteration stops when $r\boldsymbol{v}_{n+1} \approx \boldsymbol{v}_n$, where $r \equiv \boldsymbol{v}_n^{\mathrm{T}}\boldsymbol{v}_n/\boldsymbol{v}_n^{\mathrm{T}}\boldsymbol{v}_{n+1}$. If we premultiply both sides of $r\boldsymbol{v}_{n+1} \approx \boldsymbol{v}_n$ by $A - p_n I$, we see that this stopping criterion is equivalent to $A\boldsymbol{v}_n \approx (p_n + r)\boldsymbol{v}_n$, so that $\lambda = p_n + r$ and $\boldsymbol{v}_n$ are an (approximate) eigenvalue–eigenvector pair.

Let us experiment with DPOWER, by applying it to the 5 by 5 non-symmetric matrix 3.3.12. We tried various values for p_0, with and without updating p, and the results are reported in Table 3.5.1. In every case, we input $\boldsymbol{v}_0 = \mathbf{0}$, so that a random starting vector would be used. The location of the eigenvalues of this matrix are shown in Figure 3.5.1, along with the starting approximations p_0 used.

Table 3.5.1

Shifted Inverse Power Method Example

p_o	Updating Frequency	Converged to	Iterations to Convergence
$10 + 10\,i$	Never	$13.14 + 4.94\,i$	25
$13 + 5\,i$	Never	$13.14 + 4.94\,i$	5
$7\,i$	Never	$-4.58 + 6.94\,i$	30
4.8798093	Never	4.8798093	1
13	Never	$\longleftarrow$ did not converge	$\longrightarrow$
$10+ 10\,i$	Each step	$13.14 + 4.94\,i$	7
$13 + 5\,i$	Each step	$13.14 + 4.94\,i$	4
$7\,i$	Each step	$13.14 + 4.94\,i$	12

These results confirm our expectations: Without updating, the shifted inverse power method always converges to the eigenvalue nearest p and, the closer p is chosen to an eigenvalue, the faster it converges. With updating, convergence is generally more rapid, but not always to the eigenvalue nearest p_0. With no updating, when we set $p = 13$, which is equally close to the two eigenvalues $(13.14 \pm 4.94i)$, $\boldsymbol{v}_n$ oscillated indefinitely, and our estimate did not converge to either eigenvalue.

When p is very close to an eigenvalue, $A - pI$ is nearly singular; so it might seem that the inverse power method would be plagued by roundoff error then, but such is not the case. $\boldsymbol{v}_{n+1}$ may be calculated inaccurately, but it is sure to be large; so $r \equiv \boldsymbol{v}_n^{\mathrm{T}}\boldsymbol{v}_n/\boldsymbol{v}_n^{\mathrm{T}}\boldsymbol{v}_{n+1}$ will be small, and $\lambda = p + r$ will be close to the true eigenvalue. Note in Table 3.5.1 that when we set $p = 4.8798093$, which is one of the eigenvalues as calculated by DEGNON earlier (Section 3.3), DPOWER converges in a single iteration. The routine also returns a good estimate of the corresponding eigenvector, namely, $(-0.29464, -0.63515, -0.26891, 0.66035, 0.03745)$. This illustrates the usefulness of DPOWER in calculating eigenvectors, once the eigenvalues are known.

With $p = 13 + 5i$, and with no updating (3.5.3), when our starting vector $\boldsymbol{v}_0$ was chosen exactly equal to this eigenvector of 4.8798093, we obtained the sequence of eigenvalue approximations shown in Table 3.5.2.

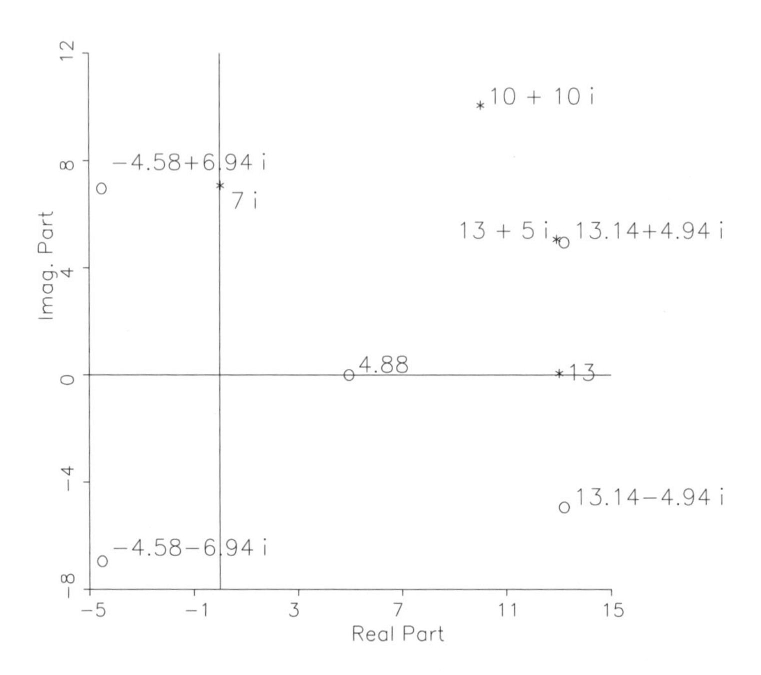

Figure 3.5.1
Locations of Eigenvalues (o) and Starting Points (*)

Since $\boldsymbol{v}_0$ is an eigenvector of A corresponding to $\lambda = 4.8798093$, it is not surprising that the first few eigenvalue estimates are close to 4.8798093, even though this is *not* the eigenvalue closest to $p = 13{+}5i$. We can show that this is one of the unlucky choices of $\boldsymbol{v}_0$ which have $G\boldsymbol{v}_0 = \mathbf{0}$ (Problem 16). However, note that after a few iterations the closest eigenvalue $13.14066 + 4.93688\ i$ takes over, and the iteration eventually converges to this eigenvalue, despite the unlucky choice of $\boldsymbol{v}_0$. $G\boldsymbol{v}_n$ will not remain exactly zero, owing to roundoff errors, and eventually the dominant eigenvalue triumphs, as it always will when inexact arithmetic is done. Here is a case where the unstable growth of small errors works to our advantage!

```
      SUBROUTINE DPOWER(A,N,EIG,V,IUPDAT)
      IMPLICIT DOUBLE PRECISION (A-H,O-Z)
C                                 DECLARATIONS FOR ARGUMENTS
      COMPLEX*16 EIG,V(N)
      DOUBLE PRECISION A(N,N)
```

Table 3.5.2

Iteration	Eigenvalue Estimate
0	4.87980 + 0.00000 i
1	4.88013 + 0.00024 i
2	4.88091 + 0.02503 i
3	3.65013 + 1.08108 i
4	12.31745 + 5.11298 i
5	13.14311 + 4.93595 i
6	13.14065 + 4.93692 i
7	13.14066 + 4.93688 i
8	13.14066 + 4.93688 i
9	13.14066 + 4.93688 i

```
      INTEGER N,IUPDAT
C                                DECLARATIONS FOR LOCAL VARIABLES
      COMPLEX*16 VN(N),VNP1(N),B(N,N),PN,R,RNUM,RDEN
      INTEGER IPERM(N)
C
C  SUBROUTINE DPOWER FINDS ONE EIGENVALUE OF A, AND A CORRESPONDING
C    EIGENVECTOR, USING THE SHIFTED INVERSE POWER METHOD.
C
C  ARGUMENTS
C
C             ON INPUT                          ON OUTPUT
C             --------                          ---------
C
C    A      - THE N BY N MATRIX.
C
C    N      - THE SIZE OF MATRIX A.
C
C    EIG    - A (COMPLEX) INITIAL GUESS AT      AN EIGENVALUE OF A,
C             AN EIGENVALUE.                    NORMALLY THE ONE CLOSEST
C                                               TO THE INITIAL GUESS.
C
C    V      - A (COMPLEX) STARTING VECTOR       AN EIGENVECTOR OF A,
C             FOR THE SHIFTED INVERSE POWER     CORRESPONDING TO THE
C             METHOD.  IF ALL COMPONENTS OF     COMPUTED EIGENVALUE.
C             V ARE ZERO ON INPUT, A RANDOM
C             STARTING VECTOR WILL BE USED.
C
C    IUPDAT - THE NUMBER OF SHIFTED INVERSE
C             POWER ITERATIONS TO BE DONE
C             BETWEEN UPDATES OF P.  IF
C             IUPDAT=1, P WILL BE UPDATED EVERY
C             ITERATION.  IF IUPDAT > 1000,
```

```
C           P WILL NEVER BE UPDATED.
C
C-----------------------------------------------------------------------
C                               EPS = MACHINE FLOATING POINT RELATIVE
C                                     PRECISION
C ***************************
      DATA EPS/2.D-16/
C ***************************
      DO 5 I=1,N
         IF (V(I).NE.0.0) GO TO 15
    5 CONTINUE
C                              IF V = 0, GENERATE A RANDOM STARTING
C                              VECTOR
      SEED = N+10000
      DEN = 2.0**31-1.0
      DO 10 I=1,N
         SEED = MOD(7**5*SEED,DEN)
         V(I) = SEED/(DEN+1.0)
   10 CONTINUE
   15 CONTINUE
C                               NORMALIZE V, AND SET VN=V
      VNORM = 0.0
      DO 20 I=1,N
         VNORM = VNORM + ABS(V(I))**2
   20 CONTINUE
      VNORM = SQRT(VNORM)
      DO 25 I=1,N
         V(I) = V(I)/VNORM
         VN(I) = V(I)
   25 CONTINUE
C                               BEGIN SHIFTED INVERSE POWER ITERATION
      NITER = 1000
      DO 60 ITER=0,NITER
         IF (MOD(ITER,IUPDAT).EQ.0) THEN
C                               EVERY IUPDAT ITERATIONS, UPDATE PN
C                               AND SOLVE (A-PN*I)*VNP1 = VN
            PN = EIG
            DO 30 I=1,N
            DO 30 J=1,N
               IF (I.EQ.J) THEN
                  B(I,J) = A(I,J) - PN
               ELSE
                  B(I,J) = A(I,J)
               ENDIF
   30       CONTINUE
            CALL CLINEQ(B,N,VNP1,V,IPERM)
         ELSE
C                               BETWEEN UPDATES, WE CAN USE THE LU
```

```
C                             DECOMPOSITION OF B=A-PN*I CALCULATED
C                             EARLIER, TO SOLVE B*VNP1=VN FASTER
            CALL CRESLV(B,N,VNP1,V,IPERM)
         ENDIF
C                             CALCULATE NEW EIGENVALUE ESTIMATE,
C                             PN + (VN*VN)/(VN*VNP1)
         RNUM = 0.0
         RDEN = 0.0
         DO 35 I=1,N
            RNUM = RNUM + VN(I)*VN(I)
            RDEN = RDEN + VN(I)*VNP1(I)
   35    CONTINUE
         R = RNUM/RDEN
         EIG = PN + R
C                             SET V = NORMALIZED VNP1
         VNORM = 0.0
         DO 40 I=1,N
            VNORM = VNORM + ABS(VNP1(I))**2
   40    CONTINUE
         VNORM = SQRT(VNORM)
         DO 45 I=1,N
            V(I) = VNP1(I)/VNORM
   45    CONTINUE
C                             IF R*VNP1 = VN  (R = (VN*VN)/(VN*VNP1) ),
C                             ITERATION HAS CONVERGED.
         ERRMAX = 0.0
         DO 50 I=1,N
            ERRMAX = MAX(ERRMAX,ABS(R*VNP1(I)-VN(I)))
   50    CONTINUE
         IF (ERRMAX.LE.SQRT(EPS)) RETURN
C                             SET VN = V = NORMALIZED VNP1
         DO 55 I=1,N
            VN(I) = V(I)
   55    CONTINUE
   60 CONTINUE
      PRINT 65
   65 FORMAT (' ***** INVERSE POWER METHOD DOES NOT CONVERGE *****')
      RETURN
      END

      SUBROUTINE CLINEQ(A,N,X,B,IPERM)
      IMPLICIT DOUBLE PRECISION (A-H,O-Z)
C                             DECLARATIONS FOR ARGUMENTS
      COMPLEX*16 A(N,N),X(N),B(N)
      INTEGER N,IPERM(N)
C                             DECLARATIONS FOR LOCAL VARIABLES
      COMPLEX*16 B_(N),LJI,TEMP,SUM
```

```
       .
  (rest of CLINEQ identical to DLINEQ, Figure 1.2.1)
       .
       .
      SUBROUTINE CRESLV(A,N,X,C,IPERM)
      IMPLICIT DOUBLE PRECISION (A-H,O-Z)
C                                DECLARATIONS FOR ARGUMENTS
      COMPLEX*16 A(N,N),X(N),C(N)
      INTEGER N,IPERM(N)
C                                DECLARATIONS FOR LOCAL VARIABLES
      COMPLEX*16 C_(N),LJI,SUM
       .
       .
  (rest of CRESLV identical to DRESLV, Figure 1.3.1)
       .
       .
```

Figure 3.5.2

3.6 The Generalized Eigenvalue Problem

The matrix eigenvalue problem corresponding to Figure 3.1.1 was derived using finite differences to approximate the partial differential equation eigenproblem 3.1.2. An alternative method for reducing a PDE eigenproblem to a matrix eigenvalue problem is the finite element method. To solve 3.1.2 using the finite element method, we start by assuming that the eigenfunction U can be written as a linear combination $U \approx \sum_j z_j \Phi_j(x,y)$ of M given linearly independent basis functions Φ_j. The Φ_j are almost always piecewise polynomial functions of some sort, and they all satisfy $\Phi_j = 0$ on the boundary, so that U will satisfy the boundary condition regardless of how the coefficients z_j are chosen. Then (in the Galerkin version of the finite element method) instead of requiring that $U_{xx} + U_{yy} + \lambda U$ be equal to zero identically, we only ask that it be "orthogonal" to each basis function Φ_i, in the sense that

$$\int_0^1 \int_0^1 \Phi_i(U_{xx} + U_{yy} + \lambda U)\, dx\, dy = 0 \qquad \text{for } i = 1, \ldots, M,$$

or

$$-\sum_{j=1}^{M} \int_0^1 \int_0^1 \Phi_i[(\Phi_j)_{xx} + (\Phi_j)_{yy}] z_j\, dx\, dy = \lambda \sum_{j=1}^{M} \int_0^1 \int_0^1 \Phi_i \Phi_j z_j\, dx\, dy.$$

This can be written as

$$A\boldsymbol{z} = \lambda B\boldsymbol{z}, \tag{3.6.1}$$

where $\boldsymbol{z} = (z_1, \ldots, z_M)$, and the components of matrices A and B are

$$A_{ij} = -\int_0^1 \int_0^1 \Phi_i[(\Phi_j)_{xx} + (\Phi_j)_{yy}]\, dx\, dy$$

$$= \int_0^1 \int_0^1 (\Phi_i)_x(\Phi_j)_x + (\Phi_i)_y(\Phi_j)_y\, dx\, dy, \qquad (3.6.2a)$$

$$B_{ij} = \int_0^1 \int_0^1 \Phi_i \Phi_j\, dx\, dy. \qquad (3.6.2b)$$

The second formula for A_{ij} was obtained from the first by integration by parts, and using the fact that each Φ_i is zero on the boundaries $x = 0, x = 1, y = 0$, and $y = 1$.

Problem 3.6.1 is called a generalized eigenvalue problem. If B^{-1} exists, 3.6.1 is obviously equivalent to $B^{-1}A\boldsymbol{z} = \lambda\boldsymbol{z}$ and could be solved by applying the techniques discussed earlier to the matrix $B^{-1}A$. On the other hand, if B^{-1} does not exist, 3.6.1 may not be equivalent to any matrix eigenvalue problem. Consider, for example, the eigenvalue problem

$$\begin{bmatrix} 1 & 1 \\ 1 & 1 \end{bmatrix} \begin{bmatrix} x \\ y \end{bmatrix} = \lambda \begin{bmatrix} 1 & 1 \\ 1 & 1 \end{bmatrix} \begin{bmatrix} x \\ y \end{bmatrix}.$$

This is equivalent to

$$0 = \det \begin{bmatrix} 1-\lambda & 1-\lambda \\ 1-\lambda & 1-\lambda \end{bmatrix} = (1-\lambda)^2 - (1-\lambda)^2 \equiv 0.$$

Since this equation is satisfied for any λ, all complex numbers are eigenvalues of $A\boldsymbol{z} = \lambda B\boldsymbol{z}$, for this choice of A and B! On the other hand, the following generalized problem has **no** eigenvalues:

$$\begin{bmatrix} 1 & 0 \\ 0 & 1 \end{bmatrix} \begin{bmatrix} x \\ y \end{bmatrix} = \lambda \begin{bmatrix} 0 & 1 \\ 0 & 0 \end{bmatrix} \begin{bmatrix} x \\ y \end{bmatrix}.$$

These examples show that the generalized eigenvalue problem is fundamentally different from the usual matrix eigenvalue problem, when B^{-1} does not exist.

It might be thought that, if A and B are both symmetric, the eigenvalues of the generalized problem have to be real, but this is not true. For example, the reader can easily verify that the problem

$$\begin{bmatrix} 0 & 1 \\ 1 & 1 \end{bmatrix} \begin{bmatrix} x \\ y \end{bmatrix} = \lambda \begin{bmatrix} 1 & 1 \\ 1 & 0 \end{bmatrix} \begin{bmatrix} x \\ y \end{bmatrix}$$

has complex eigenvalues, even though A and B are symmetric ($B^{-1}A$ is not symmetric, however). However, if, in addition to requiring that A and B be symmetric, we require that B also be positive-definite, then the eigenvalues of the generalized problem *are* guaranteed to be real. For, if B is positive-definite, we can find a nonsingular lower triangular matrix L such that $B = LL^{\mathrm{T}}$ (see 1.4.10), and then the problem $A\boldsymbol{z} = \lambda B\boldsymbol{z} = \lambda LL^{\mathrm{T}}\boldsymbol{z}$ is equivalent to

$$L^{-1}AL^{-\mathrm{T}}(L^{\mathrm{T}}\boldsymbol{z}) = \lambda(L^{\mathrm{T}}\boldsymbol{z})$$

or

$$L^{-1}AL^{-\mathrm{T}}\boldsymbol{y} = \lambda\boldsymbol{y}, \qquad (3.6.3)$$

where $\boldsymbol{y} \equiv L^{\mathrm{T}}\boldsymbol{z}$. Since $A = A^{\mathrm{T}}$, $(L^{-1}AL^{-\mathrm{T}})^{\mathrm{T}} = L^{-1}AL^{-\mathrm{T}}$, and so the matrix in 3.6.3 is symmetric, and its eigenvalues λ must be real.

In our example, A and B given by 3.6.2 are clearly symmetric, and B is also positive-definite (Problem 18) and therefore nonsingular. Thus we can solve $A\boldsymbol{z} = \lambda B\boldsymbol{z}$ either as the matrix eigenvalue problem $B^{-1}A\boldsymbol{z} = \lambda\boldsymbol{z}$ or in the form 3.6.3. Since $L^{-1}AL^{-\mathrm{T}}$ is symmetric and $B^{-1}A$ is generally nonsymmetric, 3.6.3 is the preferred form, because methods designed for general nonsymmetric problems are less efficient and must assume the eigenvalues and eigenvectors are complex, even though in this case they are not.

Thus we could, for example, use DEGSYM (Figure 3.2.1) to solve 3.6.3. Then the eigenvalues returned by DEGSYM are the eigenvalues of the original problem $A\boldsymbol{z} = \lambda B\boldsymbol{z}$, but the eigenvectors $\boldsymbol{y}_i$ returned must be multiplied by $L^{-\mathrm{T}}$ ("back transformed") to give the eigenvectors $\boldsymbol{z}_i$ of the original problem.

A better choice for our problem might be the shifted inverse power method (3.5.3 or 3.5.4), because (for reasons discussed in the last section) only the first few eigenvalues are of interest and, in addition, A and B will generally be band matrices, if the basis functions $\Phi_i(x, y)$ are chosen cleverly [Sewell 2005, Section 5.11]. According to Theorem 3.5.2, applied to the equivalent problem $B^{-1}A\boldsymbol{z} = \lambda\boldsymbol{z}$, if

$$(B^{-1}A - pI)\boldsymbol{v}_{n+1} = \boldsymbol{v}_n$$

or

$$(A - pB)\boldsymbol{v}_{n+1} = B\boldsymbol{v}_n,$$

then

$$\lambda = p + \frac{\boldsymbol{u}^{\mathrm{T}}\boldsymbol{v}_n}{\boldsymbol{u}^{\mathrm{T}}\boldsymbol{v}_{n+1}}$$

will converge to the eigenvalue λ_p of the generalized problem closest to p. It can be shown that, when A and B are symmetric, the best choice for $\boldsymbol{u}$ is $\boldsymbol{u} = B\boldsymbol{v}_n$ [Sewell 2005, Section 4.11]. The fact that A and B, and thus $A - pB$, are band matrices obviously makes it possible to calculate $\boldsymbol{v}_{n+1}$ more efficiently.

The QZ method, an alternative algorithm similar to the QR method, does simultaneous reduction of A and B, where A and B need not be symmetric or invertible. That is, it uses the fact that $Az = \lambda Bz$ and $QAZz = \lambda QBZz$ have the same eigenvalues, if Q and Z are invertible matrices, which we will take to be orthogonal. Note that every premultiplication by Q, and every postmultiplication by Z, must be applied to both matrices; however, Z does not need to be the inverse of Q, it can be chosen independently.

A subroutine HESSQZ is shown in Figure 3.6.1 which first premultiplies A and B by orthogonal Given's matrices chosen to reduce B to upper triangular form, just as done in Section 2.2; A is still full after this step. Then a series of pre- and postmultiplications are done which reduce A to upper Hessenberg form, while preserving the upper triangular structure of B. Each premultiplication eliminates one element below the first subdiagonal in A, but introduces one nonzero into the lower triangle of B; thus we follow this by a postmultiplication designed to eliminate the new nonzero in B. When the elements of A are eliminated in the right order (see HESSQZ for details), the postmultiplation will not introduce any new nonzeros into A.

The actual QZ iteration would begin after HESSQZ has finished its reduction. Each iteration, A is replaced by QAZ and B by QBZ, where Q and Z are orthogonal matrices chosen so that QAZ is still Hessenberg and QBZ is still triangular, but such that, in the limit of many iterations, A becomes upper triangular, or at least (under the usual assumptions on the eigenvalues) quasitriangular. Unfortunately, the process is not as simple to implement as the QR iteration, because reducing A to upper triangular form in the usual way (as done in Section 2.2) would result in a full B matrix. The details on how the QZ iteration is implemented can be found in Stewart [1973], Section 7.6, and an EISPACK routine QZIT [Smith et al. 1976] is available which implements the algorithm efficiently, with shifts. After convergence, the eigenvalues can be determined by looking at the determinant of $A - \lambda B$, which (hopefully) is quasitriangular, or triangular, and thus easy to expand. If the original B matrix is nonsingular, all diagonal elements of the final B will be nonzero, and the determinant of the quasitriangular matrix $A - \lambda B$ will be a polynomial of degree N, which will have N roots, counting multiple eigenvalues multiply. If B is singular, some diagonal elements of the final B will be zero, and the degree of the characteristic polynomial, and the number of eigenvalues, will be less than N.

The FORTRAN copy of Figure 3.6.1 available at the book web page (see Preface) includes a driver subroutine DEGEN which calls HESSQZ to reduce A to upper Hessenberg form and B to upper triangular form, then calls EISPACK routine QZIT to reduce A to quasitriangular form, then EISPACK routine QZVAL to extract the eigenvalues. The public domain EISPACK codes QZIT and QZVAL are also included.

A subroutine DEGENP is also provided in Figure 3.6.1 which first calls HESSQZ to transform A to upper Hessenberg and B to upper triangular form (if A is already upper Hessenberg and B is already upper triangular, neither matrix will be altered by HESSQZ). Then DEGENP uses a recurrence relation to calculate the characteristic polynomial $f(\lambda)$ and its N derivatives at $\lambda = \lambda_0$, and uses these derivatives to reconstruct the coefficients of the polynomial, in $O(N^3)$ operations. Since HESSQZ also requires $O(N^3)$ operations to reduce A and B, it is thus possible to get the characteristic polynomial of any generalized eigenvalue problem in a total of $O(N^3)$ operations. If A is tridiagonal and B is diagonal, the polynomial coefficients will be computed in $O(N^2)$ operations. It is recommended that the matrices A and B be divided by $\max(\|A\|, \|B\|)$ before calling DEGENP since the danger of overflow or underflow is significant for this problem.

Note that the degree of the characteristic polynomial may be less than N, in which case there are fewer than N eigenvalues. If the characteristic polynomial is a constant, c, then either there are no eigenvalues (if $c \neq 0$) or all complex numbers are eigenvalues (if $c = 0$).

```
      SUBROUTINE DEGENP(A,B,N,LAM0,POLY)
      IMPLICIT DOUBLE PRECISION (A-H,O-Z)
C                                  DECLARATIONS FOR ARGUMENTS
      DOUBLE PRECISION A(N,N),B(N,N),LAM0,POLY(N+1)
      INTEGER N
C                                  DECLARATIONS FOR LOCAL VARIABLES
      DOUBLE PRECISION DET(N+1,N+1)
C
C  SUBROUTINE DEGENP CALLS HESSQZ TO REDUCE THE GENERALIZED EIGENVALUE
C     PROBLEM
C
C                 A*X = LAMBDA*B*X
C
C  TO A SIMILAR PROBLEM WITH THE SAME EIGENVALUES, WHERE A IS UPPER
C  HESSENBERG AND B IS UPPER TRIANGULAR, THEN CALCULATES THE COEFFICIENTS
C  OF THE POLYNOMIAL DET(A-LAMBDA*B).  THE ROOTS OF THIS POLYNOMIAL WILL
C  BE THE EIGENVALUES OF THE GENERALIZED PROBLEM.
C
C  ARGUMENTS
C
C             ON INPUT                          ON OUTPUT
C             --------                          ---------
C
C    A      - THE N BY N A MATRIX.              A IS UPPER HESSENBERG.
C
C    B      - THE N BY N B MATRIX.              B IS UPPER TRIANGULAR.
C
C    N      - THE SIZE OF MATRICES A AND B.
C
C    LAM0   - A SCALAR, SEE POLY. (SET
C             LAM0=0 TO GET THE USUAL
C             POLYNOMIAL COEFFICIENTS)
C
```

```
C     POLY   -                                          VECTOR OF LENGTH N+1,
C                                                       CONTAINING POLYNOMIAL
C                                                       COEFFICIENTS.
C                                                       DET(A-LAMBDA*B)=
C                                                       SUM FROM I=0 TO N OF
C                                                       POLY(I+1)*
C                                                           (LAMBDA-LAM0)**I
C
C-----------------------------------------------------------------------
C
C                             CALL HESSQZ TO REDUCE A TO UPPER HESSENBERG AND
C                             B TO UPPER TRIANGULAR FORM
      CALL HESSQZ(A,B,N)
C                             DEGENP USES A RECURRENCE RELATION TO CALCULATE
C                             DET(A-LAMBDA*B) AND ALL N DERIVATIVES AT
C                             LAMBDA = LAM0, FROM WHICH THE POLYNOMIAL
C                             COEFFICIENTS CAN BE FOUND.
      DET(1,N+1) = 1.0
      DO 5 I=2,N+1
         DET(I,N+1) = 0.0
    5 CONTINUE
      DET(1,N) = A(N,N)-LAM0*B(N,N)
      DET(2,N) = -B(N,N)
      DO 10 I=3,N+1
        DET(I,N) = 0.0
   10 CONTINUE
      DO 30 K=N-1,1,-1
         DET(1,K) = (A(K,K)-LAM0*B(K,K))*DET(1,K+1)
         DO 15 I=1,N
            DET(I+1,K) = (A(K,K)-LAM0*B(K,K))*DET(I+1,K+1)
     &                              - B(K,K)*DET(I,K+1)
   15    CONTINUE
         FACT = 1.0
         DO 25 J=K+1,N
            FACT = -FACT*A(J,J-1)
            IF (A(K,J).EQ.0.0 .AND. B(K,J).EQ.0.0) GO TO 25
            DET(1,K) = DET(1,K) + FACT*(A(K,J)-LAM0*B(K,J))*DET(1,J+1)
            DO 20 I=1,N
               DET(I+1,K) = DET(I+1,K) + FACT*
     &         ((A(K,J)-LAM0*B(K,J))*DET(I+1,J+1) - B(K,J)*DET(I,J+1))
   20       CONTINUE
   25    CONTINUE
   30 CONTINUE
      DO 35 I=1,N+1
         POLY(I) = DET(I,1)
   35 CONTINUE
      RETURN
      END
```

```
      SUBROUTINE HESSQZ(A,B,N)
      IMPLICIT DOUBLE PRECISION (A-H,O-Z)
C                                 DECLARATIONS FOR ARGUMENTS
      DOUBLE PRECISION A(N,N),B(N,N)
      INTEGER N
C
C  SUBROUTINE HESSQZ REDUCES THE GENERALIZED EIGENVALUE PROBLEM
C
C              A*X = LAMBDA*B*X
C
C  TO A SIMILAR PROBLEM WITH THE SAME EIGENVALUES, WHERE A IS
C  UPPER HESSENBERG AND B IS UPPER TRIANGULAR.
C
C  ARGUMENTS
C
C             ON INPUT                          ON OUTPUT
C             --------                          ---------
C
C    A      - THE N BY N A MATRIX.              A IS UPPER HESSENBERG.
C
C    B      - THE N BY N B MATRIX.              B IS UPPER TRIANGULAR.
C
C    N      - THE SIZE OF MATRICES A AND B.
C
C-----------------------------------------------------------------------
C                                 PREMULTIPLY A AND B BY ORTHOGONAL MATRIX
C                                 (PRODUCT OF GIVENS MATRICES) Q, SUCH
C                                 THAT QB IS UPPER TRIANGULAR.
      DO 20 I=1,N-1
         DO 15 J=I+1,N
            IF (B(J,I).EQ.0.0) GO TO 15
            DEN = SQRT(B(I,I)**2+B(J,I)**2)
            S = -B(J,I)/DEN
            C =  B(I,I)/DEN
            DO 5 K=I,N
               BIK = C*B(I,K)-S*B(J,K)
               BJK = S*B(I,K)+C*B(J,K)
               B(I,K) = BIK
               B(J,K) = BJK
    5       CONTINUE
            DO 10 K=1,N
               AIK = C*A(I,K)-S*A(J,K)
               AJK = S*A(I,K)+C*A(J,K)
               A(I,K) = AIK
               A(J,K) = AJK
   10       CONTINUE
```

```
   15    CONTINUE
   20 CONTINUE
C                               PREMULTIPLY A AND B BY ORTHOGONAL MATRIX
C                               Q, AND POSTMULTIPLY BY ORTHOGONAL MATRIX
C                               Z, SUCH THAT QAZ IS UPPER HESSENBERG AND
C                               QBZ IS STILL UPPER TRIANGULAR
      DO 50 I=1,N-2
         DO 45 J=N,I+2,-1
            IF (A(J,I).EQ.0.0) GO TO 45
C                               PREMULTIPLY A TO ZERO A(J,I)
            DEN = SQRT(A(J-1,I)**2+A(J,I)**2)
            S = -A(J,I)/DEN
            C =  A(J-1,I)/DEN
            DO 25 K=I,N
               A1K = C*A(J-1,K) - S*A(J,K)
               A2K = S*A(J-1,K) + C*A(J,K)
               A(J-1,K) = A1K
               A(J,K) = A2K
   25       CONTINUE
C                               PREMULTIPLY B BY SAME MATRIX, CREATING
C                               NEW NONZERO B(J,J-1)
            DO 30 K=J-1,N
               B1K = C*B(J-1,K) - S*B(J,K)
               B2K = S*B(J-1,K) + C*B(J,K)
               B(J-1,K) = B1K
               B(J,K) = B2K
   30       CONTINUE
            IF (B(J,J-1).EQ.0.0) GO TO 45
C                               POSTMULTIPLY B TO ZERO B(J,J-1)
            DEN = SQRT(B(J,J-1)**2+B(J,J)**2)
            S = -B(J,J-1)/DEN
            C =  B(J,J)/DEN
            DO 35 K=1,J
               BK1 = C*B(K,J-1) + S*B(K,J)
               BK2 = -S*B(K,J-1) + C*B(K,J)
               B(K,J-1) = BK1
               B(K,J) = BK2
   35       CONTINUE
C                               POSTMULTIPLY A BY SAME MATRIX
            DO 40 K=1,N
               AK1 = C*A(K,J-1) + S*A(K,J)
               AK2 = -S*A(K,J-1) + C*A(K,J)
               A(K,J-1) = AK1
               A(K,J) = AK2
   40       CONTINUE
   45    CONTINUE
   50 CONTINUE
      RETURN
```

```
END
```

Figure 3.6.1

In theory, a good polynomial root finder can now be used to compute all the eigenvalues. Unfortunately, finding the roots of a high degree polynomial is known to be an unstable process numerically, so in practice this may be possible only for small or moderate size problems.

3.7 The Singular Value Decomposition

The "singular value decomposition" (SVD) of an arbitrary M by N matrix is $A = UDV^T$, where U is an M by M orthogonal matrix, V is an N by N orthogonal matrix, and D is an M by N diagonal matrix with real, nonnegative elements. The diagonal elements of D are called the singular values of A. Note that while the singular value decomposition of a matrix is not unique, the singular values are unique: if $A = U_1D_1V_1^T = U_2D_2V_2^T$, it is easy to show that $D_1^TD_1$ and $D_2^TD_2$ are similar matrices, and thus have the same eigenvalues, i.e., the same diagonal entries, not necessarily in the same order.

First, suppose A is a nonsingular square (N by N) matrix. Even though A may be nonsymmetric, A^TA is symmetric and positive-definite, so there exists an orthogonal matrix V such that $V^TA^TAV = D^2$, where D^2 is diagonal with the (real and positive) eigenvalues of A^TA. (Theorem 3.2.2 is essentially a constructive proof of the well-known result that a symmetric matrix is orthogonally diagonalizable.) If we define $U = AVD^{-1}$, then $UDV^T = (AVD^{-1})DV^T = A$, and U is orthogonal because $U^TU = (AVD^{-1})^T(AVD^{-1}) = D^{-1}V^TA^TAVD^{-1} = D^{-1}D^2D^{-1} = I$. So in this case it is easy to find a singular value decomposition of A.

If A is not square, or square and singular, A^TA is still symmetric and square, and nonnegative definite, so there still exists an orthogonal matrix V such that $V^TA^TAV = E^2$. But now the N by N diagonal matrix E will not be the right size (if $M \neq N$) or will not have an inverse (if A is square and singular), so the above approach for finding a singular value decomposition will no longer work. If E has only K nonzero (positive) elements, which we take to be the first K for convenience, then let us define an M by N diagonal matrix D that has all zero elements except that $D_{ii} = E_{ii} > 0$ for $i = 1, ..., K$. (Since it can be shown that K is the rank of A, we have $K \leq min(M, N)$.) Then the idea is still to find an orthogonal matrix U such that $UD = AV$, so that $UDV^T = A$.

If $\boldsymbol{u}_i$ is the ith column of U, the first K columns of UD will be $D_{ii}\boldsymbol{u}_i, i = 1, ..., K$, the remaining $N-K$ columns will be zero. Thus if we take $\boldsymbol{u}_i = \boldsymbol{z}_i/D_{ii}$ for $i \leq K$, where $\boldsymbol{z}_i$ is the ith column of AV, the first K columns of UD will equal the first K columns of AV. Furthermore, $V^TA^TAV = E^2$, or $(AV)^T(AV) = E^2$, so $\boldsymbol{z}_i^T\boldsymbol{z}_j = E_{ij}^2$, and taking $i = j > K$ we see that $\|\boldsymbol{z}_i\|_2^2 =$

$$\begin{bmatrix} \uparrow & \uparrow & \uparrow & \uparrow & \uparrow \\ | & | & | & | & | \\ u_1 & .. & u_K & .. & u_M \\ | & | & | & | & | \\ \downarrow & \downarrow & \downarrow & \downarrow & \downarrow \end{bmatrix} \begin{bmatrix} D_{11} & 0 & 0 & 0 & 0 & 0 \\ 0 & .. & 0 & 0 & 0 & 0 \\ 0 & 0 & D_{KK} & 0 & 0 & 0 \\ 0 & 0 & 0 & 0 & 0 & 0 \\ 0 & 0 & 0 & 0 & 0 & 0 \end{bmatrix} = \begin{bmatrix} \uparrow & \uparrow & \uparrow & 0 & 0 & 0 \\ | & | & | & 0 & 0 & 0 \\ z_1 & .. & z_K & 0 & 0 & 0 \\ | & | & | & 0 & 0 & 0 \\ \downarrow & \downarrow & \downarrow & 0 & 0 & 0 \end{bmatrix}$$

$$\text{UD} = \text{AV}$$

$E_{ii}^2 = 0$ and so the last $N-K$ columns of AV are also zero. So no matter what we take for $\boldsymbol{u}_{K+1}, ...\boldsymbol{u}_M$, the last $N-K$ columns of UD will equal the last $N-K$ columns of AV (all being **0**!). In this way we can construct a matrix U which satisfies $UD = AV$, but we also need U to be orthogonal. But note that if $i \leq K$ and $j \leq K$, $\boldsymbol{u}_i^T\boldsymbol{u}_j = \boldsymbol{z}_i^T\boldsymbol{z}_j/(D_{ii}D_{jj}) = E_{ij}^2/(D_{ii}D_{jj}) = \delta_{ij}$, so the first K columns of U form an orthonormal set. Since the remaining columns can be chosen arbitrarily, we just choose them to be an orthonormal basis for the orthogonal compliment of the K-dimensional space spanned by the first K columns, and U will be an orthogonal matrix. Thus we have constructed a singular value decomposition $A = UDV^T$ for an arbitrary rectangular matrix A.

Notice that the nonzero singular values are just the square roots of the eigenvalues of A^TA, which are positive if A is square and nonsingular, and nonnegative for any A. From Theorem 0.3.1 we see that the largest singular value is the 2-norm of A.

If A is a nonsingular square matrix, using the definition of condition number in Section 1.7, and the fact that the eigenvalues of $A^{-T}A^{-1}$ are the inverses of the eigenvalues of A^TA:

$$cond(A) = \|A\|_2\|A^{-1}\|_2 = \sqrt{\lambda_{max}(A^TA)}\sqrt{\lambda_{max}(A^{-T}A^{-1})}$$
$$= \sqrt{\lambda_{max}(A^TA)}/\sqrt{\lambda_{min}(A^TA)} = \sigma_{max}/\sigma_{min}$$

where σ_i are the singular values of A. Thus the ratio of the largest singular value to the smallest gives the (2-norm) condition number of A. If A is singular, its minimum singular value is 0 and the condition number is infinite.

One application of the singular value decomposition is data compression. If $A = UDV^T$, and $\boldsymbol{u}_i$ is the ith column of U, $\boldsymbol{v}_i^T$ is the ith row of V^T, and D_{ii} is the ith diagonal element of the diagonal matrix D, then $A = \sum_{i=1}^K D_{ii}\boldsymbol{u}_i\boldsymbol{v}_i^T$. To see this, define $B \equiv \sum_{i=1}^K D_{ii}\boldsymbol{u}_i\boldsymbol{v}_i^T$. Then element (j,k) of the rank one matrix $\boldsymbol{u}_i\boldsymbol{v}_i^T$ is the jth element of column $\boldsymbol{u}_i$ times the kth element of row $\boldsymbol{v}_i^T$, that is $U_{ji}V_{ik}^T$. Thus $B_{jk} = \sum_i U_{ji}D_{ii}V_{ik}^T = \sum_i U_{ji}(DV^T)_{ik} = (UDV^T)_{jk} = A_{jk}$, and so $A = B$. If A is a large matrix with many zero or nearly zero singular values D_{ii}, we can throw away the corresponding small terms in this series, leaving us with a more compact approximate representation of A. If there are only L terms left after we discard the smaller terms, then we have an approximation of the M by N matrix A which requires only $L(M+N)$ words of memory, which may be small compared to MN. This technique for data compression can be used to store pictures compactly; for example, the use of SVD to compress fingerprints is illustrated in Kahaner et al. [1989, p. 224]. The reason it is important that U and V be orthogonal here is so that $\boldsymbol{u}_i$ and $\boldsymbol{v}_i$ are unit vectors. If U, V were arbitrary matrices, the fact that D_{ii} is small would not necessarily imply that $D_{ii}\boldsymbol{u}_i\boldsymbol{v}_i^T$ is small.

Another application is in solving least squares problems. If D is an M by N diagonal matrix, its *pseudoinverse* D^+ is found by taking its transpose, then inverting the *nonzero* elements on the diagonal. A pseudoinverse of a general M by N matrix is defined by $A^+ = VD^+U^T$, where $A = UDV^T$ is a singular value decomposition of A. Note that if D is a square nonsingular matrix, $D^+ = D^{-1}$ and also $A^+ = A^{-1}$. Now, as before, assume only the first K elements of D are nonzero. Then DD^+ is an M by M diagonal matrix with ones in the first K diagonals and zeros thereafter, and then $DD^+ - I$ has zeros in the first K diagonal positions. Since D^T has zeros in diagonal positions $K+1, ...,$ the product $D^T(DD^+ - I) = 0$, so $D^TDD^+ = D^T$. Hence, for any vector $\boldsymbol{b}$:

$$A^TA(A^+\boldsymbol{b}) = (UDV^T)^T(UDV^T)(VD^+U^T)\boldsymbol{b} = VD^TDD^+U^T\boldsymbol{b} = VD^TU^T\boldsymbol{b} = (UDV^T)^T\boldsymbol{b} = A^T\boldsymbol{b}$$

and thus $\boldsymbol{x} = A^+\boldsymbol{b}$ is a solution of the normal equations, and therefore a solution to the least squares problem, min $\|A\boldsymbol{x} - \boldsymbol{b}\|_2$ (see Theorem 2.1.1). In fact, it is the *minimum norm* least squares solution, because, assume $\boldsymbol{x} = A^+\boldsymbol{b}$ and $\boldsymbol{x} + \boldsymbol{e}$ are both solutions of the normal equations, $A^TA(\boldsymbol{x} + \boldsymbol{e}) = A^T\boldsymbol{b}$ and $A^TA\boldsymbol{x} = A^T\boldsymbol{b}$, then by subtracting, $A^TA\boldsymbol{e} = \boldsymbol{0}$, or $(UDV^T)^T(UDV^T)\boldsymbol{e} = VD^TDV^T\boldsymbol{e} = \boldsymbol{0}$, and since V is invertible, $D^TD(V^T\boldsymbol{e}) = \boldsymbol{0}$. Since the first K diagonal elements of D^TD are nonzero, the first K components of the vector $V^T\boldsymbol{e}$ must be 0, otherwise the result could not be the zero vector. Then we see that

$$\boldsymbol{x}^T\boldsymbol{e} = (A^+\boldsymbol{b})^T\boldsymbol{e} = \boldsymbol{b}^T(A^+)^T\boldsymbol{e} = \boldsymbol{b}^T(VD^+U^T)^T\boldsymbol{e} = \boldsymbol{b}^TU(D^+)^TV^T\boldsymbol{e} = 0$$

since the first K components of $V^T\boldsymbol{e}$ are zero and diagonal elements $K+1, ...$ of $(D^+)^T$ are zero. So

$$\|\boldsymbol{x} + \boldsymbol{e}\|_2^2 = (\boldsymbol{x} + \boldsymbol{e})^T(\boldsymbol{x} + \boldsymbol{e}) = \|\boldsymbol{x}\|_2^2 + \|\boldsymbol{e}\|_2^2 + 2\boldsymbol{x}^T\boldsymbol{e} = \|\boldsymbol{x}\|_2^2 + \|\boldsymbol{e}\|_2^2 \geq \|\boldsymbol{x}\|_2^2$$

showing that $\boldsymbol{x} = A^+\boldsymbol{b}$ is the minimum norm least squares solution.

One further application: suppose we want to find the general solution to $A\boldsymbol{x} = \boldsymbol{0}$, and we have a singular value decomposition $A = UDV^T$. Since U is nonsingular, $A\boldsymbol{x} = \boldsymbol{0}$ means $DV^T\boldsymbol{x} = \boldsymbol{0}$, or $D\boldsymbol{y} = \boldsymbol{0}$, where $\boldsymbol{x} = V\boldsymbol{y}$. The components y_i of $\boldsymbol{y}$ corresponding to zero columns (if any) of D can thus have arbitrary values, the other components must be 0. So the general solution is the set of all linear combinations of the columns of V corresponding to columns of D which are zero. Note that since D has $N - K$ zero columns, the null space of A has dimension $N - K$, so the rank of A is K.

3.8 Problems

1. Verify the assertion made in the proof of Theorem 3.2.1 that

$$2b_{ji}^2 + b_{ii}^2 + b_{jj}^2 = 2a_{ji}^2 + a_{ii}^2 + a_{jj}^2,$$

where b_{ji}, b_{ii}, and b_{jj} are given by 3.2.3. You can show this directly using relations 3.2.3, or you can show that if you reduce the matrices A, B and Q_{ij} in the equation $B = Q_{ij}^T A Q_{ij}$ (3.2.2) to 2 by 2 matrices by throwing away everything except elements $(i,i), (i,j), (j,i)$ and (j,j), then $B = Q_{ij}^T A Q_{ij}$, still holds, and since Q_{ij} is still orthogonal, B and A have the same sum of squares, as shown in the proof of Theorem 3.2.1.

2. Verify the assertion made in the proof of Theorem 3.2.2 that $1 - h \leq \exp(-h)$ for any positive h. (Hint: The two functions $1-h$ and $\exp(-h)$ are equal at $h = 0$; compare their slopes for $h \geq 0$.)

3. DEGSYM (Figure 3.2.1) uses the formula $e_{k+1} = e_k - 2a_{ji}^2$ to keep up with the sum of squares of the off-diagonal elements. Explain why this idea is somewhat dangerous from a numerical point of view. It is not feasible to recalculate e_k directly (by summing the squares) after each individual transformation, but it might be reasonable to recalculate e_k directly after each complete cycle of transformations. DEGSYM (FORTRAN version) has been written so that this can be accomplished by moving a single statement. Which one and where should it be moved?

4. a. Because of the way DEGSYM chooses c and s, s may be close to one even when the matrix has almost reached diagonal form, which means that eigenvalues on the diagonal may be permuted each iteration up to the very end. Modify DEGSYM so that this cannot happen, and test your new routine on the matrix of Figure 3.1.1.

   b Modify loop 30 of DEGSYM to take advantage of the symmetry of A, and retest on the matrix of Figure 3.1.1.

5. Verify the assertions made in the proof of Theorem 3.3.2, namely, that

   a. the nonzero structure of a quasitriangular matrix is preserved when it is multiplied by another quasitriangular matrix of the same nonzero structure (once you have established this, it follows that its structure is preserved when multiplied by an upper triangular matrix—why?);

b. the nonzero structure of a quasitriangular matrix is preserved when it is inverted (assuming that the inverse exists).

6. The assumptions that A is nonsingular and diagonalizable were useful to simplify the proofs of Theorems 3.3.1 and 3.3.2. However, if

$$A = \begin{bmatrix} -1 & 0 & 0 & 1 \\ 0 & 3 & -1 & 0 \\ 0 & 1 & 1 & 0 \\ -2 & 0 & 0 & 2 \end{bmatrix}, \quad P = \begin{bmatrix} 1 & 1 & 0 & 0 \\ 0 & 0 & 1 & 2 \\ 0 & 0 & 1 & 1 \\ 1 & 2 & 0 & 0 \end{bmatrix},$$

$$J = \begin{bmatrix} 0 & 0 & 0 & 0 \\ 0 & 1 & 0 & 0 \\ 0 & 0 & 2 & 1 \\ 0 & 0 & 0 & 2 \end{bmatrix},$$

verify that $P^{-1}AP = J$, and explain why this shows that A is neither nonsingular nor diagonalizable (Hint: the Jordan canonical form is unique except for the order of the blocks). Nevertheless, verify that the QR method (use DEGNON in Figure 3.3.4) and the LR method (DEGNON + LR in Figure 3.4.5) both converge when applied to this matrix.

7. Modify DEGNON so that the QR transformations take advantage of the fact that A is symmetric (the required modifications are already suggested in the comments). Then apply this modified version to the matrix of Figure 3.1.1, which is symmetric. Compare the total computer time with that used by the unmodified version.

8. Expand the determinant 3.3.9 of the example quasitriangular matrix, and show that this determinant reduces to the factors shown in 3.3.10. (Hint: Expand along the first column.)

9. Modify DEGNON to implement the shifted QR iteration 3.3.11, with $\sigma_n = a_{mm}$, where $m = N - \text{mod}(n/5, N)$ (Note: $n/5$ will be an integer, if FORTRAN is used; if MATLAB is used, replace with $fix(n/5)$). In this way, the shift parameter is set cyclically to each diagonal element, and, at least if A_n is converging to triangular form, each diagonal element is close to an eigenvalue, for large n. Count the number of QR iterations required when this modified version of DEGNON is applied to the matrix in Figure 3.1.1, and compare this with the number of iterations used by the unshifted QR version.

Although for this small matrix the effect of shifts will not be spectacular, note the results in Table 6.1.1, where a similar diagonal shift strategy cut the computer time by a factor of over 25, for a random matrix of size $N = 201$.

10. a. Reduce the following matrix to a similar upper Hessenberg matrix, using elementary matrices M_{ij} (do calculations by hand):

$$\begin{bmatrix} 1 & -2 & 4 & 1 \\ 2 & 0 & 5 & 2 \\ 2 & -2 & 9 & 3 \\ -6 & -1 & -16 & -6 \end{bmatrix}.$$

    b. Transpose this Hessenberg matrix (transposing does not change the eigenvalues) and reduce *this* matrix to upper Hessenberg form, using more elementary matrix transformations. As long as you do not have to do any row switches, the zeros in the upper right-hand corner will be preserved (if orthogonal transformations are used, they will not). Thus the final result is a tridiagonal matrix with the same eigenvalues as the original matrix. (Although for this example row interchanges are not required for numerical stability, normally they will be, and this approach will not work, in general. In fact, it has been shown that it is impossible to reduce a *general* nonsymmetric matrix to a similar tridiagonal matrix in a stable manner.)

    c. Use DEGNON to calculate the eigenvalues of the original matrix A, the Hessenberg matrix, and the final tridiagonal matrix, to verify that the eigenvalues have not changed.

11. Use a random number generator to generate the coefficients of a 50 by 50 nonsymmetric matrix. Find the eigenvalues using DEGNON, first with HESSQ and QR called by DEGNON, and then with HESSM and LR. Compare the execution times, and print out the final (quasitriangular) matrix in each case. Note that the matrix produced by HESSM and LR contains some elements that are very large compared with the diagonal and subdiagonal elements, while the other matrix does not.

12. Use hand calculations on the upper Hessenberg matrix

$$A = \begin{bmatrix} 4 & 3 & 0 \\ 3 & 6 & 4 \\ 0 & 4 & 6 \end{bmatrix}.$$

    a. Do one Jacobi iteration, to zero $A_{2,1}$.

    b. Do one complete QR iteration (i.e., reduce to upper triangular form and then postmultiply).

    c. Do one complete LR iteration.

13. a. Modify subroutine LR (Figure 3.4.5) to do partial pivoting during the reduction to triangular form; only a one-line change is needed. Test the new LR on the large random matrix of Problem 11; you will find that pivoting makes the LR iteration much *less* stable, a somewhat counterintuitive result!

    b. A possible explanation for why pivoting often has a harmful effect on convergence of the LR method follows. In the convergence proof (Theorem 3.3.1) for the QR method, the assumption that Q_n is orthogonal was only needed to ensure that $H_n \equiv Z^{-1}Q_0 \ldots Q_{n-1}$ and its inverse did not grow without bounds as $n \to \infty$. For the LR method without pivoting, the Q_n are lower triangular matrices with 1's on the diagonal. Show that the inverse and product of such matrices also are lower triangular with 1's on the diagonal. This means that ZH_n and $H_n^{-1}Z^{-1}$ both have eigenvalues which are all ones, so at least their spectral radii are bounded as $n \to \infty$. If pivoting is done, this is no longer the case. Of course, if a zero or nearly zero pivot is encountered during the LU factorization, we have no choice but to pivot or give up: a subdiagonal entry of ∞ in one of the Q_n would surely be even more harmful!

14. The nth power method iterate $\boldsymbol{v}_n$ is equal to $A^n\boldsymbol{v}_0$. If A is symmetric, then $A = QDQ^{\mathrm{T}}$, where D is diagonal and Q is orthogonal, and so $\boldsymbol{v}_n = QD^nQ^{\mathrm{T}}\boldsymbol{v}_0$. Use this formula to express

$$\frac{\boldsymbol{v}_n^{\mathrm{T}}\boldsymbol{v}_{n+1}}{\boldsymbol{v}_n^{\mathrm{T}}\boldsymbol{v}_n}$$

in a form that makes it apparent why this gives a more accurate approximation to λ_1 than

$$\frac{\boldsymbol{u}^{\mathrm{T}}\boldsymbol{v}_{n+1}}{\boldsymbol{u}^{\mathrm{T}}\boldsymbol{v}_n}$$

does for arbitrary $\boldsymbol{u}$.

15. If A is the 2 by 2 matrix

$$\begin{bmatrix} 4 & 2 \\ -1 & 1 \end{bmatrix},$$

do several iterations (by hand) of the shifted inverse power method 3.5.3, to find the eigenvalue of A nearest to $p = 1$. Verify that the error is reduced each step by a factor of about $|\lambda_p - p|/|\lambda_q - p|$, where λ_p and λ_q are the eigenvalues of A closest and next closest, respectively, to p.

16. Show that if $\boldsymbol{v}_0$ is an eigenvector of A corresponding to an eigenvalue *other than* λ_1, then $G\boldsymbol{v}_0 = \mathbf{0}$, where G is defined in Section 3.5. (Hint: Show that $SJS^{-1}\boldsymbol{v}_0 = \lambda_i \boldsymbol{v}_0$ $(i > 1)$ implies that the first $L\alpha$ components of $S^{-1}\boldsymbol{v}_0$ are zero, and thus $E(S^{-1}\boldsymbol{v}_0) = \mathbf{0}$.)

    Construct a counterexample that shows that $G\boldsymbol{v}_0 = \mathbf{0}$ does *not* necessarily mean that $\boldsymbol{v}_0$ is an eigenvector of A.

17. Use DPOWER (Figure 3.5.2) to find an eigenvalue of the matrix A in Problem 6, with

    (a) $p_0 = 0.9$ and no updating of p (EIG $= 0.9$, IUPDAT > 1000);

    (b) $p_0 = 1.9$ and no updating of p (EIG $= 1.9$, IUPDAT > 1000);

    (c) $p_0 = 1.9$ and updating of p every step (EIG $=1.9$, IUPDAT $= 1$).

    In each case let DPOWER choose a random starting vector $\boldsymbol{v}_0$, and print out the eigenvalue estimate each iteration. Why does the iteration converge so slowly in test (b)? (Hint: Look at the Jordan canonical form of A given in Problem 6.)

18. Verify that the matrix B defined by 3.6.2b is positive-definite. (Hint: Show that

$$\boldsymbol{z}^{\mathrm{T}} B \boldsymbol{z} = \int_0^1 \int_0^1 \left(\sum_j z_j \Phi_j \right)^2 dx\, dy \text{)}$$

19. a. Use DEGENP (Figure 3.6.1) to find the characteristic polynomial of the N by N tridiagonal matrix with 2 in each main diagonal position and -1 in each sub- and super-diagonal position, with $N = 10$. Then find the roots of this characteristic polynomial using a good root finder (such as MATLAB's "roots"). The exact eigenvalues should be $\lambda_k = 4\ sin^2(\frac{k\pi}{2(N+1)})$. Note that DEGENP and HESSQZ check for already zero elements and will do only $O(N^2)$ work when the matrix A is tridiagonal and B is diagonal ($B = I$ here).

    b. Repeat this experiment, but changing the sub-diagonal elements to 0. Now the exact eigenvalues are obviously all $\lambda_k = 2$, and DEGENP should return the characteristic polynomial $(2 - \lambda)^N$ exactly, but the root finder will nevertheless probably return very poor estimates of the roots.

20. Use DEGENP (Figure 3.6.1) to find the characteristic polynomial of the matrix A shown in (3.3.12) (set $B = I$) and verify that A is a "root" of its own characteristic polynomial, that is, verify that $\sum_{i=0}^{N} poly(i+1)\ A^i = 0$.

21. a. Find the singular value decomposition for the 5 by 5 nonsymmetric matrix given below, by forming A^TA and using DEGSYM (Figure 3.2.1) to find D^2 and V, and thus $U = AVD^{-1}$.

$$\begin{bmatrix} -5 & 7 & 3 & 4 & -8 \\ 5 & 8 & 3 & 6 & 8 \\ 3 & -7 & -3 & -4 & 5 \\ -3 & 0 & 4 & 5 & 3 \\ 7 & 4 & 5 & 9 & 5 \end{bmatrix}.$$

b. One of the singular values for the matrix in part (a) is close to zero. Do a (slight) data compression by replacing this small value in D by 0, recompute UDV^T, and compare with A.

22. If

$$A = \begin{bmatrix} 1 & 1 \\ 1 & 1 \\ 1 & 1 \end{bmatrix},$$

a. Show that a singular value decomposition of A is UDV^T, where

$$U = \begin{bmatrix} \frac{1}{\sqrt{3}} & \frac{1}{\sqrt{2}} & \frac{1}{\sqrt{6}} \\ \frac{1}{\sqrt{3}} & 0 & \frac{-2}{\sqrt{6}} \\ \frac{1}{\sqrt{3}} & \frac{-1}{\sqrt{2}} & \frac{1}{\sqrt{6}} \end{bmatrix}, D = \begin{bmatrix} \sqrt{6} & 0 \\ 0 & 0 \\ 0 & 0 \end{bmatrix}, V = \begin{bmatrix} \frac{1}{\sqrt{2}} & \frac{1}{\sqrt{2}} \\ \frac{1}{\sqrt{2}} & \frac{-1}{\sqrt{2}} \end{bmatrix}$$

b. Find the pseudoinverse $A^+ = VD^+U^T$ of A.

c. Now find the minimum norm solution $A^+\boldsymbol{b}$ of min $\|A\boldsymbol{x}-\boldsymbol{b}\|_2$, where $\boldsymbol{b}^T = [1, 2, 3]$.

d. Form the normal equations $A^TA\boldsymbol{x} = A^T\boldsymbol{b}$, note that there are many solutions, and verify that $A^+\boldsymbol{b}$ is the minimum norm solution of the normal equations, and thus the minimum norm least squares solution.

e. While the singular value decomposition of an arbitrary matrix A is not unique, show that the pseudoinverse is unique. (Hint: show that the minimum norm solution, if there are solutions, of *any* linear system is unique, so the minimum norm solution of $A^TA\boldsymbol{x} = A^T\boldsymbol{b}$ is unique. Thus if B and C are both pseudoinverses of A, $B\boldsymbol{b} = C\boldsymbol{b}$ for arbitrary $\boldsymbol{b}$, and thus $B = C$.)

f. Write a subroutine to compute the N by M pseudoinverse of an arbitrary M by N matrix A. Use this subroutine to find the pseudoinverse of the above matrix, and compare with your answer in part (b). (Hint: First form A^TA, and call DEGSYM (Figure 3.2.1) to find the eigenvalues D_{ii}^2 and the matrix V whose columns contain the eigenvectors. Now if D_{ii} is zero, or nearly zero, the ith

row of D^+U^T is zero, if not, the ith column of U is the ith column of AV/D_{ii}, and thus the ith row of D^+U^T is the ith row of V^TA^T/D_{ii}^2. Finally, multiply by V to get $A^+ = VD^+U^T$.)

4

Linear Programming

4.1 Linear Programming Applications

Linear programming applications often represent problems in economics, rather than science or engineering. The following are illustrative of LP applications.

4.1.1 The Resource Allocation Problem

A factory produces two products: chairs and tables. They make a profit of \$40 on each chair produced and \$50 on each table. A chair requires the following resources to produce: 2 man-hours, 3 hours of machine time, and 1 unit of wood. The table requires 2 man-hours, 1 hour of machine time, and 4 units of wood. The factory has 60 man-hours, 75 machine hours, and 84 units of wood available each day for producing these two products. How should the resources (man-hours, machine-hours, and wood) be allocated between the two products in order to maximize the factory's profit?

If we let c be the number of chairs produced per day and t the number of tables produced per day, then this problem can be stated mathematically as

$$\text{maximize } P = 40c + 50t \tag{4.1.1}$$

with the constraints that

$$\begin{aligned} 2c + 2t &\leq 60, \\ 3c + t &\leq 75, \\ c + 4t &\leq 84, \end{aligned}$$

and the bounds

$$\begin{aligned} c &\geq 0, \\ t &\geq 0. \end{aligned}$$

The function to be maximized, $40c + 50t$, represents the profit, and the constraints state that the total man-hours, machine-hours, and wood used cannot exceed the amounts available. The bounds state the obvious facts that the factory cannot produce a negative number of chairs or tables.

4.1.2 The Blending Problem

A feed company wants each feed bag that they produce to contain a minimum of 120 units of protein and 80 units of calcium. Corn contains 10 units of protein and 5 units of calcium per pound, and bonemeal contains 2 units of protein and 5 units of calcium per pound. If corn costs 8 cents per pound and bonemeal costs 4 cents per pound, how much of each should they put in each bag, in order to minimize costs?

If we let c be the number of pounds of corn used per bag and b the number of pounds of bonemeal used per bag, then the problem can be stated mathematically as

$$\text{minimize } P = 8c + 4b \tag{4.1.2}$$

with the constraints that

$$\begin{aligned} 10c + 2b &\geq 120, \\ 5c + 5b &\geq 80, \end{aligned}$$

and the bounds

$$\begin{aligned} c &\geq 0, \\ b &\geq 0. \end{aligned}$$

The function to be minimized, $8c + 4b$, represents the cost per bag, and the two constraints ensure that the total protein and calcium per bag equal or exceed the stated minimum requirements.

4.1.3 The Transportation Problem

A bulldozer company has two warehouses and three stores. The first warehouse has 40 bulldozers in stock and the second has 20. The three stores have 25, 10, and 22 bulldozers, respectively, on order. If C_{ij} is used to represent the cost to transport a bulldozer from warehouse i to store j, we know that $C_{11} = \$550, C_{12} = \$300, C_{13} = \$400, C_{21} = \$350, C_{22} = \$300$, and $C_{23} = \$100$. Determine the routing that will satisfy the needs of the stores, at minimum cost.

If we let X_{ij} be the number of bulldozers transported from warehouse i to store j, then the problem can be stated mathematically as

$$\text{minimize } \ 550X_{11} + 300X_{12} + 400X_{13} + 350X_{21} + 300X_{22} + 100X_{23} \tag{4.1.3}$$

with the constraints that

$$\begin{array}{lllllllllll} X_{11} & + & X_{12} & + & X_{13} & & & & & & \leq 40, \\ & & & & & & X_{21} & + & X_{22} & + & X_{23} \leq 20, \\ X_{11} & & & & & + & X_{21} & & & & = 25, \\ & & X_{12} & & & & & + & X_{22} & & = 10, \\ & & & & X_{13} & & & & & + & X_{23} = 22, \end{array}$$

and the bounds

$$X_{11} \geq 0,\ X_{12} \geq 0,\ X_{13} \geq 0,\ X_{21} \geq 0,\ X_{22} \geq 0,\ X_{23} \geq 0.$$

The function to be minimized represents the total transportation cost. The first two constraints state that the number of bulldozers leaving each warehouse cannot exceed the warehouse capacity, and the last three constraints state that the number of bulldozers arriving at each of the three stores must be equal to the number ordered. (Actually, the number arriving at each store must be *at least* as many as ordered. However, the minimum cost solution will clearly not specify that we deliver more bulldozers than ordered to any store so we can write the last three constraints as equalities rather than inequalities, if we like.)

4.1.4 Curve Fitting

We want to find the straight line $y = mx + b$ that best fits the data points $(1,1)$, $(2,3)$, $(3,2)$, $(4,3)$, and $(5,4)$, in the L_∞-norm. That is, we want to find m and b such that the maximum error

$$\max_{1\leq i\leq 5} |mx_i + b - y_i|$$

is minimized, where (x_i, y_i) represent the data points.

In other words, we want to find m, b, and ϵ such that $|mx_i + b - y_i| \leq \epsilon$ for all i, with ϵ as small as possible. This problem can be posed mathematically as

$$\text{minimize } \epsilon$$

with the constraints that

$$\begin{aligned} mx_i + b - y_i &\leq \epsilon \qquad (i = 1, \ldots, 5), \\ mx_i + b - y_i &\geq -\epsilon \qquad (i = 1, \ldots, 5). \end{aligned}$$

The constraints are equivalent to $|mx_i + b - y_i| \leq \epsilon$.

For (x_i, y_i) as given above, the problem can be rewritten as

$$\text{minimize } 0m + 0b + \epsilon \tag{4.1.4}$$

with the constraints that

$$\begin{aligned}
-1m - b + \epsilon &\geq -1, \\
-2m - b + \epsilon &\geq -3, \\
-3m - b + \epsilon &\geq -2, \\
-4m - b + \epsilon &\geq -3, \\
-5m - b + \epsilon &\geq -4, \\
1m + b + \epsilon &\geq 1, \\
2m + b + \epsilon &\geq 3, \\
3m + b + \epsilon &\geq 2, \\
4m + b + \epsilon &\geq 3, \\
5m + b + \epsilon &\geq 4.
\end{aligned}$$

Each of the above problems will be solved later in this chapter.

4.2 The Simplex Method, with Artificial Variables

The linear programming problem that we shall solve is

$$\text{maximize } P \equiv c_1 x_1 + \ldots + c_N x_N \tag{4.2.1}$$

with constraints

$$\begin{array}{ccccccc}
a_{1,1}x_1 & + & \ldots & + & a_{1,N}x_N & = & b_1 \\
\vdots & & & & \vdots & & \vdots \\
a_{M,1}x_1 & + & \ldots & + & a_{M,N}x_N & = & b_M
\end{array} \qquad (\text{all } b_i \geq 0)$$

and bounds

$$\begin{aligned}
x_1 &\geq 0 \\
&\vdots \\
x_N &\geq 0.
\end{aligned}$$

Note that the constraint right-hand sides must be nonnegative; however, if b_i is negative, this is easily corrected by multiplying the constraint through by -1.

The linear expression to be maximized, $P = c_1 x_1 + \ldots + c_N x_N$, is called the objective function and points $(x_1, \ldots, x_N)$ that satisfy both constraints and bounds are called feasible points. We want to find the feasible point or points that maximize P.

The problem 4.2.1 is more general than it appears at first. If we want to *minimize* a linear function P, we can simply maximize $-P$. Furthermore, "less than or equal to" constraints, of the form

$$a_{i,1}x_1 + \ldots + a_{i,N}x_N \leq b_i \qquad (i = 1, \ldots),$$

can be converted to "equation" constraints, by introducing additional nonnegative "slack variables," defined by

$$x_{N+i} \equiv b_i - (a_{i,1}x_1 + \ldots + a_{i,N}x_N)$$

(so called because they "take up the slack" between $a_{i,1}x_1 + \ldots + a_{i,N}x_N$ and its upper limit b_i). The ith constraint then becomes

$$a_{i,1}x_1 + \ldots + a_{i,N}x_N + x_{N+i} = b_i.$$

Similarly, constraints of the form

$$a_{i,1}x_1 + \ldots + a_{i,N}x_N \geq b_i \qquad (i = 1, \ldots)$$

can be converted to equation constraints by introducing slack variables

$$x_{N+i} \equiv (a_{i,1}x_1 + \ldots + a_{i,N}x_N) - b_i.$$

In this case, the new equation constraints have the form

$$a_{i,1}x_1 + \ldots + a_{i,N}x_N - x_{N+i} = b_i.$$

In either case, the slack variables, like the other variables, are bounded below by zero.

What the LP problem 4.2.1 reduces to is this: Of all the solutions (and normally there are an infinite number of them) to $A\boldsymbol{x} = \boldsymbol{b}$ which have nonnegative coefficients, find the one(s) that maximize $\mathbf{c}^{\mathrm{T}}\boldsymbol{x}$.

The simplex method [Dantzig 1951,1963] solves this problem by iteratively moving from one feasible point to a better feasible point ("better" means the objective function is larger there). However, finding a feasible point from which to start the iteration is not easy, in general. Finding any solution to $A\boldsymbol{x} = \boldsymbol{b}$ with all coefficients nonnegative is not a trivial task—indeed, there may not be any such solutions.

We shall use the method of artificial variables to find a starting feasible point. We introduce M additional "artificial" variables $x_{N+1}, \ldots, x_{N+M}$, defined by

$$x_{N+i} = b_i - (a_{i,1}x_1 + \ldots + a_{i,N}x_N) \qquad (i = 1, \ldots, M).$$

Although they are defined in a way that reminds us of the slack variables, their purpose is quite different. We shall now solve the related problem

$$\text{maximize } P \equiv c_1x_1 + \ldots + c_Nx_N - \alpha x_{N+1} - \ldots - \alpha x_{N+M} \qquad (4.2.2)$$

with constraints

$$
\begin{array}{ccccccc}
a_{1,1}x_1 & + & \ldots & + & a_{1,N}x_N + x_{N+1} & & = & b_1 \\
\vdots & & \vdots & & & & & \vdots \\
a_{M,1}x_1 & + & \ldots & + & a_{M,N}x_N & + \; x_{N+M} & = & b_M
\end{array}
\quad (\text{all } b_i \geq 0)
$$

and bounds

$$
\begin{array}{rcl}
x_1 & \geq & 0 \\
\vdots & & \\
x_{N+M} & \geq & 0
\end{array}
$$

where α is a very large positive number. Now there is no problem getting started with the simplex method; we can use $(0_1, \ldots, 0_N, b_1, \ldots, b_M)$ as a beginning feasible point, since $b_i \geq 0$. But how does solving 4.2.2 help us with 4.2.1, the problem that we really want to solve?

If the optimal solution to 4.2.2 has $x_{N+1} = \ldots = x_{N+M} = 0$, then this solution is clearly also the solution to 4.2.1. (If there were a better solution to 4.2.1, by appending $x_{N+1} = \ldots = x_{N+M} = 0$ to that point we could get a better solution to 4.2.2.) On the other hand, if the best solution to 4.2.2 has one or more nonzero artificial variables, we conclude that 4.2.1 allows no feasible points. This conclusion follows from the fact that α will be taken to be such a large number (in fact we shall computationally treat α as if it were the symbol $+\infty$) that all feasible points of 4.2.2 with $x_{N+1} = \ldots = x_{N+M} = 0$ give a better (higher) objective function value than *any* with nonzero (i.e., positive) artificial variables. Therefore, if the best feasible point of 4.2.2 has some nonzero artificial variables, there must not exist any feasible points of 4.2.1. In other words, the coefficients of the artificial variables in the objective function P are so large and negative that the penalty for retaining any nonzero artificial variables will be such that they will be retained only if there is no alternative.

Thus solving 4.2.2 will either give us the solution to 4.2.1, or else tell us that 4.2.1 allows no feasible points.

Now we outline the simplex method, applied to 4.2.2. We combine the equation $P \equiv c_1x_1 + \ldots + c_Nx_N - \alpha x_{N+1} - \ldots - \alpha x_{N+M}$ and the constraint equations into one linear system as follows:

$$\begin{bmatrix} a_{1,1} & \dots & a_{1,N} & 1 & \dots & 0 & 0 \\ \vdots & & \vdots & \vdots & & \vdots & \vdots \\ a_{M,1} & \dots & a_{M,N} & 0 & \dots & 1 & 0 \\ -c_1 & \dots & -c_N & \alpha & \dots & \alpha & 1 \end{bmatrix} \begin{bmatrix} x_1 \\ \vdots \\ x_N \\ x_{N+1} \\ \vdots \\ x_{N+M} \\ P \end{bmatrix} = \begin{bmatrix} b_1 \\ \vdots \\ b_M \\ 0 \end{bmatrix}$$

$$(\text{all } b_i \geq 0). \qquad (4.2.3)$$

The $M+1$ by $N+M+1$ matrix, together with the right-hand side vector, is called the "simplex tableau."

Our job now is to find the solution(s) of the linear system 4.2.3 (normally there will be many solutions, since there are N more unknowns than equations), with $x_1 \geq 0, \dots, x_{N+M} \geq 0$, which maximizes P. Because this makes the explanation simpler, we shall assume that α is a specific large positive number, such as 10^{20}. This works fine as long as all arithmetic is done exactly, but, when we actually implement this algorithm on a computer, we shall have to be more clever to avoid numerical problems.

The first step is to clear the α's from the last row of 4.2.3. This can be done by adding $-\alpha$ times the first equation to the last, then $-\alpha$ times the second equation to the last, and so on. The resulting linear system, which is equivalent to 4.2.3, is

$$\begin{bmatrix} a_{1,1} & \dots & a_{1,N} & 1 & \dots & 0 & 0 \\ \vdots & & \vdots & \vdots & & \vdots & \vdots \\ a_{M,1} & \dots & a_{M,N} & 0 & \dots & 1 & 0 \\ d_1 & \dots & d_N & 0 & \dots & 0 & 1 \end{bmatrix} \begin{bmatrix} x_1 \\ \vdots \\ x_N \\ x_{N+1} \\ \vdots \\ x_{N+M} \\ P \end{bmatrix} = \begin{bmatrix} b_1 \\ \vdots \\ b_M \\ P_0 \end{bmatrix}$$

$$(\text{all } b_i \geq 0), \qquad (4.2.4)$$

where $P_0 = -\alpha(b_1 + \dots + b_M)$ and $d_j = -c_j - \alpha(a_{1,j} + \dots + a_{M,j})$.

Now a solution to 4.2.4 is immediately obvious: $(0_1, \dots, 0_N, b_1, \dots, b_M, P_0)$. This solution also satisfies the bounds $x_i \geq 0$ $(i = 1, \dots, N+M)$, and so $(0_1, \dots, 0_N, b_1, \dots, b_M)$ is a feasible point of 4.2.2. Feasible points with at most M nonzero x-components are called extreme feasible points; therefore this solution is also an extreme feasible point.

This particular solution, of the many that exist, was found by setting $x_1, \dots, x_N$ equal to zero and solving for $x_{N+1}, \dots, x_{N+M}, P$. The variables that are easy to solve for—because the corresponding columns of A form an identity matrix—are called the basis variables (P is normally not considered part of the basis, however), and those that are set to zero are called the nonbasis variables.

Currently, the basis variables are all artificial variables, but in each step of the simplex algorithm we shall add one new variable to the basis and delete one old variable. Then using the basic row operations described in Section 1.2 (multiply a row through by a nonzero constant, and add a multiple of one row to another) we manipulate the linear system 4.2.4 into a new equivalent form, in which a new solution can easily be extracted, by setting the nonbasis variables equal to zero and solving for the new basis variables. The variables to be added and deleted from the basis will be chosen in a manner that will ensure that the new solution will satisfy the bounds $\boldsymbol{x} \geq \mathbf{0}$, so that it is still an (extreme) feasible point, and that the value of the objective function at the new point is at least as good as at the previous point.

Concretely, the iteration proceeds as follows. The last equation in 4.2.4 can be rewritten

$$P = P_0 - d_1 x_1 - \ldots - d_N x_N.$$

At the current point, the nonbasis variables $x_1, \ldots, x_N$ are all zero, so that $P = P_0$. If the d_j are all positive or zero, we cannot improve on this value, because $x_1, \ldots, x_N$ are only allowed to increase from their current values of 0, and so P can only be decreased from its current value of P_0. Since we cannot remain within the feasible region and increase P by moving in any direction from the current point, this means that we have found a point $\boldsymbol{x}_{\mathrm{L}}$ that represents a local maximum of the objective function.

Does the objective function also attain its global maximum at $\boldsymbol{x}_{\mathrm{L}}$? The answer is yes, for suppose there were another feasible point $\boldsymbol{x}_{\mathrm{G}}$ at which the objective function is larger. Since both $\boldsymbol{x}_{\mathrm{L}}$ and $\boldsymbol{x}_{\mathrm{G}}$ are in the feasible region (i.e., satisfy the constraints $A\boldsymbol{x} = \boldsymbol{b}$ and bounds $\boldsymbol{x} \geq \mathbf{0}$), it is easy to verify that any point $\boldsymbol{x} = (1-s)\boldsymbol{x}_{\mathrm{L}} + s\boldsymbol{x}_{\mathrm{G}}$ $(0 \leq s \leq 1)$ on the straight line connecting $\boldsymbol{x}_{\mathrm{L}}$ and $\boldsymbol{x}_{\mathrm{G}}$ also satisfies the constraints and bounds and is thus also feasible. However, since the objective function is linear, it must increase linearly along this straight line as we move from $\boldsymbol{x}_{\mathrm{L}}$ to $\boldsymbol{x}_{\mathrm{G}}$; thus $\boldsymbol{x}_{\mathrm{L}}$ cannot represent a local maximum. For a general constrained optimization problem we may find local maxima that do not represent global maxima but, when the constraints and objective function are linear, this cannot happen. Thus, when all the d_j are nonnegative, we have found the global maximum (of 4.2.2).

On the other hand, if at least one d_j is negative, we can improve on our current solution. If there is more than one negative d_j, we shall choose the most negative one, d_{jp}, but we could use any one of them (see Problem 8). We can now increase P by increasing (from zero) the value of x_{jp} while holding the other nonbasis variables equal to zero. Of course, we want to remain in the feasible region as we increase x_{jp}; so the linear system 4.2.4, rewritten below in more detail, must still be enforced:

$$\begin{bmatrix} a_{1,1} & \dots & a_{1,jp} & \dots & a_{1,N} & 1 & \dots & 0 & \dots & 0 & 0 \\ \vdots & & \vdots & & \vdots & \vdots & & \vdots & & \vdots & \vdots \\ a_{i,1} & \dots & a_{i,jp} & \dots & a_{i,N} & 0 & \dots & 1 & \dots & 0 & 0 \\ \vdots & & \vdots & & \vdots & \vdots & & \vdots & & \vdots & \vdots \\ a_{M,1} & \dots & a_{M,jp} & \dots & a_{M,N} & 0 & \dots & 0 & \dots & 1 & 0 \\ d_1 & \dots & d_{jp} & \dots & d_N & 0 & \dots & 0 & \dots & 0 & 1 \end{bmatrix} \begin{bmatrix} x_1 \\ \vdots \\ x_{jp} \\ \vdots \\ x_N \\ x_{N+1} \\ \vdots \\ x_{N+i} \\ \vdots \\ x_{N+M} \\ P \end{bmatrix} = \begin{bmatrix} b_1 \\ \vdots \\ b_i \\ \vdots \\ b_M \\ P_0 \end{bmatrix}. \tag{4.2.5}$$

If the other nonbasis variables are all held equal to zero, we can see how the basis variables vary with x_{jp} by rewriting equations 4.2.5:

$$\begin{aligned} x_{N+1} &= b_1 - a_{1,jp}x_{jp} \\ &\vdots \\ x_{N+i} &= b_i - a_{i,jp}x_{jp} \\ &\vdots \\ x_{N+M} &= b_M - a_{M,jp}x_{jp}. \end{aligned}$$

To remain in the feasible region, we must also make sure that none of these basis variables goes negative as x_{jp} is increased. Now, if all the elements $a_{1,jp}, \dots, a_{M,jp}$ in column jp of A are negative or zero, we can increase x_{jp} indefinitely, and none of the basis variables will ever become negative. Since $P \to \infty$ as $x_{jp} \to \infty$, this means we can stay within the feasible region and increase the objective function to infinity. In this case we are finished and must conclude that P has an unbounded maximum. (This is only possible, of course, if the feasible region is unbounded.) Thus the original problem 4.2.1 also has an unbounded maximum, because the presence of artificial variables can only decrease the maximum.

On the other hand, if at least one of $a_{1,jp}, \dots, a_{M,jp}$ is positive, there is a limit on how far we can increase x_{jp} without permitting any basis variables to go negative. If $a_{i,jp}$ is positive, we must stop increasing x_{jp} when $x_{N+i} = 0$, that is, when $x_{jp} = b_i/a_{i,jp}$. Since we cannot allow any variables to become negative, we must stop increasing x_{jp} when the *first* basis variable reaches zero, that is, when

$$x_{jp} = \frac{b_{ip}}{a_{ip,jp}},$$

where ip is the value of i that minimizes ($b_i/a_{i,jp}$ such that $a_{i,jp} > 0$). (In case of a tie, set ip equal to any one of these i.) In other words, x_{N+ip} is the first basis variable to reach zero as x_{jp} increases, and it will be replaced in the next basis by x_{jp} (Figure 4.2.1). Note that, if b_{ip} is zero, unfortunately, the value of P will not actually increase this simplex iteration, since x_{jp} cannot increase past 0, but at least it will not decrease.

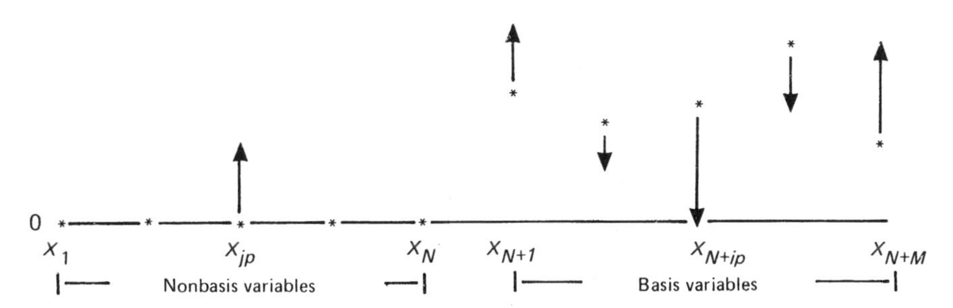

Figure 4.2.1
x_{jp} **Stops Increasing When the First Basis Variable Reaches Zero**

To calculate the new solution obtained by allowing x_{jp} to become positive (and thus to enter the basis) and forcing x_{N+ip} to zero (and thus to exit the basis), we shall perform some basic row operations on the simplex tableau 4.2.5. $a_{ip,jp}$ is by assumption positive. Therefore we can take multiples of the "pivot row" ip and add them to each of the other rows, to zero all other elements in the "pivot column" jp. Then, after we divide row ip by $a_{ip,jp}$, the system 4.2.5 has the equivalent form

$$\begin{bmatrix} a'_{1,1} & \ldots & 0 & \ldots & a'_{1,N} & 1 & \ldots & Y & \ldots & 0 & 0 \\ \vdots & & \vdots & & \vdots & \vdots & & \vdots & & \vdots & \vdots \\ a'_{ip,1} & \ldots & 1 & \ldots & a'_{ip,N} & 0 & \ldots & Y & \ldots & 0 & 0 \\ \vdots & & \vdots & & \vdots & \vdots & & \vdots & & \vdots & \vdots \\ a'_{M,1} & \ldots & 0 & \ldots & a'_{M,N} & 0 & \ldots & Y & \ldots & 1 & 0 \\ d'_1 & \ldots & 0 & \ldots & d'_N & 0 & \ldots & Y & \ldots & 0 & 1 \end{bmatrix} \begin{bmatrix} x_1 \\ \vdots \\ x_{jp} \\ \vdots \\ x_N \\ x_{N+1} \\ \vdots \\ x_{N+ip} \\ \vdots \\ x_{N+M} \\ P \end{bmatrix} = \begin{bmatrix} b'_1 \\ \vdots \\ b'_{ip} \\ \vdots \\ b'_M \\ P_1 \end{bmatrix}. \quad (4.2.6)$$

Note that the ipth column of the identity has moved from column $N + ip$ to column jp, making it now easy to solve for x_{jp} and not for x_{N+ip}. Now we *could* switch columns jp and $N + ip$, and switch the two corresponding unknowns in the "unknowns" vector, producing

$$\begin{bmatrix} a'_{1,1} & \ldots & Y & \ldots & a'_{1,N} & 1 & \ldots & 0 & \ldots & 0 & 0 \\ \vdots & & \vdots & & \vdots & \vdots & & \vdots & & \vdots & \vdots \\ a'_{ip,1} & \ldots & Y & \ldots & a'_{ip,N} & 0 & \ldots & 1 & \ldots & 0 & 0 \\ \vdots & & \vdots & & \vdots & \vdots & & \vdots & & \vdots & \vdots \\ a'_{M,1} & \ldots & Y & \ldots & a'_{M,N} & 0 & \ldots & 0 & \ldots & 1 & 0 \\ d'_1 & \ldots & Y & \ldots & d'_N & 0 & \ldots & 0 & \ldots & 0 & 1 \end{bmatrix} \begin{bmatrix} x_1 \\ \vdots \\ x_{N+ip} \\ \vdots \\ x_N \\ x_{N+1} \\ \vdots \\ x_{jp} \\ \vdots \\ x_{N+M} \\ P \end{bmatrix} = \begin{bmatrix} b'_1 \\ \vdots \\ b'_{ip} \\ \vdots \\ b'_M \\ P_1 \end{bmatrix}. \quad (4.2.7)$$

This linear system is identical in form with the linear system 4.2.5 that we started from on this step, but the names of the variables in the basis have changed. As before, we can easily extract a solution to the linear system, by setting the (new) nonbasis variables equal to zero and solving for the (new) basis variables. In fact, the values of the new basis variables (and $P = P_1$) can be read directly from the right-hand side vector, since the corresponding columns of the tableau form an identity matrix.

However, it is not necessary to switch columns physically, to keep the identity matrix in columns $N + 1$ to $N + M$. If the columns of the identity matrix are scattered throughout the tableau, that is not a problem as long as we keep up with where those columns are. We shall do this using a vector BASIS, defined so that BASIS(i) gives the tableau column number that holds the ith column of the identity. In other words. BASIS(1), ..., BASIS(M) identify the variables in the current basis. Since the initial basis consists entirely of artificial variables, we initialize BASIS(i) to $N + i$; then, when a pivot $a_{ip,jp}$ is used to knock out the other elements in column jp, column jp will take on the form of the ipth column of the identity (and another column will lose this form), so we set BASIS(ip) $= jp$.

We know that the b_i are all still nonnegative, since they are equal to the basis variables, and we took measures to ensure that no variables would become negative. We also know from our previous analysis that P_1 is at least as large as the previous objective function value P_0. We have now completed one iteration of the simplex algorithm, and the next step begins with the tableau 4.2.6.

Let us summarize (without the complicating explanations) the simplex algorithm as outlined above, starting with a tableau such as 4.2.5, with BASIS(i) initialized to $N + i$, for $i = 1, \ldots, M$.

(I) Find the most negative element in the last row, d_{jp}. The column jp is called the pivot column. If there are no negative elements, we cannot improve on the current solution; so go to step (IV).

(II) Move up the pivot column and for each *positive* element $a_{i,jp}$, calculate $b_i/a_{i,jp}$. The row on which the smallest such ratio occurs is called the pivot row, ip, and $a_{ip,jp}$ is the "pivot." If there are no positive elements in the pivot column, we conclude that the objective function (of 4.2.1) has an infinite maximum, and we are finished.

(III) By elementary row operations, use the pivot row to knock out all other elements in the pivot column, and reduce the pivot itself to 1. This transforms column jp into the ipth column of the identity matrix and destroys the ipth column that previously existed in another column; so set BASIS(ip) $= jp$. Go to step (I) and start a new iteration.

(IV) If the basis variables still include any artificial variables, that is, if BASIS(i) $> N$ for any i, we conclude that the problem 4.2.1 has no feasible solution. Otherwise, the optimal solution of 4.2.1 is obtained by setting all nonbasis variables to zero and solving (trivially) for the basis variables: if BASIS(i) $= k, x_k = b_i$.

4.3 The Dual Solution

Consider again the LP problem 4.2.1, rewritten below in matrix–vector form:

$$\text{maximize } P = \boldsymbol{c}^{\mathrm{T}}\boldsymbol{x} \tag{4.3.1}$$

with constraints

$$A\boldsymbol{x} = \boldsymbol{b}$$

and bounds

$$\boldsymbol{x} \geq \boldsymbol{0}.$$

The following problem is called the "dual" of this problem (4.3.1 is then called the "primal" problem):

$$\text{minimize } D = \boldsymbol{b}^{\mathrm{T}}\boldsymbol{y} \tag{4.3.2}$$

with constraints

$$A^{\mathrm{T}}\boldsymbol{y} \geq \boldsymbol{c}.$$

Note that there are no bounds in this dual problem.

We shall prove that the optimal objective function values for the primal problem $P_{\max}$ and the dual problem $D_{\min}$ are equal. First we prove the following theorem.

Theorem 4.3.1. *For any primal-feasible $\boldsymbol{x}$ and any dual-feasible $\boldsymbol{y}, \boldsymbol{b}^{\mathrm{T}}\boldsymbol{y} \geq \boldsymbol{c}^{\mathrm{T}}\boldsymbol{x}$.*

Proof: $\boldsymbol{b}^{\mathrm{T}}\boldsymbol{y} = (A\boldsymbol{x})^{\mathrm{T}}\boldsymbol{y} = \boldsymbol{x}^{\mathrm{T}}A^{\mathrm{T}}\boldsymbol{y} \geq \boldsymbol{x}^{\mathrm{T}}\boldsymbol{c}$. For the last step, we have used both $A^{\mathrm{T}}\boldsymbol{y} \geq \boldsymbol{c}$ and $\boldsymbol{x} \geq \boldsymbol{0}$. ■

If the dual has any feasible solution $\boldsymbol{y}$, $\boldsymbol{b}^{\mathrm{T}}\boldsymbol{y}$ will be an upper bound for the primal objective function and, similarly, if the primal problem has any feasible solution $\boldsymbol{x}$, $\boldsymbol{c}^{\mathrm{T}}\boldsymbol{x}$ will be a lower bound for the dual objective function. Hence we have the following corollary.

Corollary *If the primal problem has an unbounded maximum ($+\infty$), then the dual problem has no feasible solution and, if the dual problem has an unbounded minimum ($-\infty$), the primal problem is infeasible.*

Before proving the main theorem connecting the primal and dual problems, let us look at the simplex tableau in more detail. Let us reorder the columns of the simplex tableau 4.2.3 and the corresponding unknowns so that the *initial* tableau has the form

$$\begin{bmatrix} A_n & A_b & I & \boldsymbol{0} \\ -\boldsymbol{c}_n^{\mathrm{T}} & -\boldsymbol{c}_b^{\mathrm{T}} & \alpha\boldsymbol{1}^{\mathrm{T}} & 1 \end{bmatrix} \begin{bmatrix} \boldsymbol{x}_n \\ \boldsymbol{x}_b \\ \boldsymbol{x}_a \\ P \end{bmatrix} = \begin{bmatrix} \boldsymbol{b} \\ 0 \end{bmatrix}, \tag{4.3.3}$$

where $\boldsymbol{x}_b, \boldsymbol{x}_a$, and $\boldsymbol{x}_n$ are vectors containing the *final* basis variables, the artificial variables, and the other variables not in the *final* basis, and $\boldsymbol{1} = (1_1, \ldots, 1_M)$. We assume there are no artificial variables in the final basis (even though the initial basis consists entirely of artificial variables); otherwise the original problem 4.2.1 is infeasible. Of course, we do not know which variables will be in the final basis until we have finished the simplex algorithm (determining which variables are in the basis is the main task of the simplex method) but, for our theoretical purposes, it is convenient to assume that the variables destined to end up in the final basis are ordered after those that will not be in the final basis.

Now to transform 4.3.3 into the final tableau we can use row operations to reduce A_b to the identity matrix and $-\boldsymbol{c}_b^{\mathrm{T}}$ to zero or, equivalently, we can premultiply both sides of 4.3.3 by the $M+1$ by $M+1$ matrix

$$\begin{bmatrix} A_b^{-1} & \boldsymbol{0} \\ \boldsymbol{c}_b^{\mathrm{T}}A_b^{-1} & 1 \end{bmatrix}.$$

This produces the final simplex tableau (see Theorem 0.1.4)

$$\begin{bmatrix} A_b^{-1}A_n & I & A_b^{-1} & \boldsymbol{0} \\ \boldsymbol{c}_b^{\mathrm{T}}A_b^{-1}A_n - \boldsymbol{c}_n^{\mathrm{T}} & \boldsymbol{0}^{\mathrm{T}} & \boldsymbol{c}_b^{\mathrm{T}}A_b^{-1} + \alpha\boldsymbol{1}^{\mathrm{T}} & 1 \end{bmatrix} \begin{bmatrix} \boldsymbol{x}_n \\ \boldsymbol{x}_b \\ \boldsymbol{x}_a \\ P \end{bmatrix} = \begin{bmatrix} A_b^{-1}\boldsymbol{b} \\ \boldsymbol{c}_b^{\mathrm{T}}A_b^{-1}\boldsymbol{b} \end{bmatrix}. \tag{4.3.4}$$

Now we can prove the following theorem.

Theorem 4.3.2. *If the primal problem is feasible and has a bounded maximum, the optimal solution to the dual problem is* $\boldsymbol{y}^* = A_b^{-\mathrm{T}}\boldsymbol{c}_b$*, and the maximum primal objective function value* $P_{\max}$ *is equal to the minimum dual objective function value* $D_{\min}$.

Proof: We first show that $\boldsymbol{y}^*$ is dual feasible. We know that there can be no negative entries in the last row of the final simplex tableau; otherwise the simplex algorithm must continue. Therefore (see 4.3.4) $\boldsymbol{c}_b^{\mathrm{T}} A_b^{-1} A_n - \boldsymbol{c}_n^{\mathrm{T}} \geq \boldsymbol{0}^{\mathrm{T}}$ or, transposing both sides, $A_n^{\mathrm{T}} A_b^{-\mathrm{T}} \boldsymbol{c}_b \geq \boldsymbol{c}_n$ and so

$$A^{\mathrm{T}}\boldsymbol{y}^* = \begin{bmatrix} A_n^{\mathrm{T}} \\ A_b^{\mathrm{T}} \end{bmatrix} \left[A_b^{-\mathrm{T}}\boldsymbol{c}_b\right] \geq \begin{bmatrix} \boldsymbol{c}_n \\ \boldsymbol{c}_b \end{bmatrix} = \boldsymbol{c},$$

proving that $\boldsymbol{y}^*$ is dual feasible.

Now the solution to the primal problem is (as we have already seen) found by setting the nonbasis variables $\boldsymbol{x}_n$ and $\boldsymbol{x}_a$ to zero and solving for the basis variables and P in the final tableau 4.3.4, which gives $\boldsymbol{x}_n = \boldsymbol{0}, \boldsymbol{x}_b = A_b^{-1}\boldsymbol{b}, \boldsymbol{x}_a = \boldsymbol{0}$, $P_{\max} = \boldsymbol{c}_b^{\mathrm{T}} A_b^{-1}\boldsymbol{b}$. By Theorem 4.3.1, for any dual-feasible $\boldsymbol{y}, \boldsymbol{b}^{\mathrm{T}}\boldsymbol{y} \geq P_{\max} = \boldsymbol{c}_b^{\mathrm{T}} A_b^{-1}\boldsymbol{b} = \boldsymbol{y}^{*\mathrm{T}}\boldsymbol{b}$, and therefore $\boldsymbol{y}^*$ minimizes $\boldsymbol{b}^{\mathrm{T}}\boldsymbol{y}$ over all dual-feasible points, and $D_{\min} = \boldsymbol{b}^{\mathrm{T}}\boldsymbol{y}^* = P_{\max}$. ■

Note that, in the final tableau 4.3.4, the dual solution $\boldsymbol{y}^{*\mathrm{T}} = \boldsymbol{c}_b^{\mathrm{T}} A_b^{-1}$ appears in the last row, in the columns corresponding to artificial variables. Thus both the primal and the dual solutions are readily available from the final tableau.

As illustrated by the examples in Section 4.1, LP applications more often come with inequality constraints than equation constraints. For example, a resource allocation problem usually takes the form

$$\text{maximize } P = \boldsymbol{c}^{\mathrm{T}}\boldsymbol{x} \tag{4.3.5}$$

with constraints

$$A\boldsymbol{x} \leq \boldsymbol{b}$$

and bounds

$$\boldsymbol{x} \geq \boldsymbol{0}.$$

When slack variables s are added to convert the inequality constraints to equation constraints, the problem can be put in the form 4.3.1, as follows:

$$\text{maximize } P = [\boldsymbol{c}, \boldsymbol{0}]^{\mathrm{T}}[\boldsymbol{x}, \mathrm{s}]$$

with constraints

$$\begin{bmatrix} A & I \end{bmatrix} \begin{bmatrix} \boldsymbol{x} \\ \mathrm{s} \end{bmatrix} = \begin{bmatrix} \boldsymbol{b} \end{bmatrix}$$

and bounds

$$\begin{bmatrix} \boldsymbol{x} \\ \mathrm{s} \end{bmatrix} \geq \begin{bmatrix} \mathbf{0} \\ \mathbf{0} \end{bmatrix}.$$

The dual of this problem is, according to 4.3.2,

$$\text{minimize } D = \boldsymbol{b}^{\mathrm{T}}\boldsymbol{y}$$

with constraints

$$\begin{bmatrix} A^{\mathrm{T}} \\ I \end{bmatrix} [\, \boldsymbol{y} \,] \geq \begin{bmatrix} \boldsymbol{c} \\ \mathbf{0} \end{bmatrix}$$

or, equivalently,

$$\text{minimize } D = \boldsymbol{b}^{\mathrm{T}}\boldsymbol{y} \tag{4.3.6}$$

with constraints

$$A^{\mathrm{T}}\boldsymbol{y} \geq \boldsymbol{c}$$

and bounds

$$\boldsymbol{y} \geq \mathbf{0}.$$

Now of what use is the dual solution? First, it is sometimes more convenient to solve the dual than the primal (or vice versa). For example, if the matrix A in 4.3.5 is M by N, M slack variables and M artificial variables have to be added, and the simplex tableau 4.2.3 will be $M+1$ by $N+2M+1$. The matrix A^{T}, which appears in the dual problem, on the other hand, is N by M, and hence the simplex tableau will be $N+1$ by $M+2N+1$. If M is much larger than N, the primal tableau will be a much larger matrix than the dual tableau, and thus it is more efficient to solve the dual problem. Other LP problems that are much easier to solve in dual form are given in Problems 3 and 4.

More importantly, however, the dual solution has a physically meaningful and useful interpretation. This interpretation comes from the fact that the optimal objective function value of the primal, $P_{\max}$, is equal to $\boldsymbol{b}^{\mathrm{T}}\boldsymbol{y}^*$, or $b_1 y_1^* + \ldots + b_M y_M^*$. Hence, if b_j is changed by a small amount (small enough that the final basis does not change, and thus $\boldsymbol{y}^* = A_b^{-\mathrm{T}}\boldsymbol{c}_b$ does not change), and the other right-hand sides are held constant while the problem is re-solved, then the new optimum will change by $\Delta P_{\max} = y_j^*\ \Delta b_j$.

In other words, if we solve an LP problem such as 4.2.1 once, we shall obtain the optimal solution and the value of P (which is $P_{\max}$) at that point but, by looking at the dual solution, we can determine which right-hand side(s) should be increased if we want to obtain an even higher maximum.

It might sometimes be useful to know how the optimum depends on the coefficients $a_{i,j}$ as well as how it depends on b_j, but this type of "sensitivity analysis" is not as easy to do. In typical LP applications, however, the constraint limits b_j are more readily controlled than the matrix coefficients; so how the solution depends on the right-hand side is of more practical interest. For example, in the resource allocation problem given by 4.1.1, the number of man-hours or the quantity of wood required to build a chair is not so easily modified, but the number of man-hours available, or the quantity of wood available, can be increased simply by hiring more workers, or buying more wood.

4.4 Examples

Let us use the simplex method to solve the resource allocation problem in 4.1.1, repeated here:

$$\text{maximize } P = 40c + 50t$$

with constraints

$$\begin{aligned} 2c + 2t &\leq 60, \\ 3c + t &\leq 75, \\ c + 4t &\leq 84, \end{aligned}$$

and bounds

$$\begin{aligned} c &\geq 0, \\ t &\geq 0. \end{aligned}$$

After adding slack variables, we have

$$\text{maximize } P = 40c + 50t + 0s_1 + 0s_2 + 0s_3 \tag{4.4.1}$$

with constraints

$$\begin{array}{lllllllll} 2c & + & 2t & + & s_1 & & & & = 60, \\ 3c & + & t & & & + & s_2 & & = 75, \\ c & + & 4t & & & & & + \ s_3 & = 84, \end{array}$$

and bounds

$$\begin{aligned} c &\geq 0, \\ t &\geq 0, \\ s_1 &\geq 0, \\ s_2 &\geq 0, \\ s_3 &\geq 0. \end{aligned}$$

For resource allocation problems, there is no need for artificial variables, since the simplex tableau already contains an identity matrix, and the right-hand side components are already positive. Hence a starting feasible point, namely, $(0, 0, 60, 75, 84)$, is already available.

$$\begin{bmatrix} 2 & 2 & 1 & 0 & 0 & 0 \\ 3 & 1 & 0 & 1 & 0 & 0 \\ 1 & \textcircled{4} & 0 & 0 & 1 & 0 \\ -40 & -50 & 0 & 0 & 0 & 1 \end{bmatrix} \begin{bmatrix} c \\ t \\ s_1 \\ s_2 \\ s_3 \\ P \end{bmatrix} = \begin{bmatrix} 60 \\ 75 \\ 84 \\ 0 \end{bmatrix}. \tag{4.4.2}$$

The most negative entry in the last row is -50; so the second column will be the pivot column. Comparing $60/2, 75/1$, and $84/4$ (since 2, 1, and 4 are all positive), we see that the minimum is $84/4 = 21$; so the pivot row will be the third row, and the pivot element will be the 4 circled in 4.4.2. We shall use this pivot to knock out all other elements in the pivot column. After this is done, and the third row is divided by 4 to reduce the pivot to 1, the simplex tableau has the form shown below:

$$\begin{bmatrix} \textcircled{1.5} & 0 & 1 & 0 & -0.5 & 0 \\ 2.75 & 0 & 0 & 1 & -0.25 & 0 \\ 0.25 & 1 & 0 & 0 & 0.25 & 0 \\ -27.5 & 0 & 0 & 0 & 12.5 & 1 \end{bmatrix} \begin{bmatrix} c \\ t \\ s_1 \\ s_2 \\ s_3 \\ P \end{bmatrix} = \begin{bmatrix} 18 \\ 54 \\ 21 \\ 1050 \end{bmatrix}. \tag{4.4.3}$$

The new basis consists of (s_1, s_2, t) (BASIS$= \{3, 4, 2\}$), since the corresponding columns form an identity matrix, and we can get our new extreme feasible point by setting $c = s_3 = 0$ and solving for the basis variables. This gives $c = 0, t = 21, s_1 = 18, s_2 = 54, s_3 = 0, P = 1050$.

Now we are finished with the first simplex iteration and are ready to start a new one. The most negative entry in the last row is now in column 1, and the minimum of $18/1.5, 54/2.75, 21/0.25$ is $18/1.5$; so the new pivot element is in row 1, column 1 (circled in 4.4.3). Row operations using this pivot produced the new tableau shown below:

$$\begin{bmatrix} 1 & 0 & 0.667 & 0 & -0.333 & 0 \\ 0 & 0 & -1.833 & 1 & 0.667 & 0 \\ 0 & 1 & -0.167 & 0 & 0.333 & 0 \\ 0 & 0 & 18.333 & 0 & 3.333 & 1 \end{bmatrix} \begin{bmatrix} c \\ t \\ s_1 \\ s_2 \\ s_3 \\ P \end{bmatrix} = \begin{bmatrix} 12 \\ 21 \\ 18 \\ 1380 \end{bmatrix}.$$

The new basis consists of (c, s_2, t), and the new extreme feasible point that we have found is $c = 12$, $t = 18$, $s_1 = 0, s_2 = 21, s_3 = 0, P = 1380$. Now all entries in the last row are nonnegative, so we have finished, and we have determined that our factory should produce 12 chairs and 18 tables per day.

Furthermore, we can calculate that the dual of this problem is (18.333, 0.0, 3.333). This means that increasing the limit on the second constraint (increasing the number of machine-hours available) will not increase our profits. This is because producing 12 chairs and 18 tables per day does not exhaust the machine-hours already available. However, the dual solution tells us that profits will increase by \$18.33 for each additional man-hour, and by \$3.33 for each additional unit of wood, made available to the factory, at least until the increases force a change in the composition of the final basis. Indeed, when the above problem was re-solved with the number of units of wood available increased by 1 (to 85) the optimal profit did increase by exactly \$3.33, to \$1383.33.

It is instructive to look at this problem, and the simplex iteration, geometrically. Figure 4.4.1 shows the feasible region for the problem given by 4.1.1, which consists of those points (c, t) satisfying the three constraints and the two bounds. We started the simplex iteration at the point $c = 0$, $t = 0$ (also $s_1 = 60, s_2 = 75, s_3 = 84$). The first iteration moved us to another vertex of the feasible region, $c = 0$, $t = 21$, and the second iteration moved us to the vertex $c = 12$, $t = 18$, the optimal solution. We see here that at each iteration of the simplex method we move from one vertex (extreme feasible point) to a neighboring but better vertex.

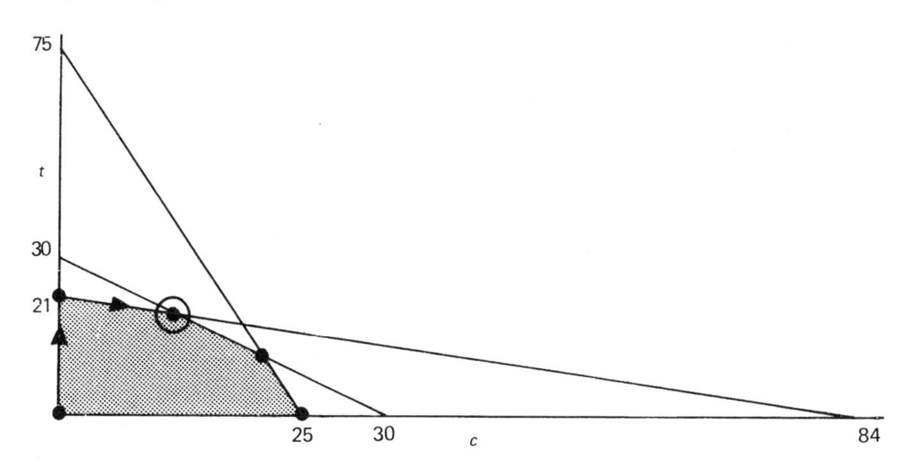

Figure 4.4.1
Feasible Region for Resource Allocation Problem

Next let us apply the simplex method to the blending problem given by 4.1.2. To convert this minimization problem to the form 4.2.1, we negate the objective function, and add slack variables to convert the inequality constraints to equation constraints:

$$\text{maximize } P = -8c - 4b + 0s_1 + 0s_2$$

with constraints

$$\begin{aligned} 10c + 2b - s_1 \qquad &= 120, \\ 5c + 5b \qquad - s_2 &= 80, \end{aligned}$$

and bounds

$$\begin{aligned} c &\geq 0, \\ b &\geq 0, \\ s_1 &\geq 0, \\ s_2 &\geq 0. \end{aligned}$$

Now we do have to use artificial variables (a_1, a_2) to get started, because no starting feasible point is readily available (setting $c = b = 0$ yields negative values for s_1, s_2). After adding artificial variables (cf. 4.2.3), the simplex tableau is

$$\begin{bmatrix} 10 & 2 & -1 & 0 & 1 & 0 & 0 \\ 5 & 5 & 0 & -1 & 0 & 1 & 0 \\ 8 & 4 & 0 & 0 & \alpha & \alpha & 1 \end{bmatrix} \begin{bmatrix} c \\ b \\ s_1 \\ s_2 \\ a_1 \\ a_2 \\ P \end{bmatrix} = \begin{bmatrix} 120 \\ 80 \\ 0 \end{bmatrix}.$$

For this problem, the artificial variables can be thought of as representing very expensive packets of pure protein and pure calcium. We start by using $a_1 = 120$ packets of pure protein and $a_2 = 80$ packets of pure calcium. If we are still using any of these expensive packets in our optimal solution, that means there is no way to fulfill the requirements with corn and bonemeal alone. If we are not, then we have solved our original problem; the expensive packets are no longer needed.

Before actually starting the first simplex iteration, we need to clear the α's from the last row. We shall not actually assign a value to α but shall treat it as an arbitrary large positive number. The starting tableau is

$$\begin{bmatrix} \textcircled{10} & 2 & -1 & 0 & 1 & 0 & 0 \\ 5 & 5 & 0 & -1 & 0 & 1 & 0 \\ 8-15\alpha & 4-7\alpha & \alpha & \alpha & 0 & 0 & 1 \end{bmatrix} \begin{bmatrix} c \\ b \\ s_1 \\ s_2 \\ a_1 \\ a_2 \\ P \end{bmatrix} = \begin{bmatrix} 120 \\ 80 \\ -200\alpha \end{bmatrix}.$$

Since α is large, $8 - 15\alpha$ is the most negative entry in the last row; so the first column is the pivot column. Comparing $120/10$ and $80/5$, we see that the first row must be chosen as the pivot row. After the row operations to knock out the other elements in the pivot column, we have the next tableau:

$$\begin{bmatrix} 1 & 0.2 & -0.1 & 0 & 0.1 & 0 & 0 \\ 0 & \textcircled{4} & 0.5 & -1 & -0.5 & 1 & 0 \\ 0 & 2.4-4\alpha & 0.8-0.5\alpha & \alpha & -0.8+1.5\alpha & 0 & 1 \end{bmatrix} \begin{bmatrix} c \\ b \\ s_1 \\ s_2 \\ a_1 \\ a_2 \\ P \end{bmatrix} = \begin{bmatrix} 12 \\ 20 \\ -96-20\alpha \end{bmatrix}.$$

After another iteration we have

$$\begin{bmatrix} 1 & 0 & -0.125 & 0.05 & 0.125 & -0.05 & 0 \\ 0 & 1 & 0.125 & -0.25 & -0.125 & 0.25 & 0 \\ 0 & 0 & 0.5 & 0.6 & -0.5+\alpha & -0.6+\alpha & 1 \end{bmatrix} \begin{bmatrix} c \\ b \\ s_1 \\ s_2 \\ a_1 \\ a_2 \\ P \end{bmatrix} = \begin{bmatrix} 11 \\ 5 \\ -108 \end{bmatrix}.$$

Since all entries on the last row are now positive, we have finished. The last basis $\{c, b\}$ no longer includes any of the artificial variables a_1, a_2; so we conclude that there is a solution, and it is $c = 11$, $b = 5$ (also $s_1 = s_2 = 0$ and $P = -108$). Thus the feed company should mix 11 pounds of corn with 5 pounds of bonemeal in each bag. Since the maximum of the negative of the cost was -108 cents, the minimum cost, corresponding to this blend, is 108 cents.

4.5 A FORTRAN90 Program

A FORTRAN90 program, DLPRG, which solves the LP problem 4.2.1, using the simplex algorithm described in Section 4.2, is shown in Figure 4.5.1. The matrix TAB holds the simplex tableau matrix but, since the last column of this matrix never changes from its initial form $(0, \ldots, 0, 1)$, there is no need to store it, and the right-hand side vector is stored in this column instead. The matrix TAB is $M + 2$ by $N + M + 1$ rather than $M + 1$ by $N + M + 1$, because the Jth element of the last row of the tableau is stored in the form TAB(M+1,J)+α*TAB(M+2,J). Thus we treat α as an arbitrary large number, rather than assigning it a specific large value. This is necessary to avoid serious accuracy problems: Adding a large number to a small one obscures the small one.

```
      SUBROUTINE DLPRG(A,B,C,N,M,P,X,Y)
      IMPLICIT DOUBLE PRECISION (A-H,O-Z)
C                              DECLARATIONS FOR ARGUMENTS
      DOUBLE PRECISION A(M,N),B(M),C(N),P,X(N),Y(M)
      INTEGER N,M
C                              DECLARATIONS FOR LOCAL VARIABLES
      DOUBLE PRECISION TAB(M+2,N+M+1)
      INTEGER BASIS(M)
C
C  SUBROUTINE DLPRG USES THE SIMPLEX METHOD TO SOLVE THE PROBLEM
C
C             MAXIMIZE      P = C(1)*X(1) + ... + C(N)*X(N)
C
C   WITH X(1),...,X(N) NONNEGATIVE, AND
C
C          A(1,1)*X(1) + ... + A(1,N)*X(N)  =  B(1)
C            .                    .              .
C            .                    .              .
C          A(M,1)*X(1) + ... + A(M,N)*X(N)  =  B(M)
C
C   WHERE B(1),...,B(M) ARE ASSUMED TO BE NONNEGATIVE.
C
C  ARGUMENTS
C
C             ON INPUT                         ON OUTPUT
C             --------                         ---------
C
C   A      - THE M BY N CONSTRAINT COEFFICIENT
C             MATRIX.
C
C   B      - A VECTOR OF LENGTH M CONTAINING
C             THE RIGHT HAND SIDES OF THE
C             CONSTRAINTS.  THE COMPONENTS OF
C             B MUST ALL BE NONNEGATIVE.
C
C   C      - A VECTOR OF LENGTH N CONTAINING
C             THE COEFFICIENTS OF THE OBJECTIVE
C             FUNCTION.
C
C   N      - THE NUMBER OF UNKNOWNS.
C
C   M      - THE NUMBER OF CONSTRAINTS.
C
C   P      -                                   THE MAXIMUM OF THE
C                                              OBJECTIVE FUNCTION.
C
C   X      -                                   A VECTOR OF LENGTH N
C                                              WHICH CONTAINS THE LP
```

```
C                                                  SOLUTION.
C
C    Y       -                                     A VECTOR OF LENGTH M
C                                                  WHICH CONTAINS THE DUAL
C                                                  SOLUTION.
C
C-----------------------------------------------------------------------
C                                  EPS = MACHINE FLOATING POINT RELATIVE
C                                        PRECISION
C ***************************
      DATA EPS/2.D-16/
C ***************************
C                                  BASIS(1),...,BASIS(M) HOLD NUMBERS OF
C                                  BASIS VARIABLES.  INITIAL BASIS CONSISTS
C                                  OF ARTIFICIAL VARIABLES ONLY
      DO 5 I=1,M
         BASIS(I) = N+I
         IF (B(I).LT.0.0) THEN
            PRINT 1
    1       FORMAT (' ***** ALL B(I) MUST BE NONNEGATIVE *****')
            RETURN
         ENDIF
    5 CONTINUE
C                                  INITIALIZE SIMPLEX TABLEAU
      DO 10 I=1,M+2
      DO 10 J=1,N+M+1
         TAB(I,J) = 0.0
   10 CONTINUE
C                                  LOAD A INTO UPPER LEFT HAND CORNER
C                                  OF TABLEAU
      DO 15 I=1,M
      DO 15 J=1,N
         TAB(I,J) = A(I,J)
   15 CONTINUE
C                                  LOAD M BY M IDENTITY TO RIGHT OF A
C                                  AND LOAD B INTO LAST COLUMN
      DO 20 I=1,M
         TAB(I,N+I) = 1.0
         TAB(I,N+M+1) = B(I)
   20 CONTINUE
C                                  ROW M+1 CONTAINS -C, INITIALLY
      DO 25 J=1,N
         TAB(M+1,J) = -C(J)
   25 CONTINUE
C                                  ROW M+2 CONTAINS COEFFICIENTS OF
C                                  "ALPHA", WHICH IS TREATED AS +INFINITY
      DO 30 I=1,M
         TAB(M+2,N+I) = 1.0
```

```
   30 CONTINUE
C                              CLEAR "ALPHAS" IN LAST ROW
      DO 35 I=1,M
      DO 35 J=1,N+M+1
         TAB(M+2,J) = TAB(M+2,J) - TAB(I,J)
   35 CONTINUE
C                              SIMPLEX METHOD CONSISTS OF TWO PHASES
      DO 90 IPHASE=1,2
         IF (IPHASE.EQ.1) THEN
C                              PHASE I:  ROW M+2 (WITH COEFFICIENTS OF
C                              ALPHA) SEARCHED FOR MOST NEGATIVE ENTRY
            MROW = M+2
            LIM = N+M
         ELSE
C                              PHASE II:  FIRST N ELEMENTS OF ROW M+1
C                              SEARCHED FOR MOST NEGATIVE ENTRY
C                              (COEFFICIENTS OF ALPHA NONNEGATIVE NOW)
            MROW = M+1
            LIM = N
C                              IF ANY ARTIFICIAL VARIABLES LEFT IN
C                              BASIS AT BEGINNING OF PHASE II, THERE
C                              IS NO FEASIBLE SOLUTION
            DO 45 I=1,M
               IF (BASIS(I).GT.LIM) THEN
                  PRINT 40
   40             FORMAT (' ***** NO FEASIBLE SOLUTION *****')
                  RETURN
               ENDIF
   45       CONTINUE
         ENDIF
C                              THRESH = SMALL NUMBER.  WE ASSUME SCALES
C                              OF A AND C ARE NOT *TOO* DIFFERENT
         THRESH = 0.0
         DO 50 J=1,LIM
            THRESH = MAX(THRESH,ABS(TAB(MROW,J)))
   50    CONTINUE
         THRESH = 1000*EPS*THRESH
C                              BEGINNING OF SIMPLEX STEP
   55    CONTINUE
C                              FIND MOST NEGATIVE ENTRY IN ROW MROW,
C                              IDENTIFYING PIVOT COLUMN JP
            CMIN = -THRESH
            JP = 0
            DO 60 J=1,LIM
               IF (TAB(MROW,J).LT.CMIN) THEN
                  CMIN = TAB(MROW,J)
                  JP = J
               ENDIF
```

```
   60       CONTINUE
C                            IF ALL ENTRIES NONNEGATIVE (ACTUALLY,
C                            IF GREATER THAN -THRESH) PHASE ENDS
            IF (JP.EQ.0) GO TO 90
C                            FIND SMALLEST POSITIVE RATIO
C                            B(*)/TAB(*,JP), IDENTIFYING PIVOT
C                            ROW IP
            RATMIN = 0.0
            IP = 0
            DO 65 I=1,M
               IF (TAB(I,JP).GT.THRESH) THEN
                  RATIO = TAB(I,N+M+1)/TAB(I,JP)
                  IF (IP.EQ.0 .OR. RATIO.LT.RATMIN) THEN
                     RATMIN = RATIO
                     IP = I
                  ENDIF
               ENDIF
   65       CONTINUE
C                            IF ALL RATIOS NONPOSITIVE, MAXIMUM
C                            IS UNBOUNDED
            IF (IP.EQ.0) THEN
               PRINT 70
   70          FORMAT (' ***** UNBOUNDED MAXIMUM *****')
               RETURN
            ENDIF
C                            ADD X(JP) TO BASIS
            BASIS(IP) = JP
C                            NORMALIZE PIVOT ROW TO MAKE TAB(IP,JP)=1
            AMULT = 1.0/TAB(IP,JP)
            DO 75 J=1,N+M+1
               TAB(IP,J) = AMULT*TAB(IP,J)
   75       CONTINUE
C                            ADD MULTIPLES OF PIVOT ROW TO OTHER
C                            ROWS, TO KNOCK OUT OTHER ELEMENTS IN
C                            PIVOT COLUMN
            DO 85 I=1,MROW
               IF (I.EQ.IP) GO TO 85
               AMULT = TAB(I,JP)
               DO 80 J=1,N+M+1
                  TAB(I,J) = TAB(I,J) - AMULT*TAB(IP,J)
   80          CONTINUE
   85       CONTINUE
         GO TO 55
C                            END OF SIMPLEX STEP
   90 CONTINUE
C                            END OF PHASE II; READ X,P,Y FROM
C                            FINAL TABLEAU
      DO 95 J=1,N
```

```
         X(J) = 0.0
 95 CONTINUE
      DO 100 I=1,M
         K = BASIS(I)
         X(K) = TAB(I,N+M+1)
100 CONTINUE
      P = TAB(M+1,N+M+1)
      DO 105 I=1,M
         Y(I) = TAB(M+1,N+I)
105 CONTINUE
      RETURN
      END
```

Figure 4.5.1

The simplex calculations are divided into two "phases." During the first phase, the most negative entry in the last row of the tableau is determined by looking only at the coefficients of α, that is, at row $M+2$ of TAB, since α is assumed to be so large we can neglect the other terms. Phase I ends when all entries in row $M+2$ are nonnegative, which means that we have solved 4.2.2 with $c_1, \ldots, c_N$ taken to be negligible (zero). If any artificial variables are still nonzero in this solution, the original problem 4.2.1 must be infeasible. Otherwise—if no artificial variables remain in the basis—this solution is a feasible point for 4.2.1 and phase II continues from this point. Throughout phase II the α-components of the artificial variables will equal one and the α-components of the other (first N) variables will equal zero (see Problem 6). Therefore, during phase II we determine the most negative entry of the last row of the tableau by looking at only the first N components of row $M+1$ of TAB.

As discussed in Section 4.2, we do not actually switch the columns of TAB every time that a new variable enters the basis, to keep the identity in the last M columns of the tableau. We simply keep up with where the columns of this identity matrix are (and hence which variables are in the basis) through the vector BASIS.

As shown in Section 4.2, the simplex method moves from one basis to another and will only stop iterating when it determines either that the maximum is unbounded or that it has reached the maximum (of 4.2.2, which is also the maximum of 4.2.1 unless the latter has no feasible points). Now there are only a finite number of possible bases, since there are only $(N+M)!/[N!M!]$ ways to choose which M of the $N+M$ variables to include in the basis. Now, as long as the value of P actually increases each simplex iteration, no basis can be repeated, and hence $(N+M)!/[N!M!]$ is an upper bound to the number of iterations required to reach the correct solution (or correctly determine that there are no feasible points, or that the maximum is unbounded). However, we noted in Section 4.2 that, if b_{ip} is zero, P will not change in that step (nor does $\boldsymbol{x}$, for that matter, although the basis does change). This raises the possibility of "cycling," where the algorithm returns to a basis that it has seen earlier, and thus cycles forever through two or more different bases corresponding to the same value of P.

In practice, cycling is a very rare phenomenon, but it can happen, unless anticycling measures are taken [Gill et al. 1991, Section 8.3]. Furthermore, the upper bound of $(N+M)!/[N!M!]$ iterations is astronomically pessimistic in real applications; the number of steps required by the simplex method to converge to the correct solution is very rarely more than a small multiple of the number M of equations. This is because the method is designed to continually move toward increasing P, and thus it never passes through the vast majority of the extreme feasible points (see Problem 7). Again, however, it is possible (with much effort) to construct examples where the number of iterations varies exponentially with the problem size.

It is easy to see by examining DLPRG that the number of multiplications done each simplex iteration is approximately $M(N+M)$. This is because, when N and M are large, nearly 100% of the computer time will be spent in loop 80, where a multiple of row IP is added to row I (this is the only doubly nested loop executed each iteration).

Now let us use DLPRG to solve the transportation problem given by 4.1.3, repeated below:

$$\text{minimize } P = 550X_{11} + 300X_{12} + 400X_{13} + 350X_{21} + 300X_{22} + 100X_{23}$$

with constraints

$$\begin{array}{lllllllllllll}
X_{11} & + & X_{12} & + & X_{13} & & & & & & & \leq 40, \\
 & & & & & & X_{21} & + & X_{22} & + & X_{23} & \leq 20, \\
X_{11} & & & & & + & X_{21} & & & & & = 25, \\
 & & X_{12} & & & & & + & X_{22} & & & = 10, \\
 & & & & X_{13} & & & & & + & X_{23} & = 22,
\end{array}$$

and the bounds

$$X_{11} \geq 0,\ X_{12} \geq 0,\ X_{13} \geq 0,\ X_{21} \geq 0,\ X_{22} \geq 0,\ X_{23} \geq 0.$$

This problem has inequality and equality constraints, but it can be put into the form 4.2.1 by adding nonnegative slack variables to the first two inequalities. Then we can call DLPRG with $N = 8, M = 5$, and

$$A = \begin{bmatrix} 1 & 1 & 1 & 0 & 0 & 0 & 1 & 0 \\ 0 & 0 & 0 & 1 & 1 & 1 & 0 & 1 \\ 1 & 0 & 0 & 1 & 0 & 0 & 0 & 0 \\ 0 & 1 & 0 & 0 & 1 & 0 & 0 & 0 \\ 0 & 0 & 1 & 0 & 0 & 1 & 0 & 0 \end{bmatrix}, \quad B = \begin{bmatrix} 40 \\ 20 \\ 25 \\ 10 \\ 22 \end{bmatrix},$$

and $C = (-550, -300, -400, -350, -300, -100, 0, 0)$.

The solution returned by DLPRG is $X_{11} = 25, X_{12} = 10, X_{13} = 2, X_{21} = 0, X_{22} = 0, X_{23} = 20, P = -19550$. This means that the optimal solution is to ship 25 bulldozers from warehouse 1 to store 1, 10 from warehouse 1 to store 2, 2 from warehouse 1 to store 3, and 20 from warehouse 2 to store 3, at a total cost of \$19,550.

The curve-fitting problem given by 4.1.4 does not require that the unknowns be nonnegative; so it does not appear at first to be solvable using DLPRG. However, note that the problem in 4.1.4 is precisely the dual 4.3.2 of the problem in 4.3.1, if we take

$$A = \begin{bmatrix} -1 & -2 & -3 & -4 & -5 & 1 & 2 & 3 & 4 & 5 \\ -1 & -1 & -1 & -1 & -1 & 1 & 1 & 1 & 1 & 1 \\ 1 & 1 & 1 & 1 & 1 & 1 & 1 & 1 & 1 & 1 \end{bmatrix}, \quad \boldsymbol{b} = \begin{bmatrix} 0 \\ 0 \\ 1 \end{bmatrix},$$

$\boldsymbol{c} = (-1, -3, -2, -3, -4, 1, 3, 2, 3, 4)$, and $\boldsymbol{y} = (m, b, \epsilon)$.

Therefore we solve 4.3.1 using DLPRG, and the dual solution returned, (0.5, 1.25, 0.75), tells us that $m = 0.5$, $b = 1.25$, and $\epsilon = 0.75$. Therefore, $y = 0.5x + 1.25$ is the best straight-line approximation to the data, in the L_∞-norm, and the maximum error is then $\epsilon = 0.75$. This line is plotted in Figure 4.5.2, along with the original data points.

4.6 The Revised Simplex Method

Consider again the LP problem

$$\text{maximize } P = \boldsymbol{c}^{\mathrm{T}}\boldsymbol{x} \tag{4.6.1}$$

with constraints

$$A\boldsymbol{x} = \boldsymbol{b} \qquad (\boldsymbol{b} \geq \boldsymbol{0})$$

and bounds

$$\boldsymbol{x} \geq \boldsymbol{0},$$

where A is an M by N matrix ($N > M$). Suppose the initial simplex tableau is written in the form

$$\begin{bmatrix} A_n & A_b & \boldsymbol{0} \\ -c_n^{\mathrm{T}} & -c_b^{\mathrm{T}} & 1 \end{bmatrix} \begin{bmatrix} \boldsymbol{x}_n \\ \boldsymbol{x}_b \\ P \end{bmatrix} = \begin{bmatrix} \boldsymbol{b} \\ 0 \end{bmatrix},$$

where $\boldsymbol{x}_b$ and $\boldsymbol{x}_n$ are vectors containing the variables that will be in the basis and will not be in the basis, respectively, at a certain stage in the simplex method computations. Then the simplex tableau at that stage will have the form (multiply the first block "row" through by A_b^{-1} and add c_b^T times the first row to the second, as in 4.3.4);

$$\begin{bmatrix} A_b^{-1}A_n & I & \boldsymbol{0} \\ c_b^{\mathrm{T}}A_b^{-1}A_n - c_n^{\mathrm{T}} & \boldsymbol{0}^{\mathrm{T}} & 1 \end{bmatrix} \begin{bmatrix} \boldsymbol{x}_n \\ \boldsymbol{x}_b \\ P \end{bmatrix} = \begin{bmatrix} A_b^{-1}\boldsymbol{b} \\ c_b^{\mathrm{T}}A_b^{-1}\boldsymbol{b} \end{bmatrix}. \tag{4.6.2}$$

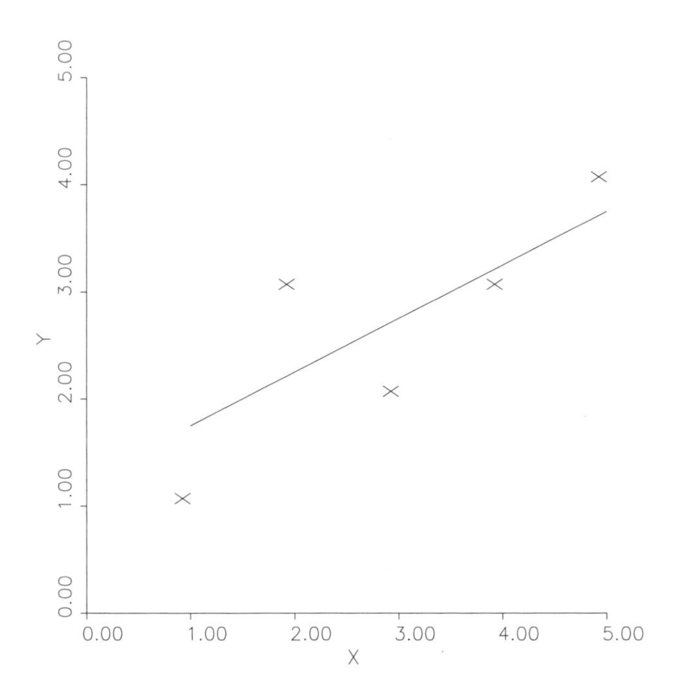

Figure 4.5.2
Best Linear Fit to Data in L_∞-norm

In the implementation of the simplex method as described in Section 4.2, the matrix A_b^{-1} is never explicitly formed or stored; the entries in the tableau are computed using elementary row operations. In the *revised* simplex method, the matrix A_b^{-1} *is* stored and updated every simplex iteration; it is the tableau itself that is not explicitly formed. However, it is easy to see from 4.6.2 that, as long as we keep up with which variables are in the basis and have A_b^{-1} available, the rest of the information contained in the tableau is computable. We shall show that it is possible to keep A_b^{-1} up to date, and to compute the information required to take one simplex step, in an efficient manner.

At each step of the simplex method we must calculate the following:

(1) $\boldsymbol{y} = A_b^{-\mathrm{T}}\boldsymbol{c}_b$ and $\boldsymbol{d}^{\mathrm{T}} = \boldsymbol{y}^{\mathrm{T}}A - \boldsymbol{c}^{\mathrm{T}}$. Then we inspect $\boldsymbol{d}^{\mathrm{T}}$ (the last row of the tableau) to determine which entry, d_{jp}, is most negative.

(2) $\boldsymbol{v} = A_b^{-1}\boldsymbol{a}_{jp}$ and $\boldsymbol{x}_b = A_b^{-1}\boldsymbol{b}$, where $\boldsymbol{a}_{jp}$ represents the jpth column of A. Then we compare the ratios of the components of these two vectors

to determine the pivot row, ip.

(3) Finally, we must update A_b^{-1}. The new A_b (which we shall call A_b') and the previous one are identical except that the ipth column of A_b has now been replaced by $\boldsymbol{a}_{jp}$, since x_{jp} has been added to the basis. Since premultiplying $\boldsymbol{a}_{jp}$ by A_b^{-1} gives $\boldsymbol{v}$, and premultiplying each of the other (unchanged) columns of A_b' by A_b^{-1} will give a column of the identity matrix, we have $A_b^{-1}A_b' = E$, where

$$E = \begin{bmatrix} 1 & 0 & \dots & v_1 & \dots & 0 & 0 \\ 0 & 1 & \dots & v_2 & \dots & 0 & 0 \\ \vdots & \vdots & & \vdots & & \vdots & \vdots \\ 0 & 0 & \dots & v_{ip} & \dots & 0 & 0 \\ \vdots & \vdots & & \vdots & & \vdots & \vdots \\ 0 & 0 & \dots & v_{M-1} & \dots & 1 & 0 \\ 0 & 0 & \dots & v_M & \dots & 0 & 1 \end{bmatrix}.$$

The inverse of this elementary matrix is given by

$$E^{-1} = \begin{bmatrix} 1 & 0 & \dots & \frac{-v_1}{v_{ip}} & \dots & 0 & 0 \\ 0 & 1 & \dots & \frac{-v_2}{v_{ip}} & \dots & 0 & 0 \\ \vdots & \vdots & & \vdots & & \vdots & \vdots \\ 0 & 0 & \dots & \frac{1}{v_{ip}} & \dots & 0 & 0 \\ \vdots & \vdots & & \vdots & & \vdots & \vdots \\ 0 & 0 & \dots & \frac{-v_{M-1}}{v_{ip}} & \dots & 1 & 0 \\ 0 & 0 & \dots & \frac{-v_M}{v_{ip}} & \dots & 0 & 1 \end{bmatrix}.$$

(Recall that v_{ip} is the simplex pivot element; so it is nonzero.) Therefore $A_b'^{-1} = E^{-1}A_b^{-1}$.

After convergence of the simplex method, $\boldsymbol{x}_b$ contains the values of the basis variables (the nonbasis variables are zero), and $\boldsymbol{y} = A_b^{-\mathrm{T}}\boldsymbol{c}_b$ contains the dual solution (see Theorem 4.3.2).

A subroutine that solves the LP problem 4.6.1 (with artificial variables added, so the components of $\boldsymbol{c}$ have the form $c_i = cc_{i,1} + cc_{i,2}\alpha$, and thus $\boldsymbol{y}$ and $\boldsymbol{d}$ will also have α components) using the revised simplex method is given in Figure 4.6.1. DLPRV starts with a basis of all artificial variables so that, initially, $A_b^{-1} = I$. Note that $\boldsymbol{x}_b$ is not calculated directly from $\boldsymbol{x}_b = A_b^{-1}\boldsymbol{b}$ each iteration but is rather updated using the formula $\boldsymbol{x}_b' = A_b'^{-1}\boldsymbol{b} = E^{-1}A_b^{-1}\boldsymbol{b} = E^{-1}\boldsymbol{x}_b$, which requires only $O(M)$ work.

The usual simplex method applied to 4.6.1 requires about $M*N$ multiplications per iteration, because that is the operation count for the row operations required to zero everything in the pivot column, above and below the pivot. The operations per iteration done by the revised simplex code DLPRV can be counted as follows (we ignore $O(M)$ and $O(N)$ computations):

$$\begin{array}{rclll} \boldsymbol{d}^{\mathrm{T}} & = & \boldsymbol{y}^{\mathrm{T}}A - \boldsymbol{c}^{\mathrm{T}} & \text{(loop 50)} & NZ \text{ multiplications,} \\ \boldsymbol{v} & = & A_b^{-1}\boldsymbol{a}_{jp} & \text{(loop 85)} & M^2, \\ A_b^{'-1} & = & E^{-1}A_b^{-1} & \text{(loop 110)} & M^2, \\ \boldsymbol{y} & = & A_b^{-\mathrm{T}}\boldsymbol{c}_b & \text{(loop 120)} & M^2. \end{array}$$

This gives a total of $NZ + 3M^2$, where NZ is the number of nonzeros in the constraint matrix A (actually $2NZ+4M^2$, during phase I, when $MROW = 2$). If M is small compared with N (frequently the case), and A is *sparse*, so that $NZ \ll M * N$, then the revised simplex method may be much faster and may require much less memory than the ordinary simplex method. This is important, because most large LP problems have sparse constraint matrices. The usual simplex algorithm cannot take advantage of sparseness because, even if the initial tableau is sparse, it will fill in rapidly as the algorithm progresses.

```
      SUBROUTINE DLPRV(A,IROW,JCOL,NZ,B,C,N,M,P,X,Y)
      IMPLICIT DOUBLE PRECISION (A-H,O-Z)
C                                  DECLARATIONS FOR ARGUMENTS
      DOUBLE PRECISION A(NZ),B(M),C(N),P,X(N),Y(M)
      INTEGER N,M,NZ,IROW(NZ),JCOL(NZ)
C                                  DECLARATIONS FOR LOCAL VARIABLES
      DOUBLE PRECISION ABINV(M,M),CC(N+M,2),YY(M,2),D(N+M,2),
     & V(M),XB(M),AP(M)
      INTEGER BASIS(M)
C
C  SUBROUTINE DLPRV USES THE REVISED SIMPLEX METHOD TO SOLVE THE PROBLEM
C
C             MAXIMIZE     P = C(1)*X(1) + ... + C(N)*X(N)
C
C    WITH X(1),...,X(N) NONNEGATIVE, AND
C
C           A(1,1)*X(1) + ... + A(1,N)*X(N)  = B(1)
C                .                  .             .
C                .                  .             .
C           A(M,1)*X(1) + ... + A(M,N)*X(N)  = B(M)
C
C    WHERE B(1),...,B(M) ARE ASSUMED TO BE NONNEGATIVE.
C
C  ARGUMENTS
C
C             ON INPUT                          ON OUTPUT
C             --------                          ---------
C
```

```
C     A      - A(IZ) IS THE CONSTRAINT MATRIX
C              ELEMENT IN ROW IROW(IZ), COLUMN
C              JCOL(IZ), FOR IZ=1,...,NZ.
C
C     IROW   - (SEE A).
C
C     JCOL   - (SEE A).
C
C     NZ     - NUMBER OF NONZEROS IN A.
C
C     B      - A VECTOR OF LENGTH M CONTAINING
C              THE RIGHT HAND SIDES OF THE
C              CONSTRAINTS.  THE COMPONENTS OF
C              B MUST ALL BE NONNEGATIVE.
C
C     C      - A VECTOR OF LENGTH N CONTAINING
C              THE COEFFICIENTS OF THE OBJECTIVE
C              FUNCTION.
C
C     N      - THE NUMBER OF UNKNOWNS.
C
C     M      - THE NUMBER OF CONSTRAINTS.
C
C     P      -                                  THE MAXIMUM OF THE
C                                               OBJECTIVE FUNCTION.
C
C     X      -                                  A VECTOR OF LENGTH N
C                                               WHICH CONTAINS THE LP
C                                               SOLUTION.
C
C     Y      -                                  A VECTOR OF LENGTH M
C                                               WHICH CONTAINS THE DUAL
C                                               SOLUTION.
C
C-----------------------------------------------------------------------
C
C                                  EPS = MACHINE FLOATING POINT RELATIVE
C                                        PRECISION
C ****************************
      DATA EPS/2.D-16/
C ****************************
C                                  INITIALIZE Ab**(-1) TO IDENTITY
      DO 5 I=1,M
      DO 5 J=1,M
         ABINV(I,J) = 0.0
         IF (I.EQ.J) ABINV(I,J) = 1.0
    5 CONTINUE
C                                  OBJECTIVE FUNCTION COEFFICIENTS ARE
```

```
C                                    CC(I,1) + CC(I,2)*ALPHA, WHERE "ALPHA"
C                                    IS TREATED AS INFINITY
      DO 10 I=1,N+M
         CC(I,1) = 0.0
         CC(I,2) = 0.0
         IF (I.LE.N) THEN
            CC(I,1) = C(I)
         ELSE
            CC(I,2) = -1
         ENDIF
   10 CONTINUE
C                                    BASIS(1),...,BASIS(M) HOLD NUMBERS OF
C                                    BASIS VARIABLES.  INITIAL BASIS CONSISTS
C                                    OF ARTIFICIAL VARIABLES ONLY
      DO 15 I=1,M
         K = N+I
         BASIS(I) = K
C                                    INITIALIZE Y TO Ab**(-T)*Cb = Cb
         YY(I,1) = CC(K,1)
         YY(I,2) = CC(K,2)
C                                    INITIALIZE Xb TO Ab**(-1)*B = B
         XB(I) = B(I)
         IF (B(I).LT.0.0) THEN
            PRINT 12
   12       FORMAT (' ***** ALL B(I) MUST BE NONNEGATIVE *****')
            RETURN
         ENDIF
   15 CONTINUE
C                                    SIMPLEX METHOD CONSISTS OF TWO PHASES
      DO 130 IPHASE=1,2
         IF (IPHASE.EQ.1) THEN
C                                    PHASE I:  ROW 2 OF D (WITH COEFFICIENTS OF
C                                    ALPHA) SEARCHED FOR MOST NEGATIVE ENTRY
            MROW = 2
            LIM = N+M
         ELSE
C                                    PHASE II:  FIRST N ELEMENTS OF ROW 1 OF
C                                    D SEARCHED FOR MOST NEGATIVE ENTRY
C                                    (COEFFICIENTS OF ALPHA NONNEGATIVE NOW)
            MROW = 1
            LIM = N
C                                    IF ANY ARTIFICIAL VARIABLES LEFT IN
C                                    BASIS AT BEGINNING OF PHASE II, THERE
C                                    IS NO FEASIBLE SOLUTION
            DO 25 I=1,M
               IF (BASIS(I).GT.LIM) THEN
                  PRINT 20
   20             FORMAT (' ***** NO FEASIBLE SOLUTION *****')
```

```
                RETURN
             ENDIF
   25     CONTINUE
       ENDIF
C                             THRESH = SMALL NUMBER.  WE ASSUME SCALES
C                             OF A AND C ARE NOT *TOO* DIFFERENT
       THRESH = 0.0
       DO 30 J=1,LIM
          THRESH = MAX(THRESH,ABS(CC(J,MROW)))
   30  CONTINUE
       THRESH = 1000*EPS*THRESH
C                             BEGINNING OF SIMPLEX STEP
   35  CONTINUE
C                             D**T = Y**T*A - C**T
          DO 55 IR=1,MROW
             DO 40 J=1,N+M
                D(J,IR) = -CC(J,IR)
   40        CONTINUE
C                             LAST M COLUMNS OF A FORM IDENTITY MATRIX
             DO 45 J=1,M
                D(N+J,IR) = D(N+J,IR) + YY(J,IR)
   45        CONTINUE
C                             FIRST N COLUMNS STORED IN SPARSE A MATRIX
             DO 50 IZ=1,NZ
                I = IROW(IZ)
                J = JCOL(IZ)
                D(J,IR) = D(J,IR) + A(IZ)*YY(I,IR)
   50        CONTINUE
   55     CONTINUE
C                             FIND MOST NEGATIVE ENTRY OF ROW MROW
C                             OF D, IDENTIFYING PIVOT COLUMN JP
          CMIN = -THRESH
          JP = 0
          DO 60 J=1,LIM
             IF (D(J,MROW).LT.CMIN) THEN
                CMIN = D(J,MROW)
                JP = J
             ENDIF
   60     CONTINUE
C                             IF ALL ENTRIES NONNEGATIVE (ACTUALLY,
C                             IF GREATER THAN -THRESH) PHASE ENDS
          IF (JP.EQ.0) GO TO 130
C                             COPY JP-TH COLUMN OF A ONTO VECTOR AP
          IF (JP.LE.N) THEN
C                             JP-TH COLUMN IS PART OF SPARSE A MATRIX
             DO 65 I=1,M
                AP(I) = 0
   65        CONTINUE
```

```
            DO 70 IZ=1,NZ
               J = JCOL(IZ)
               IF (J.EQ.JP) THEN
                  I = IROW(IZ)
                  AP(I) = A(IZ)
               ENDIF
   70       CONTINUE
         ELSE
C                                JP-TH COLUMN IS COLUMN OF FINAL IDENTITY
            DO 75 I=1,M
               AP(I) = 0
   75       CONTINUE
            AP(JP-N) = 1
         ENDIF
C                                V = Ab**(-1)*AP
         DO 85 I=1,M
            V(I) = 0.0
            DO 80 J=1,M
               V(I) = V(I) + ABINV(I,J)*AP(J)
   80       CONTINUE
   85    CONTINUE
C                                FIND SMALLEST POSITIVE RATIO
C                                Xb(I)/V(I), IDENTIFYING PIVOT ROW IP
         RATMIN = 0.0
         IP = 0
         DO 90 I=1,M
            IF (V(I).GT.THRESH) THEN
               RATIO = XB(I)/V(I)
               IF (IP.EQ.0 .OR. RATIO.LT.RATMIN) THEN
                  RATMIN = RATIO
                  IP = I
               ENDIF
            ENDIF
   90    CONTINUE
C                                IF ALL RATIOS NONPOSITIVE, MAXIMUM
C                                IS UNBOUNDED
         IF (IP.EQ.0) THEN
            PRINT 95
   95       FORMAT (' ***** UNBOUNDED MAXIMUM *****')
            RETURN
         ENDIF
C                                ADD X(JP) TO BASIS
         BASIS(IP) = JP
C                                UPDATE Ab**(-1) = E**(-1)*Ab**(-1)
C                                Xb = E**(-1)*Xb
         DO 100 J=1,M
            ABINV(IP,J) = ABINV(IP,J)/V(IP)
  100    CONTINUE
```

```
            XB(IP) = XB(IP)/V(IP)
            DO 110 I=1,M
               IF (I.EQ.IP) GO TO 110
               DO 105 J=1,M
                  ABINV(I,J) = ABINV(I,J) - V(I)*ABINV(IP,J)
  105          CONTINUE
               XB(I) = XB(I) - V(I)*XB(IP)
  110       CONTINUE
C                                  CALCULATE Y = Ab**(-T)*Cb
            DO 125 IR=1,MROW
               DO 120 I=1,M
                  YY(I,IR) = 0.0
                  DO 115 J=1,M
                     K = BASIS(J)
                     YY(I,IR) = YY(I,IR) + ABINV(J,I)*CC(K,IR)
  115             CONTINUE
  120          CONTINUE
  125       CONTINUE
         GO TO 35
C                                  END OF SIMPLEX STEP
  130 CONTINUE
C                                  END OF PHASE II; CALCULATE X
      DO 135 J=1,N
         X(J) = 0.0
  135 CONTINUE
      DO 140 I=1,M
         K = BASIS(I)
         X(K) = XB(I)
         Y(I) = YY(I,1)
  140 CONTINUE
C                                  CALCULATE P
      P = 0.0
      DO 145 I=1,N
         P = P + C(I)*X(I)
  145 CONTINUE
      RETURN
      END
```

Figure 4.6.1

If A is sparse, but M is not small compared to N, the revised simplex method can still be useful, but now the inverse of the M by M sparse matrix A_b should never be formed, instead $\boldsymbol{x}_b, \boldsymbol{v}$ and $\boldsymbol{y}$ should be computed by solving $A_b\boldsymbol{x}_b = \boldsymbol{b}, A_b\boldsymbol{v} = \boldsymbol{a}_{jp}$ and $A_b^{\mathrm{T}}\boldsymbol{y} = \boldsymbol{c}_b$, using a sparse linear system solver. Figure 4.6.2 shows a subroutine DLPRVS which is similar to DLPRV (Figure 4.6.1) except that it calls a generic sparse linear system solver SPARSOL to solve these sparse M by M systems. The calls to SPARSOL should be replaced by calls to a specific sparse linear system solver (see Problem 13). A direct solver such as IMSL Library routine LSLXG should be used (www.roguewave.com/products/imsl-numerical-libraries.aspx), as iterative solvers may not converge.

Notice how the matrix A_b, which is stored in sparse matrix format, is updated in loops 90 and 95, when one column needs to be replaced.

```
      SUBROUTINE DLPRVS(A,IROW,JCOL,NZ,B,C,N,M,P,X,Y)
      IMPLICIT DOUBLE PRECISION (A-H,O-Z)
C                                 DECLARATIONS FOR ARGUMENTS
      DOUBLE PRECISION A(NZ),B(M),C(N),P,X(N),Y(M)
      INTEGER N,M,NZ,IROW(NZ),JCOL(NZ)
C                                 DECLARATIONS FOR LOCAL VARIABLES
      DOUBLE PRECISION AB(NZ),CC(N+M,2),YY(M,2),D(N+M,2),
     & V(M),XB(M),AP(M)
      INTEGER BASIS(M),IROWB(NZ),JCOLB(NZ)
C
C  SUBROUTINE DLPRVS USES THE REVISED SIMPLEX METHOD TO SOLVE THE PROBLEM
C
C             MAXIMIZE     P = C(1)*X(1) + ... + C(N)*X(N)
C
C    WITH X(1),...,X(N) NONNEGATIVE, AND
C
C           A(1,1)*X(1) + ... + A(1,N)*X(N)  = B(1)
C             .                   .              .
C             .                   .              .
C           A(M,1)*X(1) + ... + A(M,N)*X(N)  = B(M)
C
C    WHERE B(1),...,B(M) ARE ASSUMED TO BE NONNEGATIVE.
C
C  ARGUMENTS
C
C             ON INPUT                          ON OUTPUT
C             --------                          ---------
C
C    A      - A(IZ) IS THE CONSTRAINT MATRIX
C             ELEMENT IN ROW IROW(IZ), COLUMN
C             JCOL(IZ), FOR IZ=1,...,NZ.
C
C    IROW   - (SEE A).
C
C    JCOL   - (SEE A).
C
C    NZ     - NUMBER OF NONZEROS IN A.
C
C    B      - A VECTOR OF LENGTH M CONTAINING
C             THE RIGHT HAND SIDES OF THE
C             CONSTRAINTS.  THE COMPONENTS OF
C             B MUST ALL BE NONNEGATIVE.
```

```
C
C   C       - A VECTOR OF LENGTH N CONTAINING
C             THE COEFFICIENTS OF THE OBJECTIVE
C             FUNCTION.
C
C   N       - THE NUMBER OF UNKNOWNS.
C
C   M       - THE NUMBER OF CONSTRAINTS.
C
C   P       -                                 THE MAXIMUM OF THE
C                                             OBJECTIVE FUNCTION.
C
C   X       -                                 A VECTOR OF LENGTH N
C                                             WHICH CONTAINS THE LP
C                                             SOLUTION.
C
C   Y       -                                 A VECTOR OF LENGTH M
C                                             WHICH CONTAINS THE DUAL
C                                             SOLUTION.
C
C  NOTE: DLPRVS CALLS A SPARSE LINEAR SYSTEM SOLVER
C
C          SUBROUTINE SPARSOL(A,IROW,JCOL,NZ,X,B,N)
C          DOUBLE PRECISION A(NZ),X(N),B(N)
C          INTEGER IROW(NZ),JCOL(NZ)
C
C  TO SOLVE THE N BY N SPARSE SYSTEM AX=B, WHERE A(IZ), IZ=1,...,NZ,
C  IS THE NONZERO ELEMENT OF A IN ROW IROW(IZ), COLUMN JCOL(IZ), B IS
C  THE RIGHT HAND SIDE VECTOR AND X IS THE SOLUTION.
C
C  THE CALL TO SPARSOL SHOULD BE REPLACED BY A CALL TO A DIRECT LINEAR
C  SYSTEM SOLVER (ITERATIVE SOLVERS NOT RECOMMENDED).
C
C-----------------------------------------------------------------------
C
C                             EPS = MACHINE FLOATING POINT RELATIVE
C                                   PRECISION
C ****************************
      DATA EPS/2.D-16/
C ****************************
C                             INITIALIZE Ab TO IDENTITY
      NZB = M
      DO 5 I=1,M
         IROWB(I) = I
         JCOLB(I) = I
         AB(I) = 1.0
    5 CONTINUE
C                             OBJECTIVE FUNCTION COEFFICIENTS ARE
```

```
C                                  CC(I,1) + CC(I,2)*ALPHA, WHERE "ALPHA"
C                                  IS TREATED AS INFINITY
      DO 10 I=1,N+M
         CC(I,1) = 0.0
         CC(I,2) = 0.0
         IF (I.LE.N) THEN
            CC(I,1) = C(I)
         ELSE
            CC(I,2) = -1
         ENDIF
   10 CONTINUE
C                                  BASIS(1),...,BASIS(M) HOLD NUMBERS OF
C                                  BASIS VARIABLES.  INITIAL BASIS CONSISTS
C                                  OF ARTIFICIAL VARIABLES ONLY
      DO 15 I=1,M
         K = N+I
         BASIS(I) = K
C                                  INITIALIZE Y TO Ab**(-T)*Cb = Cb
         YY(I,1) = CC(K,1)
         YY(I,2) = CC(K,2)
C                                  INITIALIZE Xb TO Ab**(-1)*B = B
         XB(I) = B(I)
         IF (B(I).LT.0.0) THEN
            PRINT 12
   12       FORMAT (' ***** ALL B(I) MUST BE NONNEGATIVE *****')
            RETURN
         ENDIF
   15 CONTINUE
C                                  SIMPLEX METHOD CONSISTS OF TWO PHASES
      DO 130 IPHASE=1,2
         IF (IPHASE.EQ.1) THEN
C                                  PHASE I:  ROW 2 OF D (WITH COEFFICIENTS OF
C                                  ALPHA) SEARCHED FOR MOST NEGATIVE ENTRY
            MROW = 2
            LIM = N+M
         ELSE
C                                  PHASE II:  FIRST N ELEMENTS OF ROW 1 OF
C                                  D SEARCHED FOR MOST NEGATIVE ENTRY
C                                  (COEFFICIENTS OF ALPHA NONNEGATIVE NOW)
            MROW = 1
            LIM = N
C                                  IF ANY ARTIFICIAL VARIABLES LEFT IN
C                                  BASIS AT BEGINNING OF PHASE II, THERE
C                                  IS NO FEASIBLE SOLUTION
            DO 25 I=1,M
               IF (BASIS(I).GT.LIM) THEN
                  PRINT 20
   20             FORMAT (' ***** NO FEASIBLE SOLUTION *****')
```

```
                  RETURN
               ENDIF
   25       CONTINUE
         ENDIF
C                                 THRESH = SMALL NUMBER.  WE ASSUME SCALES
C                                 OF A AND C ARE NOT *TOO* DIFFERENT
         THRESH = 0.0
         DO 30 J=1,LIM
            THRESH = MAX(THRESH,ABS(CC(J,MROW)))
   30    CONTINUE
         THRESH = 1000*EPS*THRESH
C                                 BEGINNING OF SIMPLEX STEP
   35    CONTINUE
C                                 D**T = Y**T*A - C**T
            DO 55 IR=1,MROW
               DO 40 J=1,N+M
                  D(J,IR) = -CC(J,IR)
   40          CONTINUE
C                                 LAST M COLUMNS OF A FORM IDENTITY MATRIX
               DO 45 J=1,M
                  D(N+J,IR) = D(N+J,IR) + YY(J,IR)
   45          CONTINUE
C                                 FIRST N COLUMNS STORED IN SPARSE A MATRIX
               DO 50 IZ=1,NZ
                  I = IROW(IZ)
                  J = JCOL(IZ)
                  D(J,IR) = D(J,IR) + A(IZ)*YY(I,IR)
   50          CONTINUE
   55       CONTINUE
C                                 FIND MOST NEGATIVE ENTRY OF ROW MROW
C                                 OF D, IDENTIFYING PIVOT COLUMN JP
            CMIN = -THRESH
            JP = 0
            DO 60 J=1,LIM
               IF (D(J,MROW).LT.CMIN) THEN
                  CMIN = D(J,MROW)
                  JP = J
               ENDIF
   60       CONTINUE
C                                 IF ALL ENTRIES NONNEGATIVE (ACTUALLY,
C                                 IF GREATER THAN -THRESH) PHASE ENDS
            IF (JP.EQ.0) GO TO 130
C                                 COPY JP-TH COLUMN OF A ONTO VECTOR AP
            IF (JP.LE.N) THEN
C                                 JP-TH COLUMN IS PART OF SPARSE A MATRIX
               DO 65 I=1,M
                  AP(I) = 0
   65          CONTINUE
```

```
            DO 70 IZ=1,NZ
               J = JCOL(IZ)
               IF (J.EQ.JP) THEN
                  I = IROW(IZ)
                  AP(I) = A(IZ)
               ENDIF
   70       CONTINUE
         ELSE
C                               JP-TH COLUMN IS COLUMN OF FINAL IDENTITY
            DO 75 I=1,M
               AP(I) = 0
   75       CONTINUE
            AP(JP-N) = 1
         ENDIF
C                               SOLVE Ab*V = AP
         CALL SPARSOL(AB,IROWB,JCOLB,NZB,V,AP,M)
C                               FIND SMALLEST POSITIVE RATIO
C                               Xb(I)/V(I), IDENTIFYING PIVOT ROW IP
         RATMIN = 0.0
         IP = 0
         DO 80 I=1,M
            IF (V(I).GT.THRESH) THEN
               RATIO = XB(I)/V(I)
               IF (IP.EQ.0 .OR. RATIO.LT.RATMIN) THEN
                  RATMIN = RATIO
                  IP = I
               ENDIF
            ENDIF
   80    CONTINUE
C                               IF ALL RATIOS NONPOSITIVE, MAXIMUM
C                               IS UNBOUNDED
         IF (IP.EQ.0) THEN
            PRINT 85
   85       FORMAT (' ***** UNBOUNDED MAXIMUM *****')
            RETURN
         ENDIF
C                               ADD X(JP) TO BASIS
         BASIS(IP) = JP
C                               UPDATE Ab.  PUT NONZEROS OF JP-TH
C                               COLUMN OF A (=AP) INTO COLUMN IP
C                               OF SPARSE MATRIX Ab
         NZBOLD = NZB
         DO 90 I=1,M
            IF (AP(I).NE.0.0) THEN
               NZB = NZB+1
               IROWB(NZB) = I
               JCOLB(NZB) = IP
               AB(NZB) = AP(I)
```

```
                ENDIF
   90        CONTINUE
             NZBNEW = NZB
C                                    REMOVE ELEMENTS OF OLD COLUMN IP
             NZB = 0
             DO 95 IZ=1,NZBNEW
                IF (JCOLB(IZ).NE.IP .OR. IZ.GT.NZBOLD) THEN
                   NZB = NZB+1
                   JCOLB(NZB) = JCOLB(IZ)
                   IROWB(NZB) = IROWB(IZ)
                   AB(NZB) = AB(IZ)
                ENDIF
   95        CONTINUE
C                                    SOLVE Ab*Xb = B
             CALL SPARSOL(AB,IROWB,JCOLB,NZB,XB,B,M)
C                                    SOLVE Ab**T*Y = Cb
             DO 105 IR=1,MROW
                DO 100 J=1,M
                   K = BASIS(J)
                   V(J) = CC(K,IR)
  100           CONTINUE
                CALL SPARSOL(AB,JCOLB,IROWB,NZB,YY(1,IR),V,M)
  105        CONTINUE
          GO TO 35
C                                    END OF SIMPLEX STEP
  130 CONTINUE
C                                    END OF PHASE II; CALCULATE X
      DO 135 J=1,N
         X(J) = 0.0
  135 CONTINUE
      DO 140 I=1,M
         K = BASIS(I)
         X(K) = XB(I)
         Y(I) = YY(I,1)
  140 CONTINUE
C                                    CALCULATE P
      P = 0.0
      DO 145 I=1,N
         P = P + C(I)*X(I)
  145 CONTINUE
      RETURN
      END
```

Figure 4.6.2

Note that now all the $O(M^2)$ loops in DLPRV are gone, DLPRVS has only loops of lengths $O(N)$, $O(M)$ or $O(NZ)$ each simplex step (recall that $MROW \leq 2$), in addition to the 3 or 4 calls to SPARSOL to solve sparse M

by M systems. Thus if the constraint matrix is very sparse, so that $NZB \leq NZ \ll M * N$ and if the sparse solver SPARSOL can solve these systems in $O(NZB)$ operations, then the total work per simplex step will still compare favorably with the unrevised code DLPRG.

Table 4.6.1 compares the speeds of DLPRG, DLPRV and DLPRVS (with IMSL Library routine LSLXG used to solve the sparse systems) on some sparse problems. Problems of the form (4.3.5) were solved, with M inequality constraints, and N variables, so M additional slack variables had to be added before calling each routine. In the first series of tests, each column of A contained about 5 random nonzeros elements, and N was equal to M. Since M was not small compared to N, DLPRV was not faster than the unrevised code DLPRG, but DLPRVS was much faster, as the results on the left side of Table 4.6.1 show. The other series was similar, but this time N was set to M^2. Now, since M is small compared to N, DLPRV and DLPRVS both do much better than DLPRG.

Recall that all three routines should do exactly the same number of simplex steps, so the differences in speed are due to the work done per iteration.

Table 4.6.1

Timings for Sparse Problems

$N = M$				$N = M^2$			
M	DLPRG (sec)	DLPRV (sec)	DLPRVS (sec)	M	DLPRG (sec)	DLPRV (sec)	DLPRVS (sec)
250	4.0	5.4	0.8	50	1.3	0.3	0.3
500	41.6	50.1	3.1	100	83.9	3.5	3.2
1000	373.2	423.7	12.0	200	2122.6	35.3	31.3

Finally, although their complexity puts them beyond the scope of this text, we should mention that, in recent years, "interior" algorithms have emerged as competitors of the simplex algorithm for solving LP problems. These algorithms, which plunge through the interior of the feasible region rather than moving along the boundary, received much attention in the early 1980s when mathematicians proved that they could be used to solve arbitrary LP problems in "polynomial time", that is, in $O(N^\alpha)$ time, where α is a constant. Although the simplex method also converges in polynomial time for all practical problems, it is possible to construct examples where it requires exponential time ($O(\alpha^N)$) to converge. At first, the simplex method codes still clearly outperformed the interior codes, despite their theoretical inferiority, but now interior algorithm codes have been developed which are competitive with the best simplex implementations, or perhaps superior to them.

The reader who would like to learn more about interior algorithms is referred to Schrijver [1986] and to Problem 12.

4.7 The Transportation Problem

The transportation problem (e.g., 4.1.3) is clearly an LP problem for which the revised simplex method is ideally suited. If N_W is the number of warehouses, N_S the number of stores, WCAP(i) the capacity of warehouse i, SREQ(j) the requirement of store j, COST(i,j) the cost per unit to ship from warehouse i to store j, and $X(i,j)$ the number of units shipped from warehouse i to store j, then the transportation problem can be formulated as follows:

$$\text{minimize} \sum_{i=1}^{N_W}\sum_{j=1}^{N_S} \text{COST}(i,j)X(i,j) \tag{4.7.1}$$

with

$$\sum_{j=1}^{N_S} X(i,j) \le \text{WCAP}(i) \qquad \text{for } i = 1,\ldots,N_W$$

and

$$\sum_{i=1}^{N_W} X(i,j) = \text{SREQ}(j) \qquad \text{for } j = 1,\ldots,N_S$$

and bounds

$$X(i,j) \ge 0 \qquad \text{for } i = 1,\ldots,N_W, j = 1,\ldots,N_S.$$

The number of unknowns is $N = N_W N_S$ while the number of constraints is only $M = N_W + N_S$; so clearly $M \ll N$ for large problems. Furthermore, the constraint matrix is extremely sparse, and the revised simplex code DLPRV (Figure 4.6.1) can take good advantage of this sparsity.

Let us consider the two-warehouse three-store problem posed in Section 4.1.3. If we add slack variables to the first two (warehouse capacity) constraints, we have a problem of the form 4.6.1:

$$\max -550X_{11} - 300X_{12} - 400X_{13} - 350X_{21} - 300X_{22} - 100X_{23} + 0S_1 + 0S_2$$

with

$$\begin{array}{lllllllll}
X_{11} + X_{12} + X_{13} & & & & & & + S_1 & & = 40, \\
& & & X_{21} + X_{22} + X_{23} & & & & + S_2 & = 20, \\
X_{11} & & + X_{21} & & & & & & = 25, \\
& X_{12} & & + X_{22} & & & & & = 10, \\
& & X_{13} & & + X_{23} & & & & = 22,
\end{array}$$

and

$$X_{11}, X_{12}, X_{13}, X_{21}, X_{22}, X_{23}, S_1, S_2 \ge 0.$$

Although DLPRV, like DLPRG, actually adds artificial variables to all constraints to create the required identity matrix in the last columns, it is really only necessary to add three artificial variables for this problem, one for each of the store constraints (see Problem 11). It is possible to view these three artificial variables as resulting from adding an "artificial warehouse" A, which has infinite capacity but is very distant from the stores (the per unit cost, α, to transport from A to any store is higher than the cost of any "real" route). If the minimum cost solution specifies that we ship anything from the distant warehouse (i.e., if any artificial variables are nonzero), then it must not be possible to fill all the stores' orders from the closer "real" warehouses. On the other hand, if the artificial variables are all zero, we have found the minimum cost solution of the original problem; the fictitious warehouse A has served its purpose—providing an initial feasible point—and we no longer need it.

The transportation problem given by 4.1.3 was solved using the FORTRAN subroutine DTRAN shown in Figure 4.7.1, which calls the revised simplex code DLPRV to solve a general transportation problem of the form 4.7.1. The solution returned was (as in Section 4.5) $C_{\text{MIN}} = 19550$ and $X_{11} = 25, X_{12} = 10, X_{13} = 2, X_{21} = 0, X_{22} = 0, X_{23} = 20$.

For this problem, the optimal X_{ij} are all integers; this will always be the case when the warehouse capacities and store requirements are integers. The dual solution returned by DTRAN provides information on how increasing warehouse capacities and decreasing store requirements could be used to cut the total cost (see Problem 9).

```
      SUBROUTINE DTRAN(WCAP,SREQ,COST,NW,NS,CMIN,X,Y)
      IMPLICIT DOUBLE PRECISION (A-H,O-Z)
C                                 DECLARATIONS FOR ARGUMENTS
      DOUBLE PRECISION WCAP(NW),SREQ(NS),COST(NW,NS),CMIN,X(NW,NS),
     & Y(NW+NS)
      INTEGER NW,NS
C                                 DECLARATIONS FOR LOCAL VARIABLES
      DOUBLE PRECISION A(2*NW*NS+NW),B(NW+NS),C(NW*NS+NW),
     & XSOL(NW*NS+NW)
      INTEGER IROW(2*NW*NS+NW),JCOL(2*NW*NS+NW)
C
C  SUBROUTINE DTRAN SOLVES THE TRANSPORTATION PROBLEM
C
C    MINIMIZE    CMIN = COST(1,1)*X(1,1) + ... + COST(NW,NS)*X(NW,NS)
C
C    WITH X(1,1),...,X(NW,NS) NONNEGATIVE, AND
C
C          X(1,1) + ... + X(1,NS)   .LE. WCAP(1)
C             .             .             .
C             .             .             .
C          X(NW,1)+ ... + X(NW,NS) .LE. WCAP(NW)
```

```
C          X(1,1) + ... + X(NW,1)     =  SREQ(1)
C             .              .             .
C             .              .             .
C          X(1,NS)+ ... + X(NW,NS)    =  SREQ(NS)
C
C  CAUTION: IF TOTAL STORE REQUIREMENTS EXACTLY EQUAL TOTAL WAREHOUSE
C           CAPACITIES, ALTER ONE WCAP(I) OR SREQ(I) SLIGHTLY, SO THAT
C           WAREHOUSE CAPACITIES SLIGHTLY EXCEED STORE REQUIREMENTS.
C
C  ARGUMENTS
C
C             ON INPUT                          ON OUTPUT
C             --------                          ---------
C
C    WCAP   - A VECTOR OF LENGTH NW CONTAINING
C             THE WAREHOUSE CAPACITIES.
C
C    SREQ   - A VECTOR OF LENGTH NS CONTAINING
C             THE STORE REQUIREMENTS.
C
C    COST   - THE NW BY NS COST MATRIX. COST(I,J)
C             IS THE PER UNIT COST TO SHIP FROM
C             WAREHOUSE I TO STORE J.
C
C    NW     - THE NUMBER OF WAREHOUSES.
C
C    NS     - THE NUMBER OF STORES.
C
C    CMIN   -                                   THE TOTAL COST OF THE
C                                               OPTIMAL ROUTING.
C
C    X      -                                   AN NW BY NS MATRIX
C                                               CONTAINING THE OPTIMAL
C                                               ROUTING.  X(I,J) UNITS
C                                               SHOULD BE SHIPPED FROM
C                                               WAREHOUSE I TO STORE J.
C
C    Y      -                                   A VECTOR OF LENGTH NW+NS
C                                               CONTAINING THE DUAL. Y(I)
C                                               GIVES THE DECREASE IN
C                                               TOTAL COST PER UNIT
C                                               INCREASE IN WCAP(I), FOR
C                                               SMALL INCREASES, AND
C                                               -Y(NW+J) GIVES THE INCREASE
C                                               IN TOTAL COST PER UNIT
C                                               INCREASE IN SREQ(J).
C
C-----------------------------------------------------------------------
```

```
      M = NW+NS
      N = NW*NS+NW
C                                  SET UP SPARSE CONSTRAINT MATRIX
      NZ = 0
      DO 10 I=1,NW
         DO 5 J=1,NS
            NZ = NZ+1
            IROW(NZ) = I
            JCOL(NZ) = (I-1)*NS + J
            A(NZ) = 1.0
    5    CONTINUE
         NZ = NZ+1
         IROW(NZ) = I
         JCOL(NZ) = NW*NS+I
         A(NZ) = 1.0
C                                  LOAD WAREHOUSE CAPACITIES INTO B
         B(I) = WCAP(I)
   10 CONTINUE
      DO 20 J=1,NS
         DO 15 I=1,NW
            NZ = NZ+1
            IROW(NZ) = NW+J
            JCOL(NZ) = J + (I-1)*NS
            A(NZ) = 1.0
   15    CONTINUE
C                                  LOAD STORE REQUIREMENTS INTO B
         B(NW+J) = SREQ(J)
   20 CONTINUE
C                                  FIRST NW*NS ENTRIES IN C ARE
C                                  -COST(I,J).  NEGATIVE SIGN USED
C                                  BECAUSE WE WANT TO MINIMIZE COST
      K = 0
      DO 30 I=1,NW
         DO 25 J=1,NS
            K = K+1
            C(K) = -COST(I,J)
   25    CONTINUE
   30 CONTINUE
C                                  NEXT NW COSTS ARE ZERO, CORRESPONDING
C                                  TO WAREHOUSE CAPACITY SLACK VARIABLES
      DO 35 I=1,NW
         K = K+1
         C(K) = 0.0
   35 CONTINUE
C                                  USE REVISED SIMPLEX METHOD TO SOLVE
C                                  TRANSPORTATION PROBLEM
      CALL DLPRV(A,IROW,JCOL,NZ,B,C,N,M,P,XSOL,Y)
C                                  FORM OPTIMAL ROUTING MATRIX, X
```

```
      CMIN = -P
      K = 0
      DO 45 I=1,NW
         DO 40 J=1,NS
            K = K+1
            X(I,J) = XSOL(K)
   40    CONTINUE
   45 CONTINUE
      RETURN
      END
```

Figure 4.7.1

IMSL Library transportation code TRAN (www.roguewave.com/products/imsl-numerical-libraries.aspx) is based on DTRAN.

4.8 Problems

1. Solve the following LP problems using the simplex method and hand calculations. Use artificial variables only when necessary. Graph the feasible regions in the x-y plane,

   a. Maximize $P = 3x + 4y$
   with constraints

$$\begin{aligned} x + y &\leq 3, \\ 2x + y &\leq 4, \end{aligned}$$

   and bounds

$$\begin{aligned} x &\geq 0, \\ y &\geq 0. \end{aligned}$$

   b. Minimize $P = 3x + 4y$
   with constraints

$$\begin{aligned} x + y &\geq 3, \\ 2x + y &\geq 4, \end{aligned}$$

   and bounds

$$\begin{aligned} x &\geq 0, \\ y &\geq 0. \end{aligned}$$

   c. Maximize $P = 3x + 4y$
   with constraints

$$\begin{aligned} x + y &\geq 5, \\ 2x + y &\leq 4, \end{aligned}$$

and bounds

$$\begin{aligned} x &\geq 0, \\ y &\geq 0. \end{aligned}$$

d. Maximize $P = 3x + 4y$
with constraints

$$\begin{aligned} x + y &\geq 3, \\ 2x + y &\geq 4, \end{aligned}$$

and bounds

$$\begin{aligned} x &\geq 0, \\ y &\geq 0. \end{aligned}$$

2. Write the dual problem for

$$\text{maximize } P = c_1 x_1 + \ldots + c_N x_N$$

with constraints

$$\begin{array}{ccccccc} a_{1,1}x_1 & + & \ldots & + & a_{1,N}x_N & \leq & b_1, \\ \vdots & & & & \vdots & & \vdots \\ a_{k,1}x_1 & + & \ldots & + & a_{k,N}x_N & \leq & b_k, \\ a_{k+1,1}x_1 & + & \ldots & + & a_{k+1,N}x_N & = & b_{k+1}, \\ \vdots & & & & \vdots & & \vdots \\ a_{M,1}x_1 & + & \ldots & + & a_{M,N}x_N & = & b_M, \end{array}$$

and bounds

$$\begin{aligned} x_1 &\geq 0, \\ &\vdots \\ x_N &\geq 0. \end{aligned}$$

3. The problem of minimizing $\|A\boldsymbol{x} - \boldsymbol{b}\|_\infty$, where A is an M by N matrix, can be posed as an LP problem:

$$\text{minimize } P = \epsilon$$

with constraints

$$\left| \sum_{j=1}^{N} a_{i,j} x_j - b_i \right| \leq \epsilon \qquad (i = 1, \ldots, M).$$

Show that this LP problem can be put into the form

$$\text{minimize } P = (\mathbf{0}, 1)^{\mathrm{T}}(\boldsymbol{x}, \epsilon)$$

with constraints

$$\begin{bmatrix} -A & \mathbf{1} \\ A & \mathbf{1} \end{bmatrix} \begin{bmatrix} \boldsymbol{x} \\ \epsilon \end{bmatrix} \geq \begin{bmatrix} -\boldsymbol{b} \\ \boldsymbol{b} \end{bmatrix},$$

where $\mathbf{1} = (1_1, \ldots, 1_M)$. Since the variables are not required to be nonnegative, we cannot solve this LP problem as it stands. However, find another problem whose dual solution is the solution to this problem.

4. The problem of minimizing $\|A\boldsymbol{x} - \boldsymbol{b}\|_1$, where A is an M by N matrix, can also be posed as an LP problem

$$\text{minimize } P = \epsilon_1 + \ldots + \epsilon_M$$

with constraints

$$\left| \sum_{j=1}^{N} a_{i,j} x_j - b_i \right| \leq \epsilon_i \qquad (i = 1, \ldots, M).$$

Show that this LP problem can be put into the form

$$\text{minimize } P = (\mathbf{0}, \mathbf{1})^{\mathrm{T}}(\boldsymbol{x}, \boldsymbol{\epsilon})$$

with constraints

$$\begin{bmatrix} -A & I \\ A & I \end{bmatrix} \begin{bmatrix} \boldsymbol{x} \\ \boldsymbol{\epsilon} \end{bmatrix} \geq \begin{bmatrix} -\boldsymbol{b} \\ \boldsymbol{b} \end{bmatrix}.$$

Find a problem whose dual solution is the solution to this problem.

5. a. Write a function which finds the N-vector $\boldsymbol{x}$ which minimizes $\|A\boldsymbol{x} - \boldsymbol{b}\|_\infty$, where A is an M by N matrix and $\boldsymbol{b}$ is an M-vector. Your function should call DLPRG (Figure 4.5.1) to solve the primal problem whose dual is described in Problem 3. $\boldsymbol{x}$ should return the first N components of the dual solution Y calculated by DLPRG, and the function value should return $Y(N+1)$ or P, which will be the minimal norm.

   To test your function, compute the polynomial of degree two, $p_2(x) = a_1 + a_2 x + a_3 x^2$, which minimizes the L_∞ error in fitting the data points $(z_i, f_i) = (0,1), (1,2), (2,4), (3,4), (4,5), (5,6)$. This means you should call the function with $A(i,j) = z_i^{j-1}, b(i) = f_i$, and $x(j) = a_j$, where $i = 1, ..., M$ and $j = 1, ..., N$, and $M = 6, N = 3$.

b. Write a function which finds the N-vector $\boldsymbol{x}$ which minimizes $\|A\boldsymbol{x} - \boldsymbol{b}\|_1$, where A is an M by N matrix and $\boldsymbol{b}$ is an M-vector. Your function should solve the primal problem whose dual is described in Problem 4. $\boldsymbol{x}$ should return the first N components of the dual solution Y calculated by DLPRG, and the function value should return P, which will be the minimal norm.

   To test your function, compute the polynomial of degree two which minimizes the L_1 error in fitting the data points given in part (a). Notice that the L_1 polynomial is not as affected as the L_∞ polynomial by the "outlier" point $(2, 4)$.

c. For comparison, also compute the polynomial of degree two which minimizes the L_2 error in fitting the data points of part (a). You may call DLLSQR (Figure 2.2.2) to minimize $\|A\boldsymbol{x} - \boldsymbol{b}\|_2$.

6. Verify the assertion made in Section 4.5 that, during phase II, the α-components of the artificial variables will equal one, while the α-components of the other variables will equal zero. (Hint: Argue that the tableau will have the form shown in Figure 4.3.4 throughout phase II.)

7. Set $\boldsymbol{b} = (1, \ldots, 1)$ and use a random number generator with output in $(0, 1)$ (e.g., the FORTRAN90 routine RANDOM_NUMBER) to generate the coefficients of an M by M matrix A, and the M-vector $\mathbf{c}$ in the resource allocation problem

$$\text{maximize } \mathbf{c}^{\mathrm{T}}\boldsymbol{x}$$

with constraints and bounds

$$A\boldsymbol{x} \le \boldsymbol{b}, \qquad \boldsymbol{x} \ge \mathbf{0}.$$

Modify DLPRG to output the total number of simplex iterations to convergence, and use this program to solve the problems generated with $M = 10, 20, 30, 40, 50$. What is the experimental dependence on M of the number of iterations? For each M, compute $(3M)!/[M!(2M)!]$, the number of possible bases, and notice that the simplex method only examines an extremely small percentage of them. (There will be $3M$ unknowns in the tableau, because slack variables must be added, and DLPRG will use artificial variables even though they could be avoided for this problem.)

8. Using the most negative element in the last row to choose the pivot column is a reasonable strategy, but as the following example illustrates, it is not always the best strategy.

Maximize $P = 3x + 2y$

with constraints

$$2x + y \leq 2$$

and bounds

$$x \geq 0,$$
$$y \geq 0.$$

Solve this problem twice using the simplex method. The first time pick the most negative element (–3) in the last row on the first iteration; the second time, try choosing the other negative element (–2). You will find the same solution each time, of course, but which requires fewer iterations?

This example suggests an alternative strategy for picking the pivot column. One could inspect each column jp for which $d_{jp} < 0$, find the corresponding pivot row ip in the usual way, and compute the resulting increase in P, $-d_{jp}b_{ip}/a_{ip,jp}$, and choose the pivot column that maximizes this projected increase. The increase in work associated with this strategy is about M times the number of negative elements in the last row, so it could be nearly as much as doing an extra iteration, and it is not likely that it will decrease the number of iterations enough to pay for the extra cost. Furthermore, even if we use this strategy there is no guarantee that the total number of iterations will be the minimum possible.

9. The dual solution for the transportation problem 4.1.3 is found to be $y = (0, 300, -550, -300, -400)$. Why is it obvious from the "physical" interpretation of the dual solution, given the solution reported in Section 4.5, that y_1 would be 0? Why could you also have predicted that y_3 would be -550?

10. The curve-fitting problem 4.1.4 had to be solved in Section 4.5 as the dual of another problem, because it does not have zero bounds on the variables and thus cannot be solved by DLPRG as it stands. An alternative approach is to define $m' = m + d$, $b' = b + d$, where d is large enough to ensure that m' and b' are positive at the optimal feasible point, so that the bounds $m' \geq 0$, $b' \geq 0$ can be added without changing the solution. Rewrite 4.1.4 in terms of the new variables m', b', ϵ, with zero bounds on all three variables, and solve the modified problem directly using DLPRG (Figure 4.5.1), then subtract d from m' and b' to

obtain the optimal m, b. For this problem, $d = 10$ will be large enough. Remember that DLPRG requires that the components of the B vector be nonnegative.

Notice, however, that this approach requires solving a problem with a 10 by 13 coefficient matrix, while in Section 4.5, A was only 3 by 10.

11. Modify DLPRV (Figure 4.6.1) so that the first MLE constraints ($0 \le MLE \le M$, where MLE is a new argument) are assumed to be $\le$, rather than $=$, constraints. This means that the first MLE artificial variables can be replaced by slack variables, since there will still be an identity matrix in the last M columns, and the total number of variables will decrease. (Hint: change c_i from $-\alpha$ to 0 for each new slack variable; then only one additional change to the code will be required, for the MATLAB version.) Now modify DTRAN (Figure 4.7.1) to call the modified DLPRV. Set $MLE = NW$, and since slack variables are now added in DLPRV instead of DTRAN, N will decrease to $NW * NS$. Use the new DTRAN/DLPRV to solve the transportation problem 4.1.3.

12. a. If we can find an $\boldsymbol{x}$ that is primal feasible ($A\boldsymbol{x} = \boldsymbol{b}, \boldsymbol{x} \ge \boldsymbol{0}$), and a $\boldsymbol{y}$ that is dual feasible ($A^{\mathrm{T}}\boldsymbol{y} - \boldsymbol{z} = \boldsymbol{c}, \boldsymbol{z} \ge \boldsymbol{0}$), such that $\boldsymbol{c}^{\mathrm{T}}\boldsymbol{x} = \boldsymbol{b}^{\mathrm{T}}\boldsymbol{y}$, explain why Theorem 4.3.1 means that $\boldsymbol{x}$ must be the primal solution and $\boldsymbol{y}$ must be the dual solution. Show that for a primal feasible $\boldsymbol{x}$ and dual feasible $\boldsymbol{y}$, $\boldsymbol{c}^{\mathrm{T}}\boldsymbol{x} = \boldsymbol{b}^{\mathrm{T}}\boldsymbol{y}$ is equivalent to $\boldsymbol{x}^{\mathrm{T}}\boldsymbol{z} = 0$, and thus (since $\boldsymbol{x}$ and $\boldsymbol{z}$ are nonnegative) $x_i z_i = 0$, for $i = 1, \ldots, N$. So the linear programming problem 4.3.1 can be reduced to finding a solution of the nonlinear system:

$$\begin{aligned} A\boldsymbol{x} - \boldsymbol{b} &= \boldsymbol{0}, \\ A^{\mathrm{T}}\boldsymbol{y} - \boldsymbol{z} - \boldsymbol{c} &= \boldsymbol{0}, \\ x_i z_i &= 0, \qquad i = 1, \ldots, N, \end{aligned}$$

with all components of $\boldsymbol{x}$ and $\boldsymbol{z}$ nonnegative. Show that the number of equations is the same as the number of unknowns in this system. Interior point methods attempt to solve this nonlinear system, while keeping $\boldsymbol{x}$ and $\boldsymbol{z}$ nonnegative. Notice that solving a nonlinear system of K equations and K unknowns normally takes $O(K^3)$ operations (what is K here?), so it is not unreasonable to expect that a good algorithm could solve this problem–even with the constraints–with $O(K^3)$ work.

b. Solve the resource allocation problem 4.4.1 using an interior point method, as follows. Replace the equations $x_i z_i = 0$ by $x_i z_i = \mu$, and solve the system of part (a) using Newton's method, with $\mu = 1$, and $x_i = z_i = 1$ ($y_i = 0$) initially. Then re-solve this system repeatedly, with μ cut in half each time, using the solution from the previous problem to start the Newton iteration for each new μ. x_i and z_i should remain nonnegative as μ goes to 0.

For this system, Newton's method has the form:

$$\begin{bmatrix} \boldsymbol{x}^{k+1} \\ \boldsymbol{y}^{k+1} \\ \boldsymbol{z}^{k+1} \end{bmatrix} = \begin{bmatrix} \boldsymbol{x}^{k} \\ \boldsymbol{y}^{k} \\ \boldsymbol{z}^{k} \end{bmatrix} - \begin{bmatrix} A & 0 & 0 \\ 0 & A^{\mathrm{T}} & -I \\ Z^k & 0 & X^k \end{bmatrix}^{-1} \begin{bmatrix} A\boldsymbol{x}^k - \boldsymbol{b} \\ A^{\mathrm{T}}\boldsymbol{y}^k - \boldsymbol{z}^k - \boldsymbol{c} \\ X^k Z^k \mathbf{1} - \mu\mathbf{1} \end{bmatrix}$$

where X^k and Z^k are diagonal matrices with the vectors $\boldsymbol{x}^k$ and $\boldsymbol{z}^k$, respectively, along the diagonals.

13. Modify DTRAN (Figure 4.7.1) to call DLPRVS (Figure 4.6.2) instead of DLPRV (Figure 4.6.1). DLPRVS calls a subroutine SPARSOL to solve the sparse linear systems; write a simple SPARSOL which simply copies the sparse matrix A to an N by N full matrix, then calls DLINEQ (Figure 1.2.1) to solve the full linear system. Then solve the transportation problem 4.1.3 using the modified DTRAN. Calling DLINEQ to solve the linear systems is of course very inefficient, but serves to verify the correctness of DLPRVS.

14. Compare the total memory requirements and operation counts (per iteration) for DLPRG and DTRAN/DLPRV, applied to a transportation problem with N_{W} warehouses and N_{S} stores. Assume $N_{\mathrm{W}} = N_{\mathrm{S}}$, and both are large.

15. For many linear programming applications, the unknowns must be integers, for example, in the problem of Section 4.1.1, we cannot produce a fractional number of chairs or tables. If we add the additional constraint that each x_i must be integer, this is an integer programming problem, which is much harder, and beyond the scope of this book. In many resource allocation applications, particularly if the number of items produced is large, a reasonable solution is simply to round each x_i down. However, if x_i represents the number of Boeing 747s we will produce, this is not a good solution. In fact, the following problem illustrates the fact that even for resource allocation problems, simply rounding down is not guaranteed to find the best integer solution:

    Maximize $P = 5x + 4y$

    with constraints

$$\begin{aligned} 3x + 4y &\le 12, \\ 16x + 11y &\le 44, \end{aligned}$$

    and bounds

$$\begin{aligned} x &\ge 0, \\ y &\ge 0. \end{aligned}$$

Solve this problem with the simplex method (or graphically) and round x and y down to the nearest smaller integers, then show that there is an integer feasible point where P is larger.

Furthermore, show that the optimum integer solution of the following problem cannot be obtained by rounding the optimum real solution components up or down. Thus even if we round each simplex solution component up and down and evaluate P at each of these points (requiring $O(2^N)$ work) there is no guarantee that this will produce the optimum integer solution of an LP problem.

Maximize $P = x + y$

with constraints

$$\begin{aligned} 3x + y &\leq 4.5, \\ -3x + y &\leq 0, \end{aligned}$$

and bounds

$$\begin{aligned} x &\geq 0, \\ y &\geq 0. \end{aligned}$$

It can be shown that if the warehouse capacities and store needs are all integer, the transportation problem will always have an integer solution, so there is no need for integer programming for this problem, at least.

5

The Fast Fourier Transform

5.1 The Discrete Fourier Transform

Although this topic is not normally covered by texts on computational linear algebra, the fast Fourier transform is really just an algorithm that calculates a certain matrix–vector product efficiently, and thus it seems appropriate to study it here.

The discrete Fourier transform of a complex N-vector $\boldsymbol{f}$ is $\boldsymbol{y} = A\boldsymbol{f}$, where A is a complex N by N matrix with components $a_{k,j} = \exp[i2\pi(k-1)(j-1)/N] (i \equiv \sqrt{-1})$. In other words,

$$y_k = \sum_{j=1}^{N} f_j \exp\left(\frac{i2\pi(k-1)(j-1)}{N}\right).$$

For a fixed j, the vector $\exp[i2\pi(k-1)(j-1)/N]$ varies periodically with k, cycling through $j-1$ complete periods as k goes from 1 to N. Thus we can think of f_j as the amplitude of that component of $\boldsymbol{y}$ which has the frequency $j-1$, where the frequency is defined as the total number of cycles.

Suppose, for example, that the temperature at a certain weather station is recorded every hour for 100 years, with the results stored in the vector $\boldsymbol{y}$ (y_k is the temperature after k hours). If the data points (k, y_k) are plotted, a daily cycle (period = 24 points) and an annual cycle (period = 24×365 points) will be easily observed. If we transform the vector $\boldsymbol{y}$ (in the "time domain") into the vector $\boldsymbol{f}$ (in the "frequency domain") using $\boldsymbol{f} = A^{-1}\boldsymbol{y}$, $\boldsymbol{f}$ will have one sharp peak corresponding to a frequency of 100 (100 complete cycles over the course of the 100-year recording period) and another peak corresponding to a frequency of 100×365 ($36,500$ cycles over the course of the data). In other words, f_{101} and $f_{36,501}$ will be large (in magnitude—the f_i are complex numbers). There might be other smaller peaks corresponding to less apparent cycles (perhaps one coinciding with the 11-year sunspot activity cycle). These minor periodic tendencies, if they exist, would be much easier to spot in the frequency domain than in the time domain of the original data.

We can explicitly display the components of the matrix A^{-1}:

$$\alpha_{j,l} = \frac{1}{N} \exp\left(-\frac{i2\pi(j-1)(l-1)}{N}\right). \tag{5.1.1}$$

Note that $\alpha_{j,l} = (1/N)\bar{a}_{j,l}$, and $A^{-1} = \bar{A}/N$, where the overbar means complex conjugate. Since $\boldsymbol{f} = A^{-1}\boldsymbol{y}$ implies $\bar{\boldsymbol{f}} = A(\bar{\boldsymbol{y}}/N)$, this means that we can calculate the inverse discrete Fourier transform of $\boldsymbol{y}$ by taking the forward transform of $\bar{\boldsymbol{y}}/N$ and then conjugating the result. (Note that $U \equiv A/\sqrt{N}$ is a *unitary* matrix, that is, $\bar{U}^T U = I$.)

To verify 5.1.1, let us multiply A by this matrix to see whether we get the identity matrix. The (k,l)th component of the product is

$$\begin{aligned}
\sum_{j=1}^{N} a_{k,j}\alpha_{j,l} &= \sum_{j=1}^{N} \frac{1}{N} \exp\left(\frac{i2\pi(k-1)(j-1)}{N}\right) \exp\left(-\frac{i2\pi(j-1)(l-1)}{N}\right) \\
&= \sum_{j=1}^{N} \frac{1}{N} \exp\left(\frac{i2\pi(j-1)(k-l)}{N}\right) \\
&= \frac{1}{N}(1 + \beta + \beta^2 + \ldots + \beta^{N-1}),
\end{aligned}$$

where $\beta = \exp[i2\pi(k-l)/N]$. If $k = l, \beta$ will be 1, and the sum will be $N/N = 1$. Otherwise, if $k \neq l$, the sum will be $1/N[(\beta^N - 1)/(\beta - 1)] = 0$, since $\beta^N = 1$, while $\beta \neq 1$. Thus the product matrix is the identity, and so 5.1.1 gives the components of A^{-1} correctly.

The discrete Fourier transform can be used not only to identify periodic tendencies but also to filter out high-frequency noise. If we transform a data sequence $\boldsymbol{y}$ into the frequency domain by $\boldsymbol{f} = A^{-1}\boldsymbol{y}$, and set $f_j = 0$ for all j greater than some given cutoff frequency, then, when the truncated $\boldsymbol{f}'$ is transformed back into the time domain, the new data sequence $\boldsymbol{y}' = A\boldsymbol{f}'$ will be smoother than the original sequence, since its rapidly oscillating components will be gone (see Problem 8).

5.2 The Fast Fourier Transform

Forming the product of an N by N matrix with an N-vector normally requires N^2 multiplications. Hence the forward ($\boldsymbol{y} = A\boldsymbol{f}$) and reverse ($\boldsymbol{f} = A^{-1}\boldsymbol{y}$) Fourier transforms would each seem to require $O(N^2)$ work and, if N is a prime number, these transforms do indeed require $O(N^2)$ operations. However, if $N = 2^m$ for some integer m, the matrices A and A^{-1} have a special structure that can be exploited using the "fast Fourier transform" [Cooley and Tukey 1965; Brigham 1974, 1988], to reduce the work in the matrix–vector multiplications to only $O[N\log(N)]$. Since in applications N may be extremely large, this represents an important savings in effort.

If N can be factored at all, there is some potential savings to be realized by cleverly exploiting the matrix structure [Singleton 1967]. However, if we cannot control the number N of data points, this savings is unlikely to be worth the effort, whereas if we can pick N as we please (as is often the case), we might as well choose N to be a power of 2 and achieve the maximum reduction in effort.

In order to achieve this work reduction, we reorder the columns of A, and the corresponding unknowns, so that the odd columns (and unknowns) come first, ahead of the even columns (and unknowns). Then (recall that $N = 2^m$, and so N is certainly even), the equation $\boldsymbol{y} = A\boldsymbol{f}$ takes the form

$$\left[\begin{array}{c} y_1 \\ \vdots \\ y_{N/2} \\ \hline y_{N/2+1} \\ \vdots \\ y_N \end{array}\right] = \left[\begin{array}{c|c} P & Q \\ \hline R & S \end{array}\right] \left[\begin{array}{c} f_1 \\ f_3 \\ \vdots \\ f_{N-1} \\ \hline f_2 \\ f_4 \\ \vdots \\ f_N \end{array}\right]. \tag{5.2.1}$$

Now let B be the $N/2$ by $N/2$ Fourier transform matrix, so that

$$b_{k,j} = \exp\left(\frac{i2\pi(k-1)(j-1)}{N/2}\right) = \exp\left(\frac{i2\pi(k-1)(2j-2)}{N}\right) \quad (k,j = 1,\ldots,N/2).$$

Also, let us define an $N/2$ by $N/2$ *diagonal* matrix D to have diagonal components

$$d_{k,k} = \exp\left(\frac{i2\pi(k-1)}{N}\right) = h^{k-1},$$

where $h \equiv \exp(i2\pi/N)$. Recalling that A (the N by N Fourier transform matrix) has components

$$a_{k,j} = \exp\left(\frac{i2\pi(k-1)(j-1)}{N}\right) \qquad (k,j = 1,\ldots,N),$$

we can compute the elements of P, Q, R, and S in terms of those of D and B:

$$\begin{aligned} p_{k,j} = a_{k,2j-1} &= \exp\left(\frac{i2\pi(k-1)(2j-2)}{N}\right) = b_{k,j}, \\ q_{k,j} = a_{k,2j} &= \exp\left(\frac{i2\pi(k-1)(2j-1)}{N}\right) \\ &= \exp\left(\frac{i2\pi(k-1)}{N}\right) b_{k,j} = d_{k,k}b_{k,j}, \end{aligned}$$

$$\begin{aligned} r_{k,j} = a_{N/2+k,2j-1} &= \exp\left(\frac{i2\pi(k-1+N/2)(2j-2)}{N}\right) \\ &= \exp[i2\pi(j-1)]b_{k,j} \\ &= b_{k,j}, \end{aligned}$$

$$\begin{aligned} s_{k,j} = a_{N/2+k,2j} &= \exp\left(\frac{i2\pi(k-1+N/2)(2j-1)}{N}\right) \\ &= \exp\left(\frac{i2\pi(k-1+N/2)}{N}\right) r_{k,j} \\ &= -\exp\left(\frac{i2\pi(k-1)}{N}\right) r_{k,j} \\ &= -d_{k,k}b_{k,j}. \end{aligned}$$

Thus the system 5.2.1 can be rewritten in terms of D and B:

$$\begin{bmatrix} y_1 \\ \vdots \\ y_N \end{bmatrix} = \left[\begin{array}{c|c} B & DB \\ \hline B & -DB \end{array}\right] \left[\begin{array}{c} \boldsymbol{f}_{odd} \\ \hline \boldsymbol{f}_{even} \end{array}\right] = \left[\begin{array}{c} B\boldsymbol{f}_{odd} + DB\boldsymbol{f}_{even} \\ \hline B\boldsymbol{f}_{odd} - DB\boldsymbol{f}_{even} \end{array}\right]. \quad (5.2.2)$$

From 5.2.2 we can see how our work reduction is achieved. To calculate the Fourier transform of an N-vector we first need to transform two $N/2$-vectors (i.e., we calculate $\boldsymbol{f}_{odd} = B\boldsymbol{f}_{odd}$ and $\boldsymbol{f}_{even} = B\boldsymbol{f}_{even}$). Next, we calculate $\boldsymbol{u} \equiv D\boldsymbol{f}_{even}$ using the formulas (recall that $d_{k,k} = h^{k-1}$)

$$\begin{aligned} & d_{1,1} = 1 \\ k = 1, \ldots, \frac{N}{2} \quad & u_k = d_{k,k}(\boldsymbol{f}_{even})_k \\ & d_{k+1,k+1} = hd_{k,k} \end{aligned}$$

This can be done in a total of $N/2 + N/2 = N$ multiplications. Finally, the first half of the vector $\boldsymbol{y}$ is calculated by adding $\boldsymbol{f}_{odd}$ and $\boldsymbol{u}$, and the second half by adding $\boldsymbol{f}_{odd}$ and $-\boldsymbol{u}$ (a total of N additions).

Therefore, if W_m represents the total number of multiplications (or additions, for that matter) required to transform a vector of length $N = 2^m$, we have the recurrence relation

$$W_m = 2W_{m-1} + N = 2W_{m-1} + 2^m. \tag{5.2.3}$$

Since the 1 by 1 Fourier transform matrix is simply the number 1, the number of multiplications required to transform a vector of length $N = 1$ is zero, so that

$$W_0 = 0. \tag{5.2.4}$$

The solution to the recurrence relation 5.2.3, with the initial condition 5.2.4, is

$$W_m = m2^m = N \log_2(N),$$

as we can verify by substituting $W_m = m2^m$ directly into 5.2.3 and 5.2.4:

$$2W_{m-1} + 2^m = 2[(m-1)2^{m-1}] + 2^m = m2^m = W_m$$

and $W_0 = 02^0 = 0$.

This establishes the claim that, if N is a power of 2, the Fourier transform of an N-vector can be done in $O[N \log(N)]$ operations. The key to this work reduction is the fact that, as exhibited in 5.2.2, the multiplication $A\boldsymbol{f}$ requires only two matrix–vector multiplications of order $N/2$, rather than four, as would be required if A had no special structure.

5.3 FORTRAN90 Programs

The fast Fourier transform algorithm is described above in a recursive manner: To transform an N-vector we must first know how to transform an $N/2$-vector. Thus it is more easily programmed in FORTRAN90 than in FORTRAN77, which does not allow subprograms to call themselves recursively. Figure 5.3.1 gives a recursive FORTRAN90 subroutine that calculates a fast Fourier transform. Note that the subroutine DFFT calls itself twice, to transform the $N/2$-vectors $\boldsymbol{f}_{odd}$ and $\boldsymbol{f}_{even}$, as part of its computation of the transform of an N-vector. Of course the recursion must stop somewhere and so, for $N = 1$ $(M = 0)$, DFFT does nothing, because the transform of a 1-vector is itself. Note the remarkable simplicity of this subroutine.

```
      RECURSIVE SUBROUTINE DFFT(F,M)
      IMPLICIT DOUBLE PRECISION (A-H,O-Z)
C                                 DECLARATIONS FOR ARGUMENTS
      COMPLEX*16 F(2**M)
      INTEGER M
C                                 DECLARATIONS FOR LOCAL VARIABLES
```

```
      COMPLEX*16 FODD(2**M/2+1),FEVEN(2**M/2+1),D,H,U
C
C  SUBROUTINE DFFT PERFORMS A FAST FOURIER TRANSFORM ON THE COMPLEX
C     VECTOR F, OF LENGTH N=2**M.  THE FOURIER TRANSFORM IS DEFINED BY
C       Y(K) = SUM FROM J=1 TO N OF: EXP[I*2*PI*(K-1)*(J-1)/N]*F(J)
C     WHERE I = SQRT(-1).
C
C  ARGUMENTS
C
C             ON INPUT                          ON OUTPUT
C             --------                          ---------
C
C    F      - THE COMPLEX VECTOR OF LENGTH      THE TRANSFORMED VECTOR
C             2**M TO BE TRANSFORMED.           Y.
C
C    M      - THE LENGTH OF THE VECTOR F
C             IS ASSUMED TO BE 2**M.
C
C-----------------------------------------------------------------------
C                                 FOURIER TRANSFORM OF A 1-VECTOR IS
C                                 UNCHANGED
      IF (M.EQ.0) RETURN
      N = 2**M
      PI = 3.141592653589793 D0
      H = COS(2*PI/N) + SIN(2*PI/N)*CMPLX(0.0,1.0)
      N2 = N/2
C                                 COPY ODD COMPONENTS OF F TO FODD
C                                 AND EVEN COMPONENTS TO FEVEN
      DO 5 K=1,N2
         FODD(K) = F(2*K-1)
         FEVEN(K)= F(2*K)
    5 CONTINUE
C                                 TRANSFORM N/2-VECTORS FODD AND FEVEN
      CALL DFFT(FODD ,M-1)
      CALL DFFT(FEVEN,M-1)
      D = 1.0
C                                 Y = (FODD+D*FEVEN , FODD-D*FEVEN)
      DO 10 K=1,N2
         U = D*FEVEN(K)
         F(K)    = FODD(K) + U
         F(N2+K) = FODD(K) - U
         D = D*H
   10 CONTINUE
      RETURN
      END
```

Figure 5.3.1

As mentioned in Section 5.1 (see 5.1.1), $A^{-1} = \bar{A}/N$, and thus the inverse Fourier transform $\boldsymbol{f} \equiv A^{-1}\boldsymbol{y}$ can be calculated as $\bar{\boldsymbol{f}} = A(\bar{\boldsymbol{y}}/N)$. Hence DFFT can be used to calculate an inverse transform as follows:

```
      N = 2**M
      DO 10 K=1,N
   10 Y(K) = CONJG(Y(K))/N
      CALL DFFT(Y,M)
      DO 20 J=1,N
   20 F(J) = CONJG(Y(J))
```

DFFT was used in this manner to calculate the inverse Fourier transform $\boldsymbol{f} = A^{-1}\boldsymbol{y}$ of the vector $\boldsymbol{y}$ of length $N = 2^7$ defined by

$$y_k = 30 + 20\sin\left(\frac{2\pi k}{30}\right) + 8\cos\left(\frac{2\pi k}{7}\right), \qquad k = 1, \ldots, 128. \tag{5.3.1}$$

y_k is plotted as a function of k in Figure 5.3.2; $\boldsymbol{y}$ has a constant component, a periodic component of period 30 (frequency$= N/30 = 4.27$), and another periodic component, of smaller amplitude, of period 7 (frequency$= N/7 = 18.29$). These periodic components are evident in Figure 5.3.2.

The absolute values of the components of the complex vector $\boldsymbol{f}$ are plotted as a function of frequency; that is, $|\boldsymbol{f}_j|$ is plotted as a function of $j - 1$, in Figure 5.3.3. There are peaks near frequencies of 4 and 18, as expected, plus another peak at zero frequency corresponding to the constant component of $\boldsymbol{y}$. There are two additional peaks (not shown in Figure 5.3.3) near frequencies of 124 and 110, whose presence is at first sight inexplicable. However, if we expand

$$\begin{aligned}
\sin\left(\frac{2\pi k}{P}\right) &= \frac{\exp(i2\pi k/P) - \exp(-i2\pi k/P)}{2i} \\
&= \frac{\exp(i2\pi k/P) - \exp(i2\pi k(P-1)/P)}{2i}, \\
\cos\left(\frac{2\pi k}{P}\right) &= \frac{\exp(i2\pi k/P) + \exp(-i2\pi k/P)}{2} \\
&= \frac{\exp(i2\pi k/P) + \exp(i2\pi k(P-1)/P)}{2},
\end{aligned}$$

we see that, when a sine or cosine term of period P appears in y_k, peaks in the frequency domain will occur at both N/P and $N(P-1)/P = N - N/P$.

Finally, a *non-recursive* fast Fourier transform routine is shown in Figure 5.3.4. All the re-ordering is done at the beginning (loop 40); then we make m passes over $\boldsymbol{f}$, the first time we "process" the 2-vectors $(f_1, f_2), (f_3, f_4), \ldots$, the second time we process the 4-vectors $(f_1, f_2, f_3, f_4), \ldots)$ and so on. Here to "process" a vector means to apply loop 10 of DFFT (Figure 5.3.1). However,

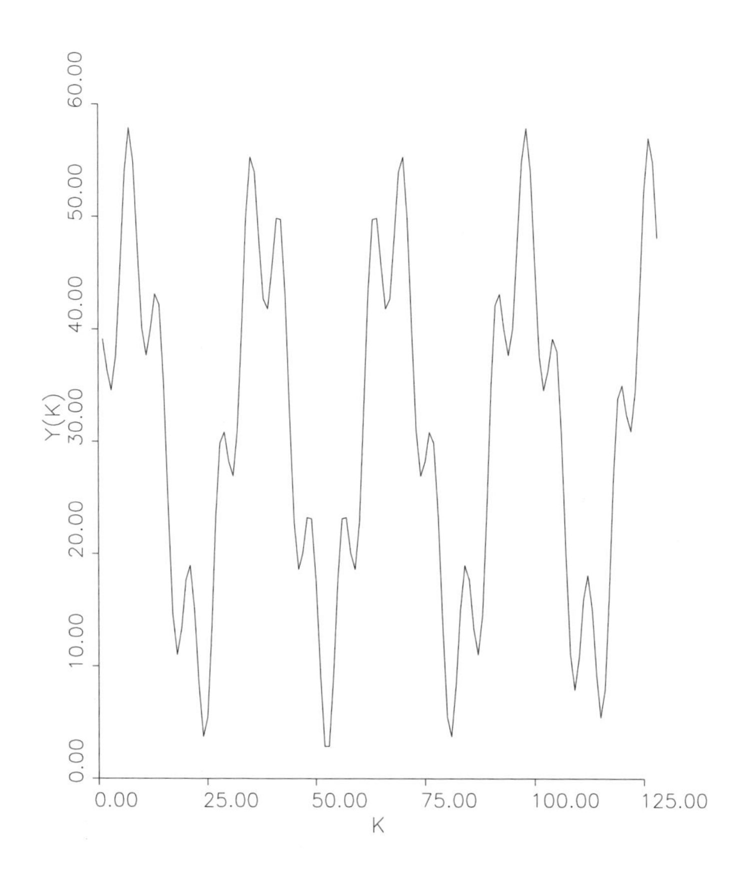

Figure 5.3.2
The Time Domain

now FODD and FEVEN in this loop are just the first and last halves of the vector, since the re-ordering has already been done.

Now this routine clearly uses only $2N$ words of memory, and it is also easy to count that there are approximately mN floating point multiplications done, or $N\log_2(N)$, the same as done by DFFT.

```
      SUBROUTINE NRFFT(F,M)
      IMPLICIT DOUBLE PRECISION(A-H,O-Z)
      COMPLEX*16 F(2**M),FODD(2**M/2),FEVEN(2**M/2)
      COMPLEX*16 D,H,U
C
C  SUBROUTINE NRFFT PERFORMS A FAST FOURIER TRANSFORM ON THE COMPLEX
C     VECTOR F, OF LENGTH N=2**M.  THE FOURIER TRANSFORM IS DEFINED BY
```

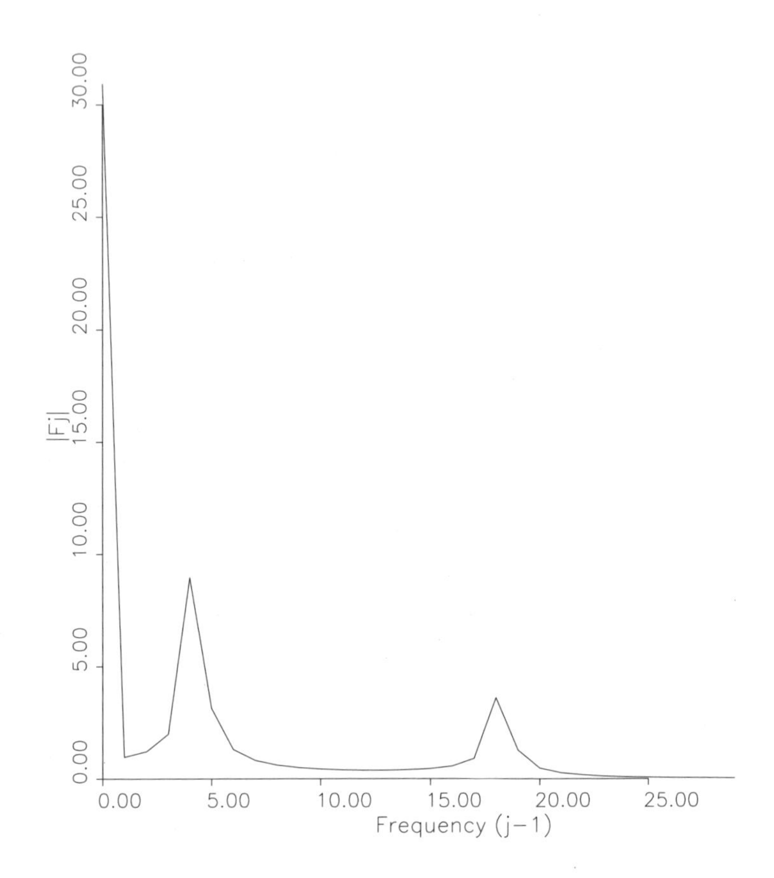

Figure 5.3.3
The Frequency Domain

```
C       Y(K) = SUM FROM J=1 TO N OF: EXP[I*2*PI*(K-1)*(J-1)/N]*F(J)
C     WHERE I = SQRT(-1).
C
C  ARGUMENTS
C
C              ON INPUT                          ON OUTPUT
C              --------                          ---------
C
C    F       - THE COMPLEX VECTOR OF LENGTH      THE TRANSFORMED VECTOR
C              2**M TO BE TRANSFORMED.           Y.
C
C    M       - THE LENGTH OF THE VECTOR F
C              IS ASSUMED TO BE 2**M.
```

```
C
C-----------------------------------------------------------------------
      N = 2**M
      PI = 3.141592653589793 D0
C                                   REORDER ELEMENTS OF F AS THEY WILL BE
C                                   IN TRANSFORMED VECTOR
      DO 40 I=M,2,-1
         NI = 2**I
         N2 = NI/2
         DO 30 J=0,N-1,NI
            DO 10 K=1,N2
               FODD(K) = F(J+2*K-1)
               FEVEN(K) = F(J+2*K)
   10       CONTINUE
            DO 20 K=1,N2
               F(J+K) = FODD(K)
               F(J+N2+K) = FEVEN(K)
   20       CONTINUE
   30    CONTINUE
   40 CONTINUE
C                                   PROCESS ELEMENTS BY GROUPS OF 2,
C                                   THEN BY GROUPS OF 4,8,...,2**M
      DO 70 I=1,M
         NI = 2**I
         N2 = NI/2
         H = COS(2*PI/NI) + SIN(2*PI/NI)*CMPLX(0.0,1.0)
         DO 60 J=0,N-1,NI
            D = 1.0
            DO 50 K=1,N2
               FODD(K) = F(J+K)
               FEVEN(K) = F(J+N2+K)
               U = D*FEVEN(K)
               F(J+K) = FODD(K) + U
               F(J+N2+K) = FODD(K) - U
               D = D*H
   50       CONTINUE
   60    CONTINUE
   70 CONTINUE
      RETURN
      END
```

Figure 5.3.4

5.4 Problems

1. Compute the vector $\boldsymbol{y}$ using

$$y_k = 60 + 20\sin\left(\frac{2\pi k}{24 \times 365}\right) + 15\sin\left(\frac{2\pi k}{24}\right) + 5(r_k - 0.5),$$

where r_k is a random variable on $(0, 1)$ generated by a random-number generator (e.g., IMSL's RNUN). Choose $N = 2^m$, where m is 17. y_k might represent the temperature after k hours recorded at a weather station; it has an annual cycle and a daily cycle, plus a random component. Perform an inverse discrete Fourier transform on $\boldsymbol{y}$, using the fast Fourier transform routine DFFT (Figure 5.3.1). Print out the absolute values of some of the complex coefficients f_k to verify that, despite the random component, $\boldsymbol{f} = A^{-1}\boldsymbol{y}$ has peaks corresponding to frequencies of $N/(24 \times 365), N/24$, and 0. (Why 0?)

2. Run Problem 1 with three consecutive values of m (preferably 15, 16, and 17) and record the execution time, to verify that the time is nearly proportional to N.

3. Consider the following "slow Fourier transform" subroutine, which computes $\boldsymbol{y} = A\boldsymbol{f}$ by ordinary matrix multiplication:

```
      SUBROUTINE SFT(F,Y,N)
      IMPLICIT DOUBLE PRECISION (A-H,O-Z)
      COMPLEX*16 F(*),Y(*),H,HK,AKJ
      PI = 3.141592653589793D0
C                H = EXP(I*2*PI/N)
      H = COS(2*PI/N) + SIN(2*PI/N)*CMPLX(0.0,1.0)
C                HK = H**(K-1)
      HK = 1.0
      DO 10 K=1,N
         Y(K) = 0.0
C                AKJ = HK**(J-1) = H**((J-1)*(K-1))
         AKJ = 1.0
         DO 5 J=1,N
            Y(K) = Y(K) + AKJ*F(J)
            AKJ = AKJ*HK
    5    CONTINUE
         HK = HK*H
   10 CONTINUE
      RETURN
      END
```

Transform the N-vector $\boldsymbol{f}$, with $f_k \equiv 1$, first using this slow routine and then using DFFT (Figure 5.3.1), with $N = 2^{14}$. Record the execution time for each and compare.

4. Another important application of the fast Fourier transform is the fast solution of the finite difference equations used to approximate certain partial differential equations. For example, the partial differential equation

$$u_{xx} + u_{yy} = u + f(x, y) \qquad \text{in } 0 \le x \le 1,\ 0 \le y \le 1,$$

with periodic boundary conditions

$$\begin{aligned} u(0, y) &= u(1, y), \qquad u(x, 0) = u(x, 1), \\ u_x(0, y) &= u_x(1, y), \qquad u_y(x, 0) = u_y(x, 1), \end{aligned}$$

can be approximated by the finite difference equations

$$\begin{aligned} u_{j+1,k} - 2u_{j,k} + u_{j-1,k} + u_{j,k+1} - 2u_{j,k} + u_{j,k-1} &= h^2(u_{j,k} + f_{j,k}), \\ \text{for } j = 1, \ldots, N, \quad k &= 1, \ldots, N, \end{aligned}$$

with

$$u_{1,k} = u_{N+1,k}, \qquad u_{j,1} = u_{j,N+1},$$

$$\frac{u_{1,k} - u_{0,k}}{h} = \frac{u_{N+1,k} - u_{N,k}}{h}, \qquad \frac{u_{j,1} - u_{j,0}}{h} = \frac{u_{j,N+1} - u_{j,N}}{h},$$

where $u_{j,k} \equiv u(x_j, y_k)$, $f_{j,k} \equiv f(x_j, y_k)$, and $x_j = (j-1)/N$, and $y_k = (k-1)/N$, and $h = 1/N$ (note that $x_1 = 0, x_{N+1} = 1, y_1 = 0, y_{N+1} = 1$). Suppose that

$$u_{j,k} = \sum_{m=1}^{N} \sum_{l=1}^{N} b_{lm} \exp\left(\frac{i2\pi(j-1)(l-1)}{N}\right) \exp\left(\frac{i2\pi(k-1)(m-1)}{N}\right)$$

and

$$f_{j,k} = \sum_{m=1}^{N} \sum_{l=1}^{N} c_{lm} \exp\left(\frac{i2\pi(j-1)(l-1)}{N}\right) \exp\left(\frac{i2\pi(k-1)(m-1)}{N}\right).$$

a. Show that, no matter how the coefficients b_{lm} are chosen, the approximate solution satisfies the periodic boundary conditions.

b. Show that, once the coefficients c_{lm} of f are known, the solution coefficients b_{lm} can be calculated from

$$b_{lm} = \frac{c_{lm}h^2}{2\cos[2\pi(l-1)/N] + 2\cos[2\pi(m-1)/N] - 4 - h^2}.$$

c. If U, F, B, and C are the N by N matrices with elements u_{jk}, f_{jk}, b_{lm}, and c_{lm}, and if A is the discrete Fourier transform matrix (defined in Section 5.1), show that

$$U^{\mathrm{T}} = A(AB)^{\mathrm{T}}$$

and

$$F^{\mathrm{T}} = A(AC)^{\mathrm{T}} \text{ or } C = A^{-1}(A^{-1}F^{\mathrm{T}})^{\mathrm{T}}.$$

d. If N is chosen to be a power of 2, the matrix–matrix multiplication AZ (or $A^{-1}Z$) can be computed in only $O[N^2 \log(N)]$ operations, because its computation requires transforming the N columns of Z, at a cost of $O[N \log(N)]$ work per column, using fast Fourier transforms. Use the results of parts (b) and (c) to show that, given the matrix F, U can be calculated in $O[N^2 \log(N)]$ operations. In other words, we can solve the above finite difference equations in $O[N^2 \log(N)]$ operations. This is much faster than even the SOR method (cf. Section 1.9), which requires $O(N^3)$ work to solve the same finite difference system. Prove that it is **impossible** to solve a 2D finite difference system in less than $O(N^2)$ operations, so this FFT algorithm is essentially impossible to beat, in terms of speed (this is also true for 3D problems, such as Problem 5d).

5. a. Generalize the formula in Problem 4b, for the partial differential equation $u_{xx}+u_{yy}+pu_x+qu_y+ru = f(x,y)$, if u_x is approximated by $(u_{j+1,k} - u_{j-1,k})/(2h)$ and similarly for u_y.

b. For the case $p = 2, q = -1, r = -20, f(x,y) = e^{2x+y}$, solve this PDE with periodic boundary conditions, using the Fourier transform method outlined in Problem 4 to exactly solve the finite difference system. Output your approximate solution at the square midpoint. The exact solution of the PDE has $u(0.5,0.5) = -0.262517$. Note that an accurate solution is found even though $f(x,y)$ is not periodic. Will this technique still work if p, q or r are functions of x and y?

c. A function $u(x,y)$ defined in $(0,1) \times (0,1)$ and extended periodically, can be expanded in a Fourier series:

$$u(x,y) = \sum_{M=-\infty}^{\infty} \sum_{L=-\infty}^{\infty} b_{LM} \exp(i2\pi Lx) \exp(i2\pi My)$$

Truncate the series, letting L and M vary from $-\frac{N}{2}$ to $\frac{N}{2}-1$, and then make the changes of variables $l = L+\frac{N}{2}+1, m = M+\frac{N}{2}+1$, to get:

$$u(x,y) \approx$$

$$\sum_{m=1}^{N}\sum_{l=1}^{N} b_{lm}\exp(i2\pi(l-1-\frac{N}{2})x)\exp(i2\pi(m-1-\frac{N}{2})y)$$

and get a similar expansion for $f(x,y)$. Plug these expansions directly into the PDE of part (a), rather than into its finite difference approximation, to find another formula for b_{lm} in terms of c_{lm}. Now if you evaluate the expansions for u and f at $x = x_j = (j-1)/N, y = y_k = (k-1)/N$ and simplify them, you will see that the solution values $u(x_j, y_k)$ can again be found from the coefficients b_{lm} using DFFT, and the values $f(x_j, y_k)$ and coefficients c_{lm} are similarly related. Rerun your program from part (b) with these minor changes, and get another estimate of $u(0.5, 0.5)$. In part (b) you were using the FFT simply to rapidly solve the finite difference equations exactly; now you are using a Fourier series method, with a truncated series.

d. Write a program which solves the 3D PDE $u_{xx} + u_{yy} + u_{zz} - u_x - 2u = f(x,y,z)$, with periodic boundary conditions on $0 \leq x \leq 2\pi, 0 \leq y \leq 2\pi, 0 \leq z \leq 2\pi$, using centered finite differences on an N by N by N grid, where $N = 2^8$, and fast Fourier transforms to solve the N^3 by N^3 linear system. You should generalize the procedure for 2D problems outlined in Problem 4. Choose $f(x,y,z)$ so that the exact solution is $u = cos(x-y+2z)$, and calculate the maximum error over the grid points.

6. The discrete sine transform of a real N-vector $\boldsymbol{f}$ is $S\boldsymbol{f}$, where S is the N by N matrix with elements $s_{k,j} = \sin(\pi kj/(N+1))$.

   a. Show that $S^{-1} = 2S/(N+1)$. (Hint: Use $\sin(\theta) = (e^{i\theta} - e^{-i\theta})/(2i)$ and use Section 5.1 to guide your calculations.)

   b. The discrete sine transform can be used to help solve some differential equations where the solution is required to be zero on the boundary. For example, consider the problem $u_{xx} = f(x)$ with $u(0) = u(1) = 0$. We approximate this differential equation using the finite difference equation $u_{j+1} - 2u_j + u_{j-1} = h^2 f_j, u_0 = u_{N+1} = 0$, where $h = 1/(N+1)$, $x_j = jh$, and u_j approximates $u(x_j)$. Now expand u and f in the form:

$$u_j = \sum_{l=1}^{N} b_l \sin(\pi jl/(N+1))$$

and

$$f_j = \sum_{l=1}^{N} c_l \sin(\pi jl/(N+1)).$$

Insert these expansions into the finite difference equation and find a formula for b_l in terms of c_l. Notice that u_0 and u_{N+1} are zero no matter what coefficients b_l are used.

c. For the case $f(x) = 2 - 6x$, solve the finite difference system, by first finding $\boldsymbol{c} = S^{-1}\boldsymbol{f}$, then $\boldsymbol{b}$, then $\boldsymbol{u} = S\boldsymbol{b}$. Although it is possible to write a "fast discrete sine transform" routine, you may do the transforms the slow way, using matrix–vector multiplications. (Even using a fast transform, this approach is not faster than using a band solver to solve these tridiagonal finite difference equations; in higher dimensions it is faster.) The exact solution of the differential equation is $u(x) = x^2(1-x)$, and this finite difference equation has zero truncation error, so u_j should exactly equal $u(x_j)$ everywhere.

7. a. Find the approximate amount of memory (number of COMPLEX*16 words) used by the fast Fourier transform routine DFFT, as a function of N. Caution: At first glance it appears that the total memory is $2N$ words. But remember that this is a RECURSIVE subroutine, which means that when DFFT calls itself, another copy of DFFT is created, with new automatic arrays FODD and FEVEN allocated when this copy is entered, and deallocated when it is exited. At any time, at most one copy of DFFT of each size $M = 0, 1, ..., M_{original}$ is open; thus what is the maximum memory allocated to automatic arrays?

   b. What minor change can be made to DFFT to cut the total memory usage to $2N$ COMPLEX*16 words? Make this change and re-run Problem 3 to verify it still works.

8. Take the sequence y_k defined in 5.3.1 and plotted in Figure 5.3.2, and calculate the inverse Fourier transform, $\boldsymbol{f}$. Then set $f_{11} = f_{12} = ... = f_{117} = 0$, and transform $\boldsymbol{f}$ back, and plot the new time sequence y_k (plot the absolute values, the y_k will actually be complex, with small imaginary parts), which should be smoother now since the high frequencies have been removed.

9. Repeat Problem 5b using the both the recursive fast Fourier transform routine DFFT (Figure 5.3.1), and the non-recursive routine NRFFT (Figure 5.3.4), and compare the computer times required for large $N = 2^m$.

6

Linear Algebra on Supercomputers

6.1 Vector Computers

For many years after the first computers were built, progress in computer arithmetic speed was steady and rapid. Finally, however, we have reached a point where it appears that further improvements in the speed of individual arithmetic operations will be harder and harder to achieve. Today's "supercomputers" are nevertheless able to solve problems more rapidly, not so much because they do individual arithmetic operations faster, but because they can do several operations simultaneously.

Consider, for example, the loop 25 in the linear system solver DLINEQ (Figure 1.2.1), where 100% of the computer time is spent in the limit of large problem size:

```
          DO 25 K=I+1,N
             A(J,K) = A(J,K) - LJI*A(I,K)
       25 CONTINUE
```

A traditional "scalar" computer executes these statements one at a time and does not begin updating $A(J,K)$ until it has finished updating $A(J,K-1)$. If the loop were, for example,

```
          DO 26 K=I+1,N
             A(J,K) = A(J,K) - LJI*A(J,K-1)
       26 CONTINUE
```

then executing these statements in sequence would be necessary, because before we update $A(J,K)$ we have to wait until the new value of $A(J,K-1)$ is available. But there is no reason, in theory, why all of the updates in loop 25 above cannot be done simultaneously, as they are completely independent of each other. (Note that trips through the *outer* loop 35 in DLINEQ can *not* be done simultaneously, we cannot start eliminating the elements below the diagonal in column $I+1$ until we have finished with column I.)

There are, in fact, many "parallel" computer systems around today which do have several processors, and we will see in the next section how such a machine can indeed be programmed to distribute the work in loop 25 over the available processors, thus doing some or all of the updates simultaneously. We will see that this is somewhat more complicated than it may sound, for the programmer. First, however, we want to talk about a simpler approach to speeding up loops such as 25, which involves doing the updates, not quite simultaneously, but rather in "assembly-line" fashion.

A "pipeline", or vector, supercomputer [Schendel 1984; Buchanan and Turner 1992] such as a Cray PVP machine, can perform the operations in loop 25 in assembly-line fashion, on a single elaborate processor, and the computer will begin updating $A(J,K)$ before it has finished updating $A(J,K-1)$. An analogy is useful here to appreciate the difference between parallel computers and pipeline computers. Suppose a radio factory initially has only one worker to assemble all the radios. If we simply increase the number of workers (processors) to P, we can produce the radios in parallel and increase the production by a factor of P. On the other hand, if construction of a radio can be divided into P subtasks, each of which requires the same time to perform, we can assign one task to each of the P workers and produce the radios in an assembly line. Once the first radio comes off the assembly line, radios will be produced at a rate P times faster than a single worker can produce them. In a similar manner, if the multiplication in loop 25 can be broken into P substeps: for example, fetch A(I,K) from memory, fetch LJI, multiply the mantissas of LJI and A(I,K), add their exponents, renormalize the result, and so on, then a pipeline computer can process the multiplication P times faster in assembly-line mode, provided the number of loop trips is sufficiently large so that the start-up time can be neglected.

The addition (subtraction) can also be done in assembly-line mode, and, in fact, most vector computers can "chain" an addition and multiplication together, so that a calculation of the form $A(K) = B(K) + C * D(K)$ (which occurs quite often in numerical linear algebra software) can be computed in one long assembly line, with the computation $B(K)+C*D(K)$ starting down the assembly line before $B(K-1)+C*D(K-1)$ has finished.

Whether the updates are done in parallel on different processors, or just in "assembly line" fashion on a single vector processor, in either case the loop can be executed more rapidly only because the updates are mutually independent, and the later updates do not have to wait for the results of the earlier calculations. For a loop such as 26, which has a "dependency", the second update must not start until the first update has passed completely through the assembly line, otherwise we will get the wrong answers.

A FORTRAN compiler designed to run on a pipeline computer will recognize that the operations of loop 25 are mutually independent and will "vectorize" this loop. That is, it will generate object code that does these operations in assembly-line form. For a loop such as 26, the compiler will recognize that vectorization would result in wrong answers and will refuse to do it. These compilers can scan very complicated DO loops and determine which involve calculations that can be done simultaneously, and thus can be vectorized, and which must be done sequentially.

Several of the FORTRAN90 codes from Chapters 1–5 were run on one processor of a Cray SV1 at the Texas Advanced Computing Center in Austin, to see the effect of vectorization, and the results are reported in Table 6.1.1. Array elements were set using a random number generator.

Table 6.1.1

Results on Cray SV1 (Single Precision)

Code	Problem Size	CPU sec. vector=on	CPU sec. vector=off	Speed up
DLINEQ	N=1001	3.2	62.1	19.4
DLLSQR (REDQ)	M=N=1001	6.9	72.0	10.4
DLLSQR (REDH)	M=N=1001	7.9	50.6	6.4
DEGSYM	N=501	56.0	310.9	5.6
DEGNON* (QR)	N=501	13.9	39.0	2.8
DEGNON (QR)	N=201	55.9	537.0	9.6
DEGNON** (QR)	N=201	2.0	20.2	10.1
DEGNON (LR)	N=201	19.4	292.6	15.1
DLPRG	M=251,N=502	8.2	169.9	20.7
DFFT	N=2^{22}	33.2	46.4	1.4

*symmetric matrix, with QR modified as suggested in comments
**with shifts, $\sigma_n = a_{mm}$, where $m = N - \text{mod}\ (n/10, N)$

When DLINEQ was used to solve a linear system with N=1001 unknowns, the Cray SV1 compiler vectorized loop 25, and the program ran in 3.2 seconds. Since solving an N by N linear system requires about $\frac{2}{3}N^3$ floating point operations, that means DLINEQ is doing about $\frac{2}{3}N^3/3.2$ calculations per second, or about 210 million floating point operations per second (210 "megaflops"). When vectorization was turned off, the program required 62.1 seconds CPU time. Thus vectorization made the entire program run about 19 times faster. Vectorization also sped up the other programs by factors of at least 5, with two exceptions. One is DEGNON in the symmetric case, where the inner DO loops in QR are shortened to take advantage of the tridiagonal structure of the reduced matrix. The other exception is DFFT; the Cray FORTRAN90 compiler was able to vectorize loop 10, even though it appears to have a dependency, because D is defined recursively by D = D*H. Apparently the compiler is able to see that D is just equal to H^{K-1}, and thus it can remove the recursion and vectorize the loop. Nevertheless, the speed-up ratio is low, perhaps because most trips through this loop are short, or because of the overhead associated with the recursive subroutine calls.

Notice that although, without vectorization, DLLSQR ran faster with Householder transformations (REDH) than with Givens transformations (REDQ) (as the operation count would predict), with vectorization REDQ was faster. Also notice that the QR method, even without shifts, was faster than the Jacobi method on the symmetric problem, but the LR method was faster than QR on the nonsymmetric problem. Implementing shifts improved the performance of the QR method dramatically.

No modifications to the original unoptimized programs were necessary to achieve these large speed-up ratios; the compiler verified that the computations in critical innermost DO loops (e.g., loop 25 of DLINEQ) were mutually independent and automatically vectorized the loops. Sometimes the programmer can improve performance by rewriting the code, for example, to reverse the order of two DO loops to make the innermost loop the longer one, or to remove a dependency that would otherwise prevent vectorization of the innermost loop. But codes for linear algebra applications usually contain inner loops that are easy for these compilers to vectorize efficiently, and the process is generally pretty automatic, in contrast to the situation in the next section.

It should be mentioned that when N is changed from 1001 to 1000, DLINEQ runs 5 times slower, with vectorization on. (The other routines exhibit similar behavior.) The reason for this strange behavior has to do with the "stride" through the matrices in the innermost DO loops. The stride is the distance between successively referenced elements of an array. For reasons beyond the scope of this book, nearly all modern computers perform best when strides are equal to 1, and worst when they are a power of 2. We will see in the next section how the performance of DLINEQ can be improved by reversing the order of loops 25 and 30, to decrease the stride through the innermost loop from N to 1.

6.2 Parallel Computers

The speed-up factor available through vectorization of a loop such as 25 is apparently limited by the number of subtasks into which the calculation $A(J,K) = A(J,K) - LJI * A(I,K)$ can be split. If we hope to get higher speed-ups we will have to take another approach, which generally involves much more programming effort, and actually run the calculations simultaneously on the different processors of a multiprocessor system. Such systems can be thought of as consisting of several autonomous computers, each with its own memory (distributed memory machine) or at least its own section of memory (shared memory machine), with the ability to pass data back and forth between computers. Although much effort has been put into maximizing the communication speed between processors, communication is still very slow compared to the speed with which the data are processed internally within a processor, so it is absolutely critical to minimize the amount of "message passing" between processors.

There are a number of libraries that provide routines that can be called by user programs to pass messages back and forth between processors. The MPI library [Pacheco 1996; Bisseling 2004] is the most widely used set of message passing routines, so we will use these routines in our examples. Figure 6.2.1 shows a routine PLINEQ, which uses Gaussian elimination with partial pivoting to solve a linear system. The program is basically a parallel version of DLINEQ (Figure 1.2.1), although it does not save the LU decomposition of A, and there are a few other minor differences.

In this program, the same code is run on each processor; however, note the call to MPI_COMM_RANK, which returns the processor number ITASK (=0,1,...,$NPES-1$), where NPES is the number of processors, so different actions can be taken on different processors. All of the examples and problems in this chapter can be done using (in addition to the initialization/finalization routines) only the two MPI routines:

MPI_BCAST(X,N,MPI_prec,JTASK,MPI_COMM_WORLD,IERR)

MPI_ALLREDUCE(XI,X,N,MPI_prec,MPI_op,MPI_COMM_WORLD,IERR)

MPI_BCAST receives the vector X, of length N, from processor JTASK (no action is taken if the local processor is JTASK). MPI_ALLREDUCE collects the vectors XI, of length N, from each processor, adds them together (if MPI_op=MPI_SUM) or finds the maximum or minimum (element by element) (if MPI_op=MPI_MAX or MPI_MIN), and returns the resulting vector X to all processors. MPI_prec can be MPI_DOUBLE_PRECISION, MPI_REAL or MPI_INTEGER. The variables MPI_DOUBLE_PRECISION, MPI_SUM, MPI_COMM_WORLD, etc. are assigned values in the include file 'mpif.h'. IERR is the output error flag.

Ideally, the columns of A should be distributed cyclically over the processors, that is, columns 1, NPES+1, 2*NPES+1,... should be stored only on processor zero, while columns 2, NPES+2, 2*NPES+2,... should be stored only on processor one, and so on. In our version, each processor holds the entire N by N matrix A in memory, but it never touches any but its "own" columns. If PLINEQ were written so that each processor stored only its own columns (Problem 2b), the program would be more efficient with regard to memory usage, but it would be more difficult to read and use. PLINEQ could be made memory-efficient by simply replacing each reference to $A(I,J)$ by $A(I,(J-1)/NPES+1)$, and dimensioning $A(N,*)$. The problem is, this means the PLINEQ "user" must dimension $A(N,(N-1)/NPES+1)$ and distribute the matrix over the available processors in the calling program; for example, he/she could define A like this:

```
            DO 10 I=1,N
               DO 5 J=ITASK+1,N,NPES
                  A(I,(J-1)/NPES+1) = [element (I,J) of matrix]
    5          CONTINUE
   10       CONTINUE
```

This is exactly what normally should be done, but the version in Figure 6.2.1 is easier for the student to follow and does not require the PLINEQ user to know anything about parallel programming. As written, the program is very wasteful in its use of memory, but it does take good advantage of multiple processors as far as computation time is concerned.

```
      SUBROUTINE PLINEQ(A,N,X,B)
      IMPLICIT DOUBLE PRECISION (A-H,O-Z)
C                                  DECLARATIONS FOR ARGUMENTS
      DOUBLE PRECISION A(N,N),X(N),B(N)
C                                  DECLARATIONS FOR LOCAL VARIABLES
      DOUBLE PRECISION B_(N),LJI,COLUMNI(N)
      INCLUDE 'mpif.h'
C
C  SUBROUTINE PLINEQ SOLVES THE LINEAR SYSTEM A*X=B
C
C  ARGUMENTS
C
C             ON INPUT                          ON OUTPUT
C             --------                          ---------
C
C    A      - THE N BY N COEFFICIENT MATRIX.    DESTROYED
C
C    N      - THE SIZE OF MATRIX A.
C
C    X      -                                   AN N-VECTOR CONTAINING
C                                               THE SOLUTION.
C
C    B      - THE RIGHT HAND SIDE N-VECTOR.
C
C-----------------------------------------------------------------------
C                                INITIALIZE MPI
      CALL MPI_INIT (IERR)
C                                NPES = NUMBER OF PROCESSORS
      CALL MPI_COMM_SIZE (MPI_COMM_WORLD,NPES,IERR)
C                                ITASK = MY PROCESSOR NUMBER (0,1,...,NPES-1)
C                                I WILL NEVER TOUCH ANY COLUMNS OF A EXCEPT
C                                MY COLUMNS, ITASK+1+ K*NPES, K=0,1,2,...
```

```
      CALL MPI_COMM_RANK (MPI_COMM_WORLD,ITASK,IERR)
C                                COPY B TO B_, SO B WILL NOT BE ALTERED
      B_(1:N) = B(1:N)
C                                BEGIN FORWARD ELIMINATION
      DO 35 I=1,N
C                                JTASK IS PROCESSOR THAT OWNS ACTIVE COLUMN
         JTASK = MOD(I-1,NPES)
         IF (ITASK.EQ.JTASK) THEN
C                                IF JTASK IS ME, SAVE ACTIVE COLUMN IN
C                                VECTOR COLUMNI
            DO 10 J=I,N
               COLUMNI(J) = A(J,I)
   10       CONTINUE
         ENDIF
C                                RECEIVE COLUMNI FROM PROCESSOR JTASK
         CALL MPI_BCAST(COLUMNI(I),N-I+1,MPI_DOUBLE_PRECISION,JTASK,
     &        MPI_COMM_WORLD,IERR)
C                                SEARCH FROM A(I,I) ON DOWN FOR LARGEST
C                                POTENTIAL PIVOT, A(L,I)
         BIG = ABS(COLUMNI(I))
         L = I
         DO 15 J=I+1,N
            IF (ABS(COLUMNI(J)).GT.BIG) THEN
               BIG = ABS(COLUMNI(J))
               L = J
            ENDIF
   15    CONTINUE
C                                IF LARGEST POTENTIAL PIVOT IS ZERO,
C                                MATRIX IS SINGULAR
         IF (BIG.EQ.0.0) GO TO 50
C                                I0 IS FIRST COLUMN >= I THAT BELONGS TO ME
         L0 = (I-1+NPES-(ITASK+1))/NPES
         I0 = ITASK+1+L0*NPES
C                                SWITCH ROW I WITH ROW L, TO BRING UP
C                                LARGEST PIVOT; BUT ONLY IN MY COLUMNS
         DO 20 K=I0,N,NPES
            TEMP = A(L,K)
            A(L,K) = A(I,K)
            A(I,K) = TEMP
   20    CONTINUE
         TEMP = COLUMNI(L)
         COLUMNI(L) = COLUMNI(I)
         COLUMNI(I) = TEMP
C                                SWITCH B_(I) AND B_(L)
         TEMP = B_(L)
         B_(L) = B_(I)
         B_(I) = TEMP
         DO 30 J=I+1,N
```

```
C                                CHOOSE MULTIPLIER TO ZERO A(J,I)
            LJI = COLUMNI(J)/COLUMNI(I)
            IF (LJI.NE.0.0) THEN
C                                SUBTRACT LJI TIMES ROW I FROM ROW J;
C                                BUT ONLY IN MY COLUMNS
               DO 25 K=I0,N,NPES
                  A(J,K) = A(J,K) - LJI*A(I,K)
   25          CONTINUE
C                                SUBTRACT LJI TIMES B_(I) FROM B_(J)
               B_(J) = B_(J) - LJI*B_(I)
            ENDIF
   30    CONTINUE
   35 CONTINUE
C                                SOLVE U*X=B_ USING BACK SUBSTITUTION.
      DO 45 I=N,1,-1
C                                I0 IS FIRST COLUMN >= I+1 THAT BELONGS
C                                TO ME
         L0 = (I+NPES-(ITASK+1))/NPES
         I0 = ITASK+1+L0*NPES
         SUMI = 0.0
         DO 40 J=I0,N,NPES
            SUMI = SUMI + A(I,J)*X(J)
   40    CONTINUE
         CALL MPI_ALLREDUCE(SUMI,SUM,1,MPI_DOUBLE_PRECISION,
     &    MPI_SUM,MPI_COMM_WORLD,IERR)
C                                JTASK IS PROCESSOR THAT OWNS A(I,I)
         JTASK = MOD(I-1,NPES)
C                                IF JTASK IS ME, CALCULATE X(I)
         IF (ITASK.EQ.JTASK) THEN
            X(I) = (B_(I)-SUM)/A(I,I)
         ENDIF
C                                RECEIVE X(I) FROM PROCESSOR JTASK
         CALL MPI_BCAST(X(I),1,MPI_DOUBLE_PRECISION,JTASK,
     &   MPI_COMM_WORLD,IERR)
   45 CONTINUE
      GO TO 60
   50 IF (ITASK.EQ.0) PRINT 55
   55 FORMAT ('***** THE MATRIX IS SINGULAR *****')
C                                CLOSE MPI
   60 CONTINUE
      CALL MPI_FINALIZE(IERR)
      RETURN
      END
```

Figure 6.2.1

Figure 6.2.2 illustrates how the forward elimination proceeds when the matrix is distributed over the processors in this way, when we have NPES=3 processors. After the first 2 columns have been zeroed, the "active" column is column $I = 3$, which "belongs" to processor 2. Now we need to switch row 3 with the row corresponding to the largest potential pivot (in absolute value) of the active column, and then take a multiple of the 3rd row and add it to the 4th row, another multiple of the 3rd row to add to the 5th row, and so on. Processor 0 can do its share of these row operations, to "its" columns 4, 7,..., and processor 1 can do its share, to its columns (5, 8,...), but each has to see the active column before it can know which row to switch with row 3, and what multiples of row 3 to add to rows 4, 5,... . So processor 2 has to "broadcast" the active column to the other processors before they can proceed.

```
                          *
processor no. -   0 1 2 0 1 2 0 1 2 ...
column no. -      1 2 3 4 5 6 7 8 9 ...

                  --------------------
                  x x x x x x x x x ...
                  0 x x x x x x x x
                * 0 0 x x x x x x x
                  0 0 x x x x x x x
                  0 0 x x x x x x x
                  0 0 x x x x x x x
                  0 0 x x x x x x x
                  0 0 x x x x x x x
                  0 0 x x x x x x x ...
                  .               .
                  .               .
```

Figure 6.2.2

When the size N of the matrix is large, each processor has many columns to process and only one to receive (from whomever has the active column); thus the amount of communication is small compared to the amount of work done between communications. Since, for large N, the process of adding multiples of one row to others consumes nearly 100% of the computer time when Gaussian elimination is done, distributing this process over the available processors, allowing all processors to work simultaneously on their parts of the task, ensures that the overall computation time is greatly diminished, compared to doing the whole elimination on one processor. In fact, each processor will have approximately the same number of columns to work on, so the total work should be decreased by approximately a factor of NPES, when N is large. Note that loop 25 in Figure 6.2.1 now looks like:

```
      DO 25 K=I0,N,NPES
         A(J,K) = A(J,K) - LJI*A(I,K)
   25 CONTINUE
```

so on any given processor, the work in this critical loop has decreased by a factor of about NPES.

After the forward elimination, back substitution is used to find X; the calculation of SUM in loop 40 is distributed over the processors, then the portions of this sum held on different processors must be collected together before the processor that owns A(I,I) can solve for X(I). This is done using MPI_ALLREDUCE.

A band solver such as DBAND (Figure 1.5.4) could be modified to take advantage of multiple processors in a very similar way. In fact, if we simply change the upper limits of loops 10,15 and 30 to MIN(I+NLD,N) and loops 20,25 and 40 to MIN(I+NUD+NLD,N), PLINEQ will solve banded systems efficiently in parallel, though it will still store the entire matrix.

When PLINEQ was used to solve linear systems of N=2000 and 4000 unknowns, with elements generated by a random number generator, on a Linux cluster of Intel Xeon processors, at the University of Texas El Paso, the results were as shown in Table 6.2.1.

Table 6.2.1
PLINEQ Results

$N = 2000$		$N = 4000$	
NPES	Time(sec)	NPES	Time(sec)
1	19.7	1	200
2	10.1	2	109
4	5.1	4	57
8	2.9	8	34
16	2.5	16	34
32	3.2	32	35

For both problems, PLINEQ "scales" well for a while; that means doubling the number of processors approximately doubles the speed. But eventually, adding more processors failed to improve the speed. This is in part because more processors means more messages between processors, but even if the communication time were negligible there would always be a limit to how fast the code could run as NPES is increased. This is because, although the most expensive section of the code is parallelized (the only part with an O(N^3) operation count, loop 25), there are other sections of the code with O(N^2) or lower operation counts that are not parallelized, so even if there are so many processors that loop 25 uses no time at all, the total time to solve the linear system will never fall below the time required to execute the scalar sections. If a code requires T_1 seconds to run using one processor, and f_p is the fraction of the calculations which can be parallelized, then it will require $f_pT_1/\text{NPES}+(1-f_p)T_1$ seconds to run on NPES processors, assuming message passing overhead to be negligible. Thus "Amdahl's law" [Leiss 1995] tells us that the best speed-up factor we can hope for is given by:

$$S = \frac{T_1}{f_pT_1/\text{NPES} + (1-f_p)T_1} = \frac{\text{NPES}}{f_p + (1-f_p)\text{NPES}}$$

Amdahl's law imposes an upper limit of $1/(1-f_p)$ as NPES $\to \infty$ on the speed-up attainable, which ensures that any program will eventually quit improving as NPES increases; however, usually the larger the size of the problem, the later this occurs, since f_p normally increases with problem size.

As mentioned in the last section, most processors, including these, perform better when DO loops operate on arrays with unit strides. Since FORTRAN matrices are stored by columns, this means that innermost DO loops should, when possible, operate on the columns, not the rows, of matrices, so the stride will be equal to 1 rather than N. In fact, when we replace loop 30 in PLINEQ with the following code:

```
      DO 30 K=I0,N,NPES
         LJI = A(I,K)/COLUMNI(I)
         IF (LJI.NE.0.0) THEN
            DO 25 J=I+1,N
               A(J,K) = A(J,K) - LJI*COLUMNI(J)
25          CONTINUE
         ENDIF
30    CONTINUE
      LJI = B_(I)/COLUMNI(I)
      DO 31 J=I+1,N
         B_(J) = B_(J) - LJI*COLUMNI(J)
31    CONTINUE
```

which basically just reverses the order of loops 25 and 30, we get the better results shown in Table 6.2.2. This time the columns of A were physically distributed over the processors, as suggested above, but this had no measurable effect on the speed, only on the memory requirements.

For the case $N = 8000, NPES = 8$, our program is now doing about $\frac{2}{3}N^3/113$ operations per second, or 3.0 gigaflops. Replacing loops 25 and 30 by a call to the assembly-language BLAS (Basic Linear Algebra Subroutines) [Dongarra et al. 1988] routine SGER gave no further improvements.

A parallel version, PLPRG, of the linear programming routine of Figure 4.5.1 is given in Figure 6.2.3. Again, each processor ITASK holds the entire tableaux, TAB, in memory, but only touches its own columns ITASK+1+ k*NPES, k=0,1,2..., except in the initializations, where speed is not important. However, the last column, B, is common to all processors. Thus, like

Table 6.2.2
PLINEQ Results with Stride=1

$N = 4000$		$N = 8000$	
NPES	Time(sec)	NPES	Time(sec)
1	103	1	812
2	53	2	419
4	27	4	214
8	14	8	113
16	13	16	105
32	17	32	103

PLINEQ, PLPRG takes advantage of multiple processors to decrease the computation time, but not to decrease the memory usage. The last row is spread out over the processors, and the portions on the different processors must be merged (using MPI_ALLREDUCE) so it can be inspected to find the pivot column, JP. The pivot column must be broadcast, by whoever owns it, to the other processors, who save it in array COLJP. Now when a multiple of the pivot row is added to the other rows, each processor processes its own columns, and so the work is almost evenly distributed over the processors. In fact, the inner loop 80, where all computer time is spent in the limit of large problem size, now has limits J=ITASK+1,N+M,NPES.

```
      SUBROUTINE PLPRG(A,B,C,N,M,P,X,Y)
      IMPLICIT DOUBLE PRECISION (A-H,O-Z)
C                                 DECLARATIONS FOR ARGUMENTS
      DOUBLE PRECISION A(M,N),B(M),C(N),P,X(N),Y(M)
      INTEGER N,M
C                                 DECLARATIONS FOR LOCAL VARIABLES
      DOUBLE PRECISION TAB(M+2,N+M+1),COLJP(M+2),LROWI(N+M),
     &  LROW(N+M)
      INTEGER BASIS(M)
      include 'mpif.h'
C
C  SUBROUTINE PLPRG USES THE SIMPLEX METHOD TO SOLVE THE PROBLEM
C
C             MAXIMIZE      P = C(1)*X(1) + ... + C(N)*X(N)
C
C   WITH X(1),...,X(N) NONNEGATIVE, AND
C
C           A(1,1)*X(1) + ... + A(1,N)*X(N)   =  B(1)
C             .                   .                .
C             .                   .                .
C           A(M,1)*X(1) + ... + A(M,N)*X(N)   =  B(M)
C
```

```
C    WHERE B(1),...,B(M) ARE ASSUMED TO BE NONNEGATIVE.
C
C  ARGUMENTS
C
C            ON INPUT                          ON OUTPUT
C            --------                          ---------
C
C    A      - THE M BY N CONSTRAINT COEFFICIENT
C            MATRIX.
C
C    B      - A VECTOR OF LENGTH M CONTAINING
C            THE RIGHT HAND SIDES OF THE
C            CONSTRAINTS.  THE COMPONENTS OF
C            B MUST ALL BE NONNEGATIVE.
C
C    C      - A VECTOR OF LENGTH N CONTAINING
C            THE COEFFICIENTS OF THE OBJECTIVE
C            FUNCTION.
C
C    N      - THE NUMBER OF UNKNOWNS.
C
C    M      - THE NUMBER OF CONSTRAINTS.
C
C    P      -                                  THE MAXIMUM OF THE
C                                              OBJECTIVE FUNCTION.
C
C    X      -                                  A VECTOR OF LENGTH N
C                                              WHICH CONTAINS THE LP
C                                              SOLUTION.
C
C    Y      -                                  A VECTOR OF LENGTH M
C                                              WHICH CONTAINS THE DUAL
C                                              SOLUTION.
C
C-----------------------------------------------------------------------
C                            EPS = MACHINE FLOATING POINT RELATIVE
C                                  PRECISION
C *****************************
      DATA EPS/2.D-16/
C *****************************
C                            INITIALIZE MPI
      CALL MPI_INIT (IERR)
C                            NPES = NUMBER OF PROCESSORS
      CALL MPI_COMM_SIZE (MPI_COMM_WORLD,NPES,IERR)
C                            ITASK = MY PROCESSOR NUMBER (0,1,...,NPES-1).
C                            I WILL NEVER TOUCH ANY COLUMNS OF TAB EXCEPT
C                            MY COLUMNS, ITASK+1+ K*NPES, K=0,1,2,...
C                            (EXCEPT IN INITIALIZATION STAGE)
```

```
      CALL MPI_COMM_RANK (MPI_COMM_WORLD,ITASK,IERR)
C                                 BASIS(1),...,BASIS(M) HOLD NUMBERS OF
C                                 BASIS VARIABLES.  INITIAL BASIS CONSISTS
C                                 OF ARTIFICIAL VARIABLES ONLY
      DO 5 I=1,M
         BASIS(I) = N+I
         IF (B(I).LT.0.0) THEN
            IF (ITASK.EQ.0) PRINT 1
    1       FORMAT (' ***** ALL B(I) MUST BE NONNEGATIVE *****')
            GO TO 120
         ENDIF
    5 CONTINUE
C                                 INITIALIZE SIMPLEX TABLEAU
      DO 10 I=1,M+2
      DO 10 J=1,N+M+1
         TAB(I,J) = 0.0
   10 CONTINUE
C                                 LOAD A INTO UPPER LEFT HAND CORNER
C                                 OF TABLEAU
      DO 15 I=1,M
      DO 15 J=1,N
         TAB(I,J) = A(I,J)
   15 CONTINUE
C                                 LOAD M BY M IDENTITY TO RIGHT OF A
C                                 AND LOAD B INTO LAST COLUMN
      DO 20 I=1,M
         TAB(I,N+I) = 1.0
         TAB(I,N+M+1) = B(I)
   20 CONTINUE
C                                 ROW M+1 CONTAINS -C, INITIALLY
      DO 25 J=1,N
         TAB(M+1,J) = -C(J)
   25 CONTINUE
C                                 ROW M+2 CONTAINS COEFFICIENTS OF
C                                 "ALPHA", WHICH IS TREATED AS +INFINITY
      DO 30 I=1,M
         TAB(M+2,N+I) = 1.0
   30 CONTINUE
C                                 CLEAR "ALPHAS" IN LAST ROW
      DO 35 I=1,M
      DO 35 J=1,N+M+1
         TAB(M+2,J) = TAB(M+2,J) - TAB(I,J)
   35 CONTINUE
C                                 SIMPLEX METHOD CONSISTS OF TWO PHASES
      DO 90 IPHASE=1,2
         IF (IPHASE.EQ.1) THEN
C                                 PHASE I:  ROW M+2 (WITH COEFFICIENTS OF
C                                 ALPHA) SEARCHED FOR MOST NEGATIVE ENTRY
```

```
            MROW = M+2
            LIM = N+M
         ELSE
C                                PHASE II:  FIRST N ELEMENTS OF ROW M+1
C                                SEARCHED FOR MOST NEGATIVE ENTRY
C                                (COEFFICIENTS OF ALPHA NONNEGATIVE NOW)
            MROW = M+1
            LIM = N
C                                IF ANY ARTIFICIAL VARIABLES LEFT IN
C                                BASIS AT BEGINNING OF PHASE II, THERE
C                                IS NO FEASIBLE SOLUTION
            DO 45 I=1,M
               IF (BASIS(I).GT.LIM) THEN
                  IF (ITASK.EQ.0) PRINT 40
   40             FORMAT (' ***** NO FEASIBLE SOLUTION *****')
                  GO TO 120
               ENDIF
   45       CONTINUE
         ENDIF
C                                THRESH = SMALL NUMBER.  WE ASSUME SCALES
C                                OF A AND C ARE NOT *TOO* DIFFERENT
         THRESHI = 0.0
         DO 50 J=ITASK+1,LIM,NPES
            THRESHI = MAX(THRESHI,ABS(TAB(MROW,J)))
   50    CONTINUE
         CALL MPI_ALLREDUCE(THRESHI,THRESH,1,MPI_DOUBLE_PRECISION,
     &   MPI_MAX,MPI_COMM_WORLD,IERR)
         THRESH = 1000*EPS*THRESH
C                                BEGINNING OF SIMPLEX STEP
   55    CONTINUE
C                                COLLECT PORTIONS (LROWI) OF LAST ROW FROM
C                                DIFFERENT PROCESSORS AND MERGE THEM
C                                INTO LROW, USING MPI_ALLREDUCE.
            DO 56 J=1,LIM
               LROWI(J) = 0
               IF (MOD(J-1,NPES).EQ.ITASK) LROWI(J) = TAB(MROW,J)
   56       CONTINUE
            CALL MPI_ALLREDUCE(LROWI,LROW,LIM,MPI_DOUBLE_PRECISION,
     &      MPI_SUM,MPI_COMM_WORLD,IERR)
C                                FIND MOST NEGATIVE ENTRY IN ROW MROW,
C                                IDENTIFYING PIVOT COLUMN JP.
            CMIN = -THRESH
            JP = 0
            DO 60 J=1,LIM
               IF (LROW(J).LT.CMIN) THEN
                  CMIN = LROW(J)
                  JP = J
               ENDIF
```

```
   60      CONTINUE
C                               IF ALL ENTRIES NONNEGATIVE (ACTUALLY,
C                               IF GREATER THAN -THRESH) PHASE ENDS
           IF (JP.EQ.0) GO TO 90
C                               IF I OWN COLUMN JP, SAVE IT IN COLJP
           JTASK = MOD(JP-1,NPES)
           IF (ITASK.EQ.JTASK) THEN
              DO 61 I=1,MROW
                 COLJP(I) = TAB(I,JP)
   61         CONTINUE
           ENDIF
C                               RECEIVE COLJP FROM PROCESSOR THAT OWNS IT
           CALL MPI_BCAST(COLJP,MROW,MPI_DOUBLE_PRECISION,
     &     JTASK,MPI_COMM_WORLD,IERR)
C                               FIND SMALLEST POSITIVE RATIO
C                               B(*)/TAB(*,JP), IDENTIFYING PIVOT
C                               ROW IP
           RATMIN = 0.0
           IP = 0
           DO 65 I=1,M
              IF (COLJP(I).GT.THRESH) THEN
                 RATIO = TAB(I,N+M+1)/COLJP(I)
                 IF (IP.EQ.0 .OR. RATIO.LT.RATMIN) THEN
                    RATMIN = RATIO
                    IP = I
                 ENDIF
              ENDIF
   65      CONTINUE
C                               IF ALL RATIOS NONPOSITIVE, MAXIMUM
C                               IS UNBOUNDED
           IF (IP.EQ.0) THEN
              IF (ITASK.EQ.0) PRINT 70
   70         FORMAT (' ***** UNBOUNDED MAXIMUM *****')
              GO TO 120
           ENDIF
C                               ADD X(JP) TO BASIS
           BASIS(IP) = JP
C                               NORMALIZE PIVOT ROW TO MAKE TAB(IP,JP)=1
           AMULT = 1.0/COLJP(IP)
           DO 75 J=ITASK+1,N+M,NPES
              TAB(IP,J) = AMULT*TAB(IP,J)
   75      CONTINUE
           TAB(IP,N+M+1) = AMULT*TAB(IP,N+M+1)
C                               ADD MULTIPLES OF PIVOT ROW TO OTHER
C                               ROWS, TO KNOCK OUT OTHER ELEMENTS IN
C                               PIVOT COLUMN
           DO 85 I=1,MROW
              IF (I.EQ.IP) GO TO 85
```

```
               AMULT = COLJP(I)
               DO 80 J=ITASK+1,N+M,NPES
                  TAB(I,J) = TAB(I,J) - AMULT*TAB(IP,J)
   80          CONTINUE
C                             MODIFY B(I)
               TAB(I,N+M+1) = TAB(I,N+M+1) - AMULT*TAB(IP,N+M+1)
   85       CONTINUE
         GO TO 55
C                             END OF SIMPLEX STEP
   90 CONTINUE
C                             END OF PHASE II; READ X,P,Y FROM
C                             FINAL TABLEAU
      DO 95 J=1,N
         X(J) = 0.0
   95 CONTINUE
      DO 100 I=1,M
         K = BASIS(I)
         X(K) = TAB(I,N+M+1)
  100 CONTINUE
      P = TAB(M+1,N+M+1)
C                             COLLECT PORTIONS (LROWI) OF LAST ROW FROM
C                             DIFFERENT PROCESSORS AND MERGE THEM
C                             INTO LROW, USING MPI_ALLREDUCE.
      DO 105 J=1,N+M
         LROWI(J) = 0.0
         IF (MOD(J-1,NPES).EQ.ITASK) LROWI(J) = TAB(M+1,J)
  105 CONTINUE
      CALL MPI_ALLREDUCE(LROWI,LROW,N+M,MPI_DOUBLE_PRECISION,
     & MPI_SUM,MPI_COMM_WORLD,IERR)
      DO 110 I=1,M
         Y(I) = LROW(N+I)
  110 CONTINUE
  120 CONTINUE
      CALL MPI_FINALIZE(IERR)
      RETURN
      END
```

Figure 6.2.3

Table 6.2.3 shows the results on the UTEP Linux cluster when PLPRG was used to solve a resource allocation problem with randomly generated coefficients, with 2000 inequalities and 2000 variables (thus $N = 4000$ after adding slack variables, and $M = 2000$). Loop 85 was replaced with the following code, which essentially reverses the order of loops 80 and 85, so that the stride through the inner loop is 1, resulting in a significant improvement in speed again:

```
          DO 85 J=ITASK+1,N+M,NPES
             AMULT = TAB(IP,J)
             DO 80 I=1,MROW
                IF (I.EQ.IP) GO TO 80
                TAB(I,J) = TAB(I,J) - AMULT*COLJP(I)
   80        CONTINUE
   85     CONTINUE
C                                  MODIFY B
          AMULT = TAB(IP,N+M+1)
          DO 86 I=1,MROW
             IF (I.EQ.IP) GO TO 86
             TAB(I,N+M+1) = TAB(I,N+M+1) - AMULT*COLJP(I)
   86     CONTINUE
```

Table 6.2.3

PLPRG Results with Stride=1

$M = 2000, N = 4000$	
NPES	Time(sec)
1	142
2	76
4	39
8	21
16	21

A parallel version of the preconditioned conjugate-gradient iterative linear system solver of Figure 1.10.1 is shown in Figure 6.2.4. The multiplication of the sparse matrix A by $\boldsymbol{p}_n$ is the most time consuming calculation, and that is very easy to parallelize, because if A_i is the part of A stored on processor i, then $A = A_0 + A_1 + ...$, and so $A\boldsymbol{p}_n = A_0\boldsymbol{p}_n + A_1\boldsymbol{p}_n + \,...$. The other calculations are also parallelized, and although all N elements of X, R, P and AP are stored on each processor, the local processor never touches any but "its" elements, ITASK+1 + k*NPES, k=0,1,2,..., except after convergence, when the parts of the solution vector X stored on the different processors are merged together.

```
      SUBROUTINE PCG(A,IROW,JCOL,NZ,X,B,N,D)
      IMPLICIT DOUBLE PRECISION (A-H,O-Z)
C                                  DECLARATIONS FOR ARGUMENTS
      DOUBLE PRECISION A(NZ),B(N),X(N),D(N)
      INTEGER IROW(NZ),JCOL(NZ)
C                                  DECLARATIONS FOR LOCAL VARIABLES
      DOUBLE PRECISION R(N),P(N),API(N),AP(N),LAMBDA
      include 'mpif.h'
```

```
C
C  SUBROUTINE PCG SOLVES THE SYMMETRIC LINEAR SYSTEM A*X=B, USING THE
C     CONJUGATE GRADIENT ITERATIVE METHOD.  THE NON-ZEROS OF A ARE STORED
C     IN SPARSE FORMAT.  THE COLUMNS OF A ARE DISTRIBUTED CYCLICALLY OVER
C     THE AVAILABLE PROCESSORS.
C
C  ARGUMENTS
C
C             ON INPUT                          ON OUTPUT
C             --------                          ---------
C
C    A      - A(IZ) IS THE MATRIX ELEMENT IN
C             ROW IROW(IZ), COLUMN JCOL(IZ),
C             FOR IZ=1,...,NZ. ELEMENTS WITH
C             MOD(JCOL(IZ)-1,NPES)=ITASK
C             ARE STORED ON PROCESSOR ITASK.
C
C    IROW   - (SEE A).
C
C    JCOL   - (SEE A).
C
C    NZ     - NUMBER OF NONZEROS STORED ON
C             THE LOCAL PROCESSOR.
C
C    X      -                                   AN N-VECTOR CONTAINING
C                                               THE SOLUTION.
C
C    B      - THE RIGHT HAND SIDE N-VECTOR.
C
C    N      - SIZE OF MATRIX A.
C
C    D      - VECTOR HOLDING A DIAGONAL
C             PRECONDITIONING MATRIX.
C             D = DIAGONAL(A) IS RECOMMENDED.
C             D(I) = 1 FOR NO PRECONDITIONING
C
C-----------------------------------------------------------------------
C                                 NPES = NUMBER OF PROCESSORS
      CALL MPI_COMM_SIZE (MPI_COMM_WORLD,NPES,IERR)
C                                 ITASK = MY PROCESSOR NUMBER
      CALL MPI_COMM_RANK (MPI_COMM_WORLD,ITASK,IERR)
C                                 X0 = 0
C                                 R0 = D**(-1)*B
C                                 P0 = R0
      R0MAX = 0
      DO 10 I=1,N
         X(I) = 0
         R(I) = B(I)/D(I)
```

```
         ROMAX = MAX(ROMAX,ABS(R(I)))
         P(I) = R(I)
   10 CONTINUE
C                                   NITER = MAX NUMBER OF ITERATIONS
      NITER = 3*N
      DO 90 ITER=1,NITER
C                                   AP = A*P
         DO 20 I=1,N
            API(I) = 0
   20    CONTINUE
         DO 30 IZ=1,NZ
            I = IROW(IZ)
            J = JCOL(IZ)
            API(I) = API(I) + A(IZ)*P(J)
   30    CONTINUE
C                                   MPI_ALLREDUCE COLLECTS THE VECTORS API
C                                   (API = LOCAL(A)*P) FROM ALL PROCESSORS
C                                   AND ADDS THEM TOGETHER, THEN SENDS
C                                   THE RESULT, AP, BACK TO ALL PROCESSORS.
         CALL MPI_ALLREDUCE(API,AP,N,MPI_DOUBLE_PRECISION,
     &    MPI_SUM,MPI_COMM_WORLD,IERR)
C                                   PAP = (P,AP)
C                                   RP = (R,D*P)
         PAPI = 0.0
         RPI = 0.0
         DO 40 I=ITASK+1,N,NPES
            PAPI = PAPI + P(I)*AP(I)
            RPI = RPI + R(I)*D(I)*P(I)
   40    CONTINUE
         CALL MPI_ALLREDUCE(PAPI,PAP,1,MPI_DOUBLE_PRECISION,
     &    MPI_SUM,MPI_COMM_WORLD,IERR)
         CALL MPI_ALLREDUCE(RPI,RP,1,MPI_DOUBLE_PRECISION,
     &    MPI_SUM,MPI_COMM_WORLD,IERR)
C                                   LAMBDA = (R,D*P)/(P,AP)
         LAMBDA = RP/PAP
C                                   X = X + LAMBDA*P
C                                   R = R - LAMBDA*D**(-1)*AP
         DO 50 I=ITASK+1,N,NPES
            X(I) = X(I) + LAMBDA*P(I)
            R(I) = R(I) - LAMBDA*AP(I)/D(I)
   50    CONTINUE
C                                   RAP = (R,AP)
         RAPI = 0.0
         DO 60 I=ITASK+1,N,NPES
            RAPI = RAPI + R(I)*AP(I)
   60    CONTINUE
         CALL MPI_ALLREDUCE(RAPI,RAP,1,MPI_DOUBLE_PRECISION,
     &    MPI_SUM,MPI_COMM_WORLD,IERR)
```

```
C                                       ALPHA = -(R,AP)/(P,AP)
         ALPHA = -RAP/PAP
C                                       P = R + ALPHA*P
         DO 70 I=ITASK+1,N,NPES
            P(I) = R(I) + ALPHA*P(I)
   70    CONTINUE
C                                       RMAX = MAX OF RESIDUAL (R)
         RMAXI = 0
         DO 80 I=ITASK+1,N,NPES
            RMAXI = MAX(RMAXI,ABS(R(I)))
   80    CONTINUE
         CALL MPI_ALLREDUCE(RMAXI,RMAX,1,MPI_DOUBLE_PRECISION,
     &    MPI_MAX,MPI_COMM_WORLD,IERR)
C                                       IF CONVERGED, MERGE PORTIONS OF X
C                                       STORED ON DIFFERENT PROCESSORS
         IF (RMAX.LE.1.D-10*R0MAX) THEN
            IF (ITASK.EQ.0) PRINT *, ' Number of iterations = ',ITER
            CALL MPI_ALLREDUCE(X,R,N,MPI_DOUBLE_PRECISION,
     &       MPI_SUM,MPI_COMM_WORLD,IERR)
            X(1:N) = R(1:N)
            RETURN
         ENDIF
   90 CONTINUE
C                                       PCG DOES NOT CONVERGE
      IF (ITASK.EQ.0) PRINT 100
  100 FORMAT('***** PCG DOES NOT CONVERGE *****')
      RETURN
      END
```

Figure 6.2.4

PCG, with $D(I) = 1$, is used in Figure 6.2.5 to solve the symmetric, positive definite, linear system 1.9.10. The columns are distributed over the processors in the main program, and the solution at the box center is output after convergence, to check the result. With $M = 100 (N = 970299)$, it takes 19.6 seconds with one processor, 12.7 seconds with two, 8.8 seconds with four, and 7.6 seconds with eight processors.

```
      PARAMETER (M=100,N=(M-1)**3,NZMAX=7*N)
      IMPLICIT DOUBLE PRECISION (A-H,O-Z)
      DIMENSION IROW(NZMAX),JCOL(NZMAX),AS(NZMAX),X(N),B(N),D(N)
      INCLUDE 'mpif.h'
C                       INITIALIZE MPI
      CALL MPI_INIT (IERR)
C                       NPES = NUMBER OF PROCESSORS
      CALL MPI_COMM_SIZE (MPI_COMM_WORLD,NPES,IERR)
C                       ITASK = MY PROCESSOR NUMBER
      CALL MPI_COMM_RANK (MPI_COMM_WORLD,ITASK,IERR)
```

```
C                    SOLVE 1.9.10 USING CONJUGATE GRADIENT ITERATION
      H = 1.d0/M
      L = 0
      NZ = 0
      DO 10 I=1,M-1
      DO 10 J=1,M-1
      DO 10 K=1,M-1
         L = L+1
         IF (MOD(L-1,NPES).EQ.ITASK) THEN
C                    DISTRIBUTE L-TH COLUMN TO PROCESSOR MOD(L-1,NPES)
            NZ = NZ + 1
            IROW(NZ) = L
            JCOL(NZ) = L
            AS(NZ) = 6
            IF (K.NE.1) THEN
               NZ = NZ + 1
               IROW(NZ) = L-1
               JCOL(NZ) = L
               AS(NZ) = -1
            ENDIF
            IF (K.NE.M-1) THEN
               NZ = NZ + 1
               IROW(NZ) = L+1
               JCOL(NZ) = L
               AS(NZ) = -1
            ENDIF
            IF (J.NE.1) THEN
               NZ = NZ + 1
               IROW(NZ) = L-(M-1)
               JCOL(NZ) = L
               AS(NZ) = -1
            ENDIF
            IF (J.NE.M-1) THEN
               NZ = NZ + 1
               IROW(NZ) = L+(M-1)
               JCOL(NZ) = L
               AS(NZ) = -1
            ENDIF
            IF (I.NE.1) THEN
               NZ = NZ + 1
               IROW(NZ) = L-(M-1)**2
               JCOL(NZ) = L
               AS(NZ) = -1
            ENDIF
            IF (I.NE.M-1) THEN
               NZ = NZ + 1
               IROW(NZ) = L+(M-1)**2
               JCOL(NZ) = L
```

```
            AS(NZ) = -1
         ENDIF
      ENDIF
      B(L) = H**2
      D(L) = 1.0
   10 CONTINUE
      CALL PCG(AS,IROW,JCOL,NZ,X,B,N,D)
C                          SOLUTION AT BOX CENTER SHOULD BE ABOUT 0.056
      IF (ITASK.EQ.0) PRINT *, ' Solution at midpoint = ',X((N+1)/2)
      CALL MPI_FINALIZE(IERR)
      STOP
      END
```

Figure 6.2.5

PCG was also used to solve a large 3D linear system generated by the finite element program PDE2D (www.pde2d.com), with N=15552 and NZ(total) = 10623956, and the results are seen in Table 6.2.4. The diagonal elements for this matrix vary over a very wide range, so D was set to the diagonal of A. With this "Jacobi" preconditioning, the method converged in 221 iterations; without preconditioning, it did not converge at all, even though the matrix was positive definite.

Table 6.2.4

PCG Results on Finite Element Linear System

$N = 15552, NZ = 10623956$	
NPES	Time(sec)
1	14.7
2	7.4
4	3.8
8	3.4

6.3 Computational Linear Algebra in a PDE Solver

In this section we will see how many of the ideas discussed in this and previous chapters are used in the solution, on a multi-processor machine, of an eigenvalue partial differential equation (PDE),

$$\frac{\partial^2 U}{\partial x^2} + \frac{\partial^2 U}{\partial y^2} + \frac{\partial^2 U}{\partial z^2} - \frac{U}{\sqrt{x^2 + y^2 + z^2}} = \lambda U$$

in part of a torus, shown in Figure 6.3.1. The boundary conditions are $\frac{\partial U}{\partial n} + U = 0$ at one flat end, and $U = 0$ on the rest of the boundary. The computations were done using PDE2D, a finite element program developed by the author over a 35 year period, which solves very general systems of nonlinear steady-state, time-dependent and eigenvalue PDE systems in 1D intervals, general 2D regions, and a wide range of simple 3D regions. Appendix A of [Sewell 2005] contains detailed documentation on the algorithms used by PDE2D. Free (limited size) versions of the program can be downloaded at www.pde2d.com, which also contains a video on PDE2D usage, and a list of over 200 scientific journal publications in which PDE2D has been used to generate numerical results. A Graphical User Interface makes PDE2D exceptionally easy to use.

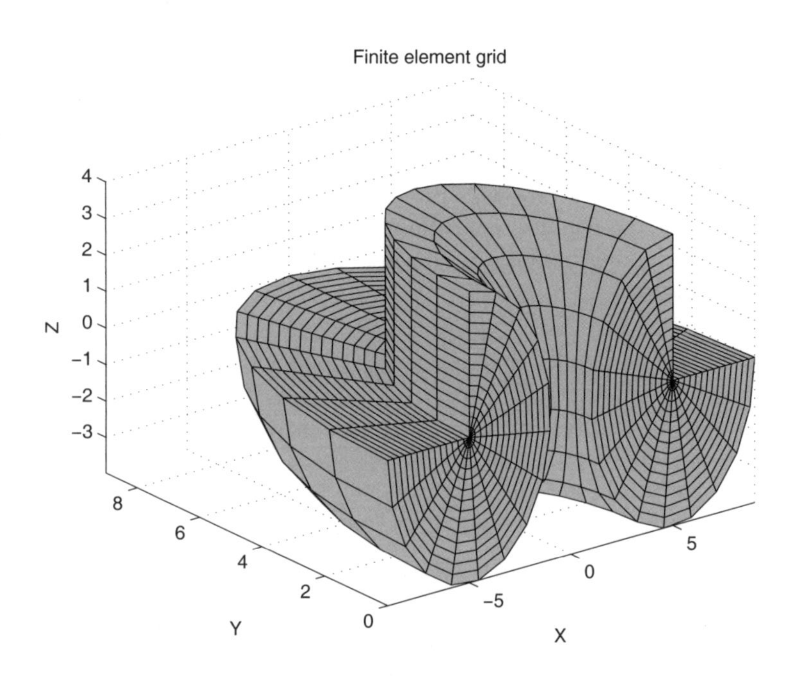

Figure 6.3.1
Domain of PDE

PDE2D uses a finite element method to solve the above eigenvalue PDE, which means (Section 3.6) that it discretizes the PDE and reduces it to a generalized eigenvalue problem $A\boldsymbol{z} = \lambda B\boldsymbol{z}$. The user can provide a number p, and PDE2D will use the shifted inverse power method (Section 3.5) to find the eigenvalue closest to p, and the corresponding eigenvector (eigenfunction). As outlined in Section 3.6, for a generalized problem this means a linear system $(A - pB)\boldsymbol{v}_{n+1} = B\boldsymbol{v}_n$ must be solved every power iteration, where the components of $\boldsymbol{v}_0$ are chosen using a random number generator (alternatively,

$\boldsymbol{v}_0$ may be supplied by the user). The matrices A and B are banded and sparse, and this can be taken advantage of by both direct and iterative methods, and the matrix is the same every power iteration, so direct solvers can save the LU decomposition (Section 1.4) of $A - pB$ calculated the first iteration and use it on subsequent iterations.

When a linear steady-state PDE system is solved, PDE2D must solve one linear system $C\boldsymbol{x} = \boldsymbol{b}$, and when a nonlinear PDE system is solved, a linear system must be solved every Newton iteration. If a time-dependent problem is solved, a linear system must be solved every time step, and for eigenvalue problems, as discussed above, a linear system must be solved every shifted inverse power iteration. For 3D problems PDE2D uses a collocation finite element method [Sewell 2010], which means that, even for symmetric PDE systems, the matrices generated are highly nonsymmetric and do not even have a symmetric nonzero structure, and sparse direct and iterative methods perform very poorly on these systems. When PDE2D uses a Galerkin finite element method for a 2D problem, on the other hand, the nonzero structure *is* symmetric, and the matrices will be symmetric if the PDE system is symmetric.

For 3D problems, then, PDE2D provides the following options to solve the linear systems which arise every time step or iteration:

IS=1 Harwell Library sparse direct solver MA27 (Section 1.6) is used to solve the normal equations, $C^TC\boldsymbol{x} = C^T\boldsymbol{b}$. Since C^TC is symmetric and positive definite, pivoting is not necessary, and the work can be cut in half by taking advantage of the symmetry (Problem 4a of Chapter 1), and sparse direct methods perform much better on this than on the original system. C^TC is still sparse, though not as sparse as C.

IS=2 A frontal method, which is just a band solver (Section 1.5) with out-of-core storage of the matrix, is used to solve the original system $C\boldsymbol{x} = \boldsymbol{b}$. This option is slow, but requires very little memory.

IS=3 Iterative methods perform *very* poorly on the original, highly nonsymmetric, system, so PDE2D uses a Jacobi conjugate-gradient method to solve the positive definite normal equations. This is essentially the same as PCG (Figure 6.2.4) applied to $C^TC\boldsymbol{x} = C^T\boldsymbol{b}$, with D set to the diagonal of C^TC. Unfortunately, C^TC is ill conditioned (Section 2.1), which tends to slow the convergence.

Note that, in this case, it is not necessary to actually form the sparse matrix C^TC, since the only place it is used (see PCG loop 30) is in the matrix-vector multiplication $C^TC\boldsymbol{p}$, and this can obviously be done without forming C^TC. Like PCG, the PDE2D preconditioned conjugate-gradient method is "MPI-enhanced," to take advantage of multiple processors.

IS=6 A parallel band solver similar to PLINEQ (Figure 6.2.1) is used to solve the original system $C\boldsymbol{x} = \boldsymbol{b}$, though the PDE2D implementation does compute and save the LU decomposition.

Table 6.3.1 shows the computer time and memory requirements when these four options were used to solve the linear systems, when 10 iterations of the shifted inverse power method were done to find the eigenvalue closest to $p = -2$ on the UTEP Linux cluster. A 15x15x15 grid of tri-cubic Hermite elements was used, resulting in matrices of rank $N = 27000$. The eigenvalue -1.9856568 was found in each case, and one of the cross-sections of the corresponding eigenfunction is shown in Figure 6.3.2, coded by color.

Table 6.3.1

Times for Linear System Solvers

$N = 27000, N_{LD} = N_{UD} = 2791$				
IS	NPES	first iteration (sec)	each iter. after first (sec)	per-processor memory (Mwords)
1	1	185	1	64.8
2	1	932	11	7.2
3	1	172	199	5.5
3	2	90	97	5.5
3	4	57	54	2.9
3	8	40	35	1.6
3	16	47	37	0.9
6	1	356	5	231.4
6	2	187	2	118.4
6	4	102	2	59.2
6	8	65	1	29.7
6	16	56	2	14.9

On the first iteration, when the LU decomposition is not yet available, the iterative solver is the fastest option, though on some other problems, it may converge very slowly or not at all, despite the fact that the matrix is positive definite. Once the LU decomposition is available, the direct solvers are much faster, as expected.

PDE2D can also be used to find all eigenvalues of the generalized problem $A\boldsymbol{z} = \lambda B\boldsymbol{z}$. This generalized problem is converted to an equivalent ordinary eigenvalue problem, $F\boldsymbol{z} \equiv (A - qB)^{-1}B\boldsymbol{z} = \frac{1}{\lambda - q}\boldsymbol{z}$, where q is any number such that $A - qB$ is nonsingular; usually q may be set to zero. This means the eigenvalues of a full matrix (F) must be found, even though A and B are sparse and banded, but there seems to be no good algorithm available for finding all eigenvalues of a nonsymmetric banded problem.

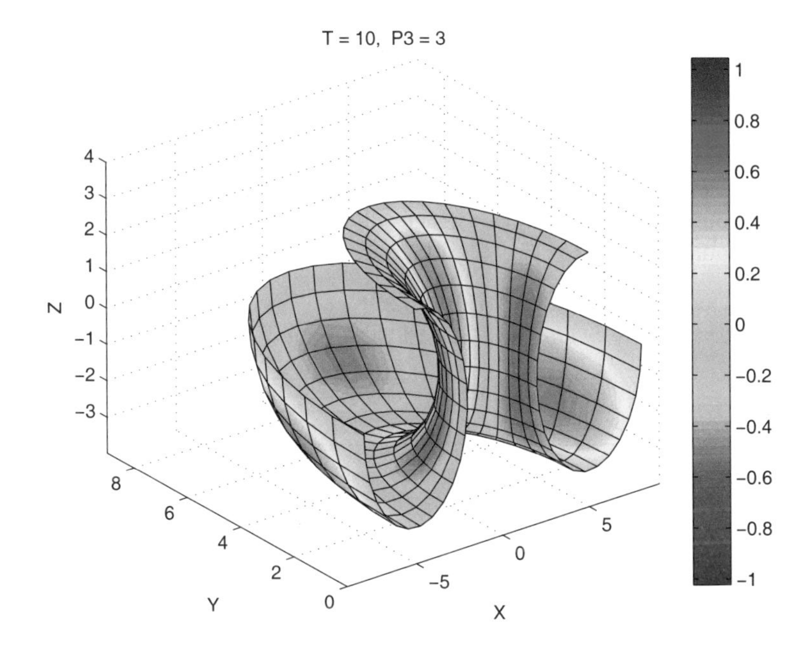

Figure 6.3.2
One Cross-section of Eigenfunction

For symmetric problems, when the Galerkin method is used and B is diagonal and positive, PDE2D employs an algorithm, BANDR from EISPACK [Smith et al. 1976], which does take advantage of the *symmetric* band structure of $B^{-\frac{1}{2}}AB^{-\frac{1}{2}}$ in the reduction of this matrix to tridiagonal form, then uses EISPACK routine TQLRAT to find all eigenvalues of the tridiagonal matrix, with a shifted QR algorithm. Otherwise, PDE2D finds all eigenvalues (without eigenvectors) of $A\boldsymbol{z} = \lambda B\boldsymbol{z}$ in three stages. First, the columns $\boldsymbol{f}_i$ of the full matrix F are found by solving $(A - qB)\boldsymbol{f}_i = \boldsymbol{b}_i$, where $\boldsymbol{b}_i$ is the ith column of B. Since $A - qB$ is banded, the first column is found using a band solver, saving the LU decomposition as outlined in Problem 8 of Chapter 1. This LU decomposition is then used to solve for the other columns, with the different $\boldsymbol{f}_i$ computed on different processors, if a multi-processor machine is used. Next, the full matrix F is reduced to a similar upper Hessenberg matrix, using a Householder orthogonal reduction routine similar to HESSH (Figure 3.4.2), but parallelized as discussed in Problem 7. The parallelized version is similar to PHESSH, displayed in Appendix B. Finally, the eigenvalues of this upper Hessenberg matrix are found using the EISPACK routine HQR which employs the shifted QR algorithm, discussed in Section 3.3. This stage is not parallelized.

Since the calculation of all eigenvalues requires much more time than the calculation of a single eigenvalue using the shifted inverse power method, a smaller (9x9x9) grid was used, and A and B are of rank $N = 5832$ now. Despite the fact that the last stage is not parallelized, the overall calculation scales fairly well, as seen by the results in Table 6.3.2. The 9th smallest eigenvalue found was -1.9855032, which differs by less than 0.01% from the eigenvalue found on the 15x15x15 grid.

Table 6.3.2

Time to Find All Eigenvalues

$N = 5832, N_{LD} = N_{UD} = 1027$	
NPES	Time(sec)
1	6446
2	2440
4	1520
8	1077
16	1270

6.4 Problems

1. The Jacobi iterative method 1.9.4 can be written in the form $\boldsymbol{x}_{n+1} = \boldsymbol{x}_n + D^{-1}(\boldsymbol{b} - A\boldsymbol{x}_n)$. Write an MPI-based subroutine JACOBI with arguments (A,IROW,JCOL,NZ,X,B,N), which iterates the Jacobi method to convergence, to solve an N by N (possibly nonsymmetric) linear system $A\boldsymbol{x} = \boldsymbol{b}$, with sparse matrix A distributed over the available processors. Solve the system 1.9.10, with $M = 40$ using a main program similar to Figure 6.2.5 to test your program. Run with 1 and 2 processors.

2. a. Use PLINEQ (Figure 6.2.1) to solve the linear system 1.9.10, and output the solution at the midpoint again, to check your answer. You can use the main program from Figure 6.2.5 to define the matrix AS in sparse format; then copy AS to an N by N full matrix A (inefficient for a sparse system such as this, of course), with columns also distributed over the available processors. You will need to remove the CALL MPI_INIT in PLINEQ since you are now initializing MPI in the main program.

   Run with 2, 4, and 8 processors, with $M = 16$.

   b. In part (a), although no processor ever touches any columns but its own, you are still storing the entire N by N matrix on every processor. This can be avoided, as suggested in the text, by replacing every reference to $A(I,J)$ by $A(I,(J-1)/NPES+1)$, in both the main program and PLINEQ. Then A can be dimensioned $A(N,(N-1)/NPES+1)$ and each processor will only store its own columns. However, NPES is not known until execution time, so you cannot dimension A in a DIMENSION statement; you should use the FORTRAN90 ALLOCATE statement to dynamically allocate space for A. This means

```
ALLOCATABLE A(:,:)
  .
CALL MPI_COMM_SIZE (MPI_COMM_WORLD,NPES,IERR)
ALLOCATE (A(N,(N-1)/NPES+1))
  .
```

Make these modifications and retest your program. You should be able to solve problems with larger M now, using many processors.

3. Write a parallel version, PLLSQR, of the linear least squares routine of Figure 2.2.2. You can follow closely the pattern used by PLINEQ (Figure 6.2.1), and distribute the columns of A cyclically over the available processors. In REDQ, whoever has the active column L should broadcast it to the other processors, who should save it in a vector COLUMNL, since all need access to this column. Each processor will need to modify COLUMNL itself, as well as their own columns, since COLUMNL(I) changes and is used, each pass through the loop 10. The back substitution can be done almost exactly as in PLINEQ. ERRLIM can be passed as an argument to PLLSQR, if desired.

   Since PLLSQR should find the exact solution of $A\boldsymbol{x} = \boldsymbol{b}$ if there is one, you can test your routine on the system 1.9.10 again, simply repeat both parts of Problem 2 using PLLSQR in place of PLINEQ. Physically distributing the columns of A over the processors (part (b)) will be easy, as in PLINEQ. However, switching the order of the REDQ loops 5 and 10, to produce a unit stride in the inner loop, would be more difficult. Why?

4. Set $\boldsymbol{b} = (1,\ldots,1)$ and use a random number generator with output in $(0,1)$ (e.g., the FORTRAN90 routine RANDOM_NUMBER) to generate the coefficients of an M by M matrix A, and the M-vector $\boldsymbol{c}$ in the resource allocation problem

   $$\text{maximize } \boldsymbol{c}^{\mathrm{T}}\boldsymbol{x}$$

   with constraints and bounds

   $$A\boldsymbol{x} \leq \boldsymbol{b}, \qquad \boldsymbol{x} \geq \boldsymbol{0}.$$

With $M = 250$, solve this problem using DLPRG (Figure 4.5.1), with one modification: increase the first dimension of TAB from M+2 to 256. Record the computer time to solve this problem, then re-run the same problem with the first dimension of TAB set to 257. On most computers—pipeline, parallel, or neither—DLPRG will run much faster when the first dimension is 257. Explain why.

5. Why is it preferable, in PLINEQ, to distribute the columns of A cyclically over the processors rather than by blocks, that is, assigning the first N/NPES columns to processor 0, and so on?

6. Repeat Problem 3a, only this time use a parallelized version, PREDH, of REDH (Figure 2.3.1) instead of REDQ. Again, distribute the columns of A cyclically over the available processors. In PREDH, whoever has the active column L should call CALW to compute the vector W, and broadcast W to the other processors. $A(I, L)$ will also need to be broadcast to the other processors after it changes.

7. Write a parallel version, PHESSH, of the subroutine HESSH (Figure 3.4.2) which reduces a matrix to a similar upper Hessenberg matrix using Householder transformations. Each processor should store the entire matrix, but never touch any columns but its "own" $(ITASK + 1 + K * NPES, K = 0, 1, 2...)$, until the end. The processor which owns column $I - 1$ should call CALW, and then broadcast the vector W to all processors, since all of them need W. At the end of PHESSH, before the QR iteration begins, the entire Hessenberg matrix should be redistributed to all processors, using MPI_BCAST.

   Test PHESSH by generating a random 1000 by 1000 symmetric matrix, which is passed to DEGNON (Figure 3.3.4), modified to call PHESSH instead of HESSQ. Since the Hessenberg matrix output by PHESSH will still be symmetric, and thus tridiagonal, modify QR as indicated in the comments, to take advantage of this. Make runs with 1 and 4 processors. Since you have only parallelized the reduction to Hessenberg form, and not the QR iteration, the speed-up observed should be substantially less than 4. Explain why it would be more difficult to parallelize QR (or HESSQ) efficiently, than HESSH.

8. Write a parallel version, PLPRV, of the revised simplex method routine DLPRV (Figure 4.6.1). Each processor should store the entire ABINV matrix, but never touch any columns but its own $(ITASK + 1 + K * NPES, K = 0, 1, 2...)$. Test PLPRV by calling it from DTRAN, with DTRAN used to solve a transportation problem with $NS = 100$ stores and $NW = 150$ warehouses, with the $SREQ, WCAP$ and $COST$ arrays generated by a random number generator. Make runs with 1 and 4 processors.

Appendix A—MATLAB Programs

All of the FORTRAN90 codes given in Chapters 1–5 have been translated into MATLAB programs, which are listed below. Both versions are available from the author (see Preface).

Since MATLAB does not allow zero or negative subscripts, Figures 1.5.4, 1.9.1, and 1.9.2 had to be rewritten slightly to keep subscripts positive; otherwise the MATLAB versions are almost line by line translations of the FORTRAN90 codes. MATLAB vector syntax (e.g., A(1:N) = B(1:N)), while popular with users, is not used here, for the same reason it is not used in the FORTRAN90 versions: It is easier for the student to analyze the computational complexity of the algorithms if all for/DO loops are shown explicitly (and indented).

There are many good MATLAB references, including Higham and Higham [2000] and Hanselman and Littlefield [2001].

```
%-----------------------------------------------------------------------
%--------------------------- FIGURE 1.2.1 ----------------------------
%-----------------------------------------------------------------------
      function [X,IPERM,A] = DLINEQ(A,N,B)
%
%  FUNCTION DLINEQ SOLVES THE LINEAR SYSTEM A*X=B
%
%  ARGUMENTS
%
%             ON INPUT                          ON OUTPUT
%             --------                          ---------
%
%    A      - THE N BY N COEFFICIENT MATRIX.    THE DIAGONAL AND UPPER
%                                               TRIANGLE OF A CONTAINS U
%                                               AND THE LOWER TRIANGLE
%                                               OF A CONTAINS THE LOWER
%                                               TRIANGLE OF L, WHERE
%                                               PA = LU, P BEING THE
%                                               PERMUTATION MATRIX
%                                               DEFINED BY IPERM.
%
```

```
%    N       - THE SIZE OF MATRIX A.
%
%    X       -                                  AN N-VECTOR CONTAINING
%                                               THE SOLUTION.
%
%    B       - THE RIGHT HAND SIDE N-VECTOR.
%
%    IPERM   -                                  AN N-VECTOR CONTAINING
%                                               A RECORD OF THE ROW
%                                               INTERCHANGES MADE.  IF
%                                               J = IPERM(K), THEN ROW
%                                               J ENDED UP AS THE K-TH
%                                               ROW.
%
%-----------------------------------------------------------------------
%                               INITIALIZE IPERM = (1,2,3,...,N)
      for K=1:N
         IPERM(K) = K;
      end
%                               BEGIN FORWARD ELIMINATION
      for I=1:N-1
%                               SEARCH FROM A(I,I) ON DOWN FOR
%                               LARGEST POTENTIAL PIVOT, A(L,I)
         BIG = abs(A(I,I));
         L = I;
         for J=I+1:N
            if (abs(A(J,I)) > BIG)
               BIG = abs(A(J,I));
               L = J;
            end
         end
%                               IF LARGEST POTENTIAL PIVOT IS ZERO,
%                               MATRIX IS SINGULAR
         if (BIG == 0.0)
            error('***** THE MATRIX IS SINGULAR *****')
         end
%                               SWITCH ROW I WITH ROW L, TO BRING
%                               UP LARGEST PIVOT
         for  K=1:N
            TEMP = A(L,K);
            A(L,K) = A(I,K);
            A(I,K) = TEMP;
         end
%                               SWITCH B(I) AND B(L)
         TEMP = B(L);
         B(L) = B(I);
         B(I) = TEMP;
%                               SWITCH IPERM(I) AND IPERM(L)
         ITEMP = IPERM(L);
         IPERM(L) = IPERM(I);
         IPERM(I) = ITEMP;
         for J=I+1:N
%                               CHOOSE MULTIPLIER TO ZERO A(J,I)
            LJI = A(J,I)/A(I,I);
            if (LJI ~= 0.0)
```

```
%                             SUBTRACT LJI TIMES ROW I FROM ROW J
               for K=I+1:N
                  A(J,K) = A(J,K) - LJI*A(I,K);
               end
%                             SUBTRACT LJI TIMES B(I) FROM B(J)
               B(J) = B(J) - LJI*B(I);
            end
%                             SAVE LJI IN A(J,I).  IT IS UNDERSTOOD,
%                             HOWEVER, THAT A(J,I) IS REALLY ZERO.
            A(J,I) = LJI;
         end
      end
      if (A(N,N) == 0.0)
         error('***** THE MATRIX IS SINGULAR *****')
      end
%                             SOLVE U*X = B USING BACK SUBSTITUTION.
      X(N) = B(N)/A(N,N);
      for I=N-1:-1:1
         SUM = 0.0;
         for J=I+1:N
            SUM = SUM + A(I,J)*X(J);
         end
         X(I) = (B(I)-SUM)/A(I,I);
      end
%-----------------------------------------------------------------------
%---------------------------- FIGURE 1.3.1 -----------------------------
%-----------------------------------------------------------------------
      function X = DRESLV(A,N,C,IPERM)
%
%  FUNCTION DRESLV SOLVES THE LINEAR SYSTEM A*X=C IN O(N**2) TIME,
%    AFTER DLINEQ HAS PRODUCED AN LU DECOMPOSITION OF PA.
%
%  ARGUMENTS
%
%             ON INPUT                          ON OUTPUT
%             --------                          ---------
%
%    A      - THE N BY N COEFFICIENT MATRIX
%             AFTER PROCESSING BY DLINEQ.
%             AS OUTPUT BY DLINEQ, A CONTAINS
%             AN LU DECOMPOSITION OF PA.
%
%    N      - THE SIZE OF MATRIX A.
%
%    X      -                                   AN N-VECTOR CONTAINING
%                                               THE SOLUTION.
%
%    C      - THE RIGHT HAND SIDE N-VECTOR.
%
%    IPERM  - THE PERMUTATION VECTOR OF
%             LENGTH N OUTPUT BY DLINEQ.
%
%-----------------------------------------------------------------------
%                             CALCULATE C=P*C, WHERE P IS PERMUTATION
%                             MATRIX DEFINED BY IPERM.
```

```
      for K=1:N
         J = IPERM(K);
         X(K) = C(J);
      end
      for K=1:N
         C(K) = X(K);
      end
%                                BEGIN FORWARD ELIMINATION, TO CALCULATE
%                                C = L^(-1)*C
      for I=1:N-1
         for J=I+1:N
%                                RETRIEVE MULTIPLIER SAVED IN A(J,I)
            LJI = A(J,I);
%                                SUBTRACT LJI TIMES C(I) FROM C(J)
            C(J) = C(J) - LJI*C(I);
         end
      end
%                                SOLVE U*X = C USING BACK SUBSTITUTION.
      X(N) = C(N)/A(N,N);
      for I=N-1:-1:1
         SUM = 0.0;
         for J=I+1:N
            SUM = SUM + A(I,J)*X(J);
         end
         X(I) = (C(I)-SUM)/A(I,I);
      end
%-----------------------------------------------------------------------
%--------------------------- FIGURE 1.5.4 ---------------------------
%-----------------------------------------------------------------------
      function X = DBAND(A,N,NLD,NUD,B)
%
%  FUNCTION DBAND SOLVES THE LINEAR SYSTEM A*X=B, WHERE A IS A
%    BAND MATRIX.
%
%  ARGUMENTS
%
%             ON INPUT                          ON OUTPUT
%             --------                          ---------
%
%    A      - THE N BY NUD+2*NLD+1 BAND MATRIX.
%             FIRST NLD COLUMNS = SUBDIAGONALS
%             NEXT  1           = MAIN DIAGONAL
%             NEXT  NUD         = SUPERDIAGONALS
%             NEXT  NLD         = WORKSPACE (FILL-IN)
%
%    N      - THE SIZE OF MATRIX A.
%
%    NLD    - NUMBER OF NONZERO LOWER DIAGONALS
%             IN A, I.E., NUMBER OF DIAGONALS
%             BELOW THE MAIN DIAGONAL.
%
%    NUD    - NUMBER OF NONZERO UPPER DIAGONALS
%             IN A, I.E., NUMBER OF DIAGONALS
%             ABOVE THE MAIN DIAGONAL.
%
```

```
%     X        -                                      AN N-VECTOR CONTAINING
%                                                     THE SOLUTION.
%
%     B        - THE RIGHT HAND SIDE N-VECTOR.
%
%
%-----------------------------------------------------------------------
      MD = NLD+1;
%                                ZERO TOP NLD DIAGONALS (WORKSPACE)
      for I=1:N
      for J=NUD+1:NUD+NLD
         A(I,MD+J) = 0.0;
      end
      end
%                                BEGIN FORWARD ELIMINATION
      for I=1:N-1
%                                SEARCH FROM AII ON DOWN FOR
%                                LARGEST POTENTIAL PIVOT, ALI
         BIG = abs(A(I,MD));
         L = I;
         for J=I+1:min(I+NLD,N);
            if (abs(A(J,MD+I-J)) > BIG)
               BIG = abs(A(J,MD+I-J));
               L = J;
            end
         end
%                                IF LARGEST POTENTIAL PIVOT IS ZERO,
%                                MATRIX IS SINGULAR
         if (BIG == 0.0)
            error('***** THE MATRIX IS SINGULAR *****')
         end
%                                SWITCH ROW I WITH ROW L, TO BRING
%                                UP LARGEST PIVOT
         for K=I:min(I+NUD+NLD,N)
            TEMP = A(L,MD+K-L);
            A(L,MD+K-L) = A(I,MD+K-I);
            A(I,MD+K-I) = TEMP;
         end
%                                SWITCH B(I) AND B(L)
         TEMP = B(L);
         B(L) = B(I);
         B(I) = TEMP;
         for J=I+1:min(I+NLD,N)
%                                CHOOSE MULTIPLIER TO ZERO AJI
            LJI = A(J,MD+I-J)/A(I,MD);
            if (LJI ~= 0.0)
%                                SUBTRACT LJI TIMES ROW I FROM ROW J
               for K=I:min(I+NUD+NLD,N)
                  A(J,MD+K-J) = A(J,MD+K-J) - LJI*A(I,MD+K-I);
               end
%                                SUBTRACT LJI TIMES B(I) FROM B(J)
               B(J) = B(J) - LJI*B(I);
            end
         end
      end
```

```
      if (A(N,MD) == 0.0)
         error('***** THE MATRIX IS SINGULAR *****')
      end
%                                SOLVE U*X = B USING BACK SUBSTITUTION.
      X(N) = B(N)/A(N,MD);
      for I=N-1:-1:1
         SUM = 0.0;
         for J=I+1:min(I+NUD+NLD,N)
            SUM = SUM + A(I,MD+J-I)*X(J);
         end
         X(I) = (B(I)-SUM)/A(I,MD);
      end
%-----------------------------------------------------------------------
%------------------------- FIGURE 1.7.2 --------------------------------
%-----------------------------------------------------------------------
      function YOUT = DSPLN(X,Y,N,YXX1,YXXN,XOUT,NOUT)
%
%  FUNCTION DSPLN FITS AN INTERPOLATORY CUBIC SPLINE THROUGH THE
%    POINTS (X(I),Y(I)), I=1,...,N, WITH SPECIFIED SECOND DERIVATIVES
%    AT THE END POINTS, AND EVALUATES THIS SPLINE AT THE OUTPUT POINTS
%    XOUT(1),...,XOUT(NOUT).
%
%  ARGUMENTS
%
%             ON INPUT                          ON OUTPUT
%             --------                          ---------
%
%    X      - A VECTOR OF LENGTH N CONTAINING
%             THE X-COORDINATES OF THE DATA
%             POINTS.
%
%    Y      - A VECTOR OF LENGTH N CONTAINING
%             THE Y-COORDINATES OF THE DATA
%             POINTS.
%
%    N      - THE NUMBER OF DATA POINTS
%             (N >= 3).
%
%    YXX1   - THE SECOND DERIVATIVE OF THE
%             CUBIC SPLINE AT X(1).
%
%    YXXN   - THE SECOND DERIVATIVE OF THE
%             CUBIC SPLINE AT X(N).  (YXX1=0
%             AND YXXN=0 GIVES A NATURAL
%             CUBIC SPLINE)
%
%    XOUT   - A VECTOR OF LENGTH NOUT CONTAINING
%             THE X-COORDINATES AT WHICH THE
%             CUBIC SPLINE IS EVALUATED.  THE
%             ELEMENTS OF XOUT MUST BE IN
%             ASCENDING ORDER.
%
%    YOUT   -                                   A VECTOR OF LENGTH NOUT.
%                                               YOUT(I) CONTAINS THE
%                                               VALUE OF THE SPLINE
```

```
%                                                   AT XOUT(I).
%
%    NOUT   - THE NUMBER OF OUTPUT POINTS.
%
%-----------------------------------------------------------------------
      SIG(1) = YXX1;
      SIG(N) = YXXN;
%                                  SET UP TRIDIAGONAL SYSTEM SATISFIED
%                                  BY SECOND DERIVATIVES (SIG(I)=SECOND
%                                  DERIVATIVE AT X(I)).
      for I=1:N-2
         HI   = X(I+1)-X(I);
         HIP1 = X(I+2)-X(I+1);
         R(I) = (Y(I+2)-Y(I+1))/HIP1 - (Y(I+1)-Y(I))/HI;
         A(I,1) = HI/6.0;
         A(I,2) = (HI + HIP1)/3.0;
         A(I,3) = HIP1/6.0;
         if (I == 1)
            R(1)   = R(1)   - HI/  6.0*SIG(1);
         end
         if (I == N-2)
            R(N-2) = R(N-2) - HIP1/6.0*SIG(N);
         end
      end
%                                  CALL DBAND TO SOLVE TRIDIAGONAL SYSTEM
      NLD = 1;
      NUD = 1;
      SIG(2:N-1) = DBAND(A,N-2,NLD,NUD,R);
%                                  CALCULATE COEFFICIENTS OF CUBIC SPLINE
%                                  IN EACH SUBINTERVAL
      for I=1:N-1
         HI = X(I+1)-X(I);
         COEFF(1,I) = Y(I);
         COEFF(2,I) = (Y(I+1)-Y(I))/HI - HI/6.0*(2*SIG(I)+SIG(I+1));
         COEFF(3,I) = SIG(I)/2.0;
         COEFF(4,I) = (SIG(I+1)-SIG(I))/(6.0*HI);
      end
      L = 1;
      for I=1:NOUT
%                                  FIND FIRST VALUE OF J FOR WHICH X(J+1) IS
%                                  GREATER THAN OR EQUAL TO XOUT(I).  SINCE
%                                  ELEMENTS OF XOUT ARE IN ASCENDING ORDER,
%                                  WE ONLY NEED CHECK THE KNOTS X(L+1)...X(N)
%                                  WHICH ARE GREATER THAN OR EQUAL TO
%                                  XOUT(I-1).
         for J=L:N-1
            JSAVE = J;
            if (X(J+1) >= XOUT(I))
               break
            end
         end
         L = JSAVE;
%                                  EVALUATE CUBIC SPLINE IN INTERVAL
%                                  (X(L),X(L+1))
         P = XOUT(I)-X(L);
```

```
         YOUT(I) = COEFF(1,L)      + COEFF(2,L)*P        ...
                 + COEFF(3,L)*P*P + COEFF(4,L)*P*P*P;
      end
%-----------------------------------------------------------------------
%------------------------ FIGURE 1.9.1 ---------------------------------
%-----------------------------------------------------------------------
%                              JACOBI METHOD
      M = 10;
      H = 1.0/M;
%                              SET BOUNDARY KNOWNS TO ZERO PERMANENTLY
%                              AND INTERIOR UNKNOWNS TO ZERO TEMPORARILY
      for I=1:M+1
      for J=1:M+1
      for K=1:M+1
         UOLD(I,J,K) = 0.0;
      end
      end
      end
%                              BEGIN JACOBI ITERATION
      NITER = (M-1)^3;
      for ITER = 1:NITER
%                              UPDATE UNKNOWNS ONLY
         for I=2:M
         for J=2:M
         for K=2:M
            UNEW(I,J,K)  = H^2/6.0 + ( UOLD(I+1,J,K) + UOLD(I-1,J,K)    ...
                                     + UOLD(I,J+1,K) + UOLD(I,J-1,K)    ...
                                     + UOLD(I,J,K+1) + UOLD(I,J,K-1))/6.0;
         end
         end
         end
%                              COPY UNEW ONTO UOLD
         for I=2:M
         for J=2:M
         for K=2:M
           UOLD(I,J,K) = UNEW(I,J,K);
         end
         end
         end
%                              EVERY 10 ITERATIONS CALCULATE MAXIMUM
%                              RESIDUAL AND CHECK FOR CONVERGENCE
         if (mod(ITER,10) ~= 0)
            continue
         end
         RMAX = 0.0;
         for I=2:M
         for J=2:M
         for K=2:M
            RESID = 6*UOLD(I,J,K) - UOLD(I+1,J,K) - UOLD(I-1,J,K)      ...
                                  - UOLD(I,J+1,K) - UOLD(I,J-1,K)      ...
                                  - UOLD(I,J,K+1) - UOLD(I,J,K-1) - H^2;
            RMAX = max(RMAX,abs(RESID));
         end
         end
         end
```

```
         ITER
         RMAX = RMAX/H^2
         if (RMAX <= 1.e-10)
            break
         end
      end
%-----------------------------------------------------------------------
%--------------------------- FIGURE 1.9.2 -----------------------------
%-----------------------------------------------------------------------
%                                GAUSS-SEIDEL METHOD
      M = 10;
      H = 1.0/M;
%                                SET BOUNDARY KNOWNS TO ZERO PERMANENTLY
%                                AND INTERIOR UNKNOWNS TO ZERO TEMPORARILY
      for I=1:M+1
      for J=1:M+1
      for K=1:M+1
         U(I,J,K) = 0.0;
      end
      end
      end
%                                BEGIN GAUSS-SEIDEL ITERATION
      NITER = (M-1)^3;
      for ITER = 1:NITER
%                                UPDATE UNKNOWNS ONLY
         for I=2:M
         for J=2:M
         for K=2:M
            GAUSS = H^2/6.0 + ( U(I+1,J,K) + U(I-1,J,K)     ...
                              + U(I,J+1,K) + U(I,J-1,K)     ...
                              + U(I,J,K+1) + U(I,J,K-1))/6.0;
            U(I,J,K) = GAUSS;
         end
         end
         end
%                                EVERY 10 ITERATIONS CALCULATE MAXIMUM
%                                RESIDUAL AND CHECK FOR CONVERGENCE
         if (mod(ITER,10) ~= 0)
            continue
         end
         RMAX = 0.0;
         for I=2:M
         for J=2:M
         for K=2:M
            RESID = 6*U(I,J,K) - U(I+1,J,K) - U(I-1,J,K)     ...
                               - U(I,J+1,K) - U(I,J-1,K)     ...
                               - U(I,J,K+1) - U(I,J,K-1) - H^2;
            RMAX = max(RMAX,abs(RESID));
         end
         end
         end
         ITER
         RMAX = RMAX/H^2
         if (RMAX <= 1.e-10)
            break
```

```
         end
      end
%-----------------------------------------------------------------------
%-------------------------- FIGURE 1.10.1 ------------------------------
%-----------------------------------------------------------------------
      function X = DCG(A,IROW,JCOL,NZ,B,N,D)
%
%  FUNCTION DCG SOLVES THE SYMMETRIC LINEAR SYSTEM A*X=B, USING THE
%     CONJUGATE GRADIENT ITERATIVE METHOD.  THE NON-ZEROS OF A ARE STORED
%     IN SPARSE FORMAT.
%
%  ARGUMENTS
%
%              ON INPUT                           ON OUTPUT
%              --------                           ---------
%
%    A      - A(IZ) IS THE MATRIX ELEMENT IN
%             ROW IROW(IZ), COLUMN JCOL(IZ),
%             FOR IZ=1,...,NZ.
%
%    IROW   - (SEE A).
%
%    JCOL   - (SEE A).
%
%    NZ     - NUMBER OF NONZEROS.
%
%    X      -                                     AN N-VECTOR CONTAINING
%                                                 THE SOLUTION.
%
%    B      - THE RIGHT HAND SIDE N-VECTOR.
%
%    N      - SIZE OF MATRIX A.
%
%    D      - VECTOR HOLDING A DIAGONAL
%             PRECONDITIONING MATRIX.
%             D = DIAGONAL(A) IS RECOMMENDED.
%             D(I) = 1 FOR NO PRECONDITIONING
%
%-----------------------------------------------------------------------
%                                  X0 = 0
%                                  R0 = D^(-1)*B
%                                  P0 = R0
      R0MAX = 0;
      for I=1:N
         X(I) = 0;
         R(I) = B(I)/D(I);
         R0MAX = max(R0MAX,abs(R(I)));
         P(I) = R(I);
      end
%                                  NITER = MAX NUMBER OF ITERATIONS
      NITER = 3*N;
      for ITER=1:NITER
%                                  AP = A*P
         for I=1:N
            AP(I) = 0;
```

```
         end
         for IZ=1:NZ
            I = IROW(IZ);
            J = JCOL(IZ);
            AP(I) = AP(I) + A(IZ)*P(J);
         end
%                                    PAP = (P,AP)
%                                    RP = (R,D*P)
         PAP = 0.0;
         RP = 0.0;
         for I=1:N
            PAP = PAP + P(I)*AP(I);
            RP = RP + R(I)*D(I)*P(I);
         end
%                                    LAMBDA = (R,D*P)/(P,AP)
         LAMBDA = RP/PAP;
%                                    X = X + LAMBDA*P
%                                    R = R - LAMBDA*D^(-1)*AP
         for I=1:N
            X(I) = X(I) + LAMBDA*P(I);
            R(I) = R(I) - LAMBDA*AP(I)/D(I);
         end
%                                    RAP = (R,AP)
         RAP = 0.0;
         for I=1:N
            RAP = RAP + R(I)*AP(I);
         end
%                                    ALPHA = -(R,AP)/(P,AP)
         ALPHA = -RAP/PAP;
%                                    P = R + ALPHA*P
         for I=1:N
            P(I) = R(I) + ALPHA*P(I);
         end
%                                    RMAX = MAX OF RESIDUAL (R)
         RMAX = 0;
         for I=1:N
            RMAX = max(RMAX,abs(R(I)));
         end
%                                    CHECK FOR CONVERGENCE
         if (RMAX <= 1.e-10*R0MAX)
            disp ([' Number of iterations = ',num2str(ITER)])
            return
         end
      end
%                                    DCG DOES NOT CONVERGE
      error ('***** DCG DOES NOT CONVERGE *****')
      end
%-----------------------------------------------------------------------
%--------------------------- FIGURE 2.2.2 ------------------------------
%-----------------------------------------------------------------------
      function X = DLLSQR(A,M,N,B)
%
%  FUNCTION DLLSQR SOLVES THE LINEAR LEAST SQUARES PROBLEM
%
%          MINIMIZE  2-NORM OF (A*X-B)
```

```
%
%
% ARGUMENTS
%
%           ON INPUT                         ON OUTPUT
%           --------                         ---------
%
%   A     - THE M BY N MATRIX.
%
%   M     - THE NUMBER OF ROWS IN A.
%
%   N     - THE NUMBER OF COLUMNS IN A.
%
%   X     -                                  AN N-VECTOR CONTAINING
%                                            THE LEAST SQUARES
%                                            SOLUTION.
%
%   B     - THE RIGHT HAND SIDE M-VECTOR.
%
%-----------------------------------------------------------------------
%                              EPS = MACHINE FLOATING POINT RELATIVE
%                                    PRECISION
% ****************************
      EPS = eps;
% ****************************
%                              AMAX = MAXIMUM ELEMENT OF A
      AMAX = 0.0;
      for I=1:M
      for J=1:N
         AMAX = max(AMAX,abs(A(I,J)));
      end
      end
      ERRLIM = 1000*EPS*AMAX;
%                              REDUCTION TO ROW ECHELON FORM
      [A,B,PIVOT,NPIVOT] = REDQ(A,M,N,B,ERRLIM);
%                              CAUTION USER IF SOLUTION NOT UNIQUE.
      if (NPIVOT ~= N)
         disp(' NOTE: SOLUTION IS NOT UNIQUE ')
      end
%                              ASSIGN VALUE OF ZERO TO NON-PIVOT
%                              VARIABLES.
      for K=1:N
         X(K) = 0.0;
      end
%                              SOLVE FOR PIVOT VARIABLES USING BACK
%                              SUBSTITUTION.
      for I=NPIVOT:-1:1
         L = PIVOT(I);
         SUM = 0.0;
         for K=L+1:N
            SUM = SUM + A(I,K)*X(K);
         end
         X(L) = (B(I)-SUM)/A(I,L);
      end
%%%%%%%%
```

```
      function [A,B,PIVOT,NPIVOT] = REDQ(A,M,N,B,ERRLIM)
%                               USE GIVENS ROTATIONS TO REDUCE A
%                               TO ROW ECHELON FORM
      I = 1;
      for L=1:N
%                               USE PIVOT A(I,L) TO KNOCK OUT ELEMENTS
%                               I+1 TO M IN COLUMN L.
         for J=I+1:M
            if (A(J,L) == 0.0)
               continue
            end
            DEN = sqrt(A(I,L)^2+A(J,L)^2);
            C = A(I,L)/DEN;
            S = A(J,L)/DEN;
%                               PREMULTIPLY A BY Qij^T
            for K=L:N
               BIK = C*A(I,K) + S*A(J,K);
               BJK =-S*A(I,K) + C*A(J,K);
               A(I,K) = BIK;
               A(J,K) = BJK;
            end
%                               PREMULTIPLY B BY Qij^T
            BI = C*B(I) + S*B(J);
            BJ =-S*B(I) + C*B(J);
            B(I) = BI;
            B(J) = BJ;
         end
%                               PIVOT A(I,L) IS NONZERO AFTER PROCESSING
%                               COLUMN L--MOVE DOWN TO NEXT ROW, I+1
         if (abs(A(I,L)) <= ERRLIM)
            A(I,L) = 0.0;
         end
         if (A(I,L) ~= 0.0)
            NPIVOT = I;
            PIVOT(NPIVOT) = L;
            I = I+1;
            if (I > M)
               return
            end
         end
      end
%-----------------------------------------------------------------------
%------------------------- FIGURE 2.3.1 --------------------------------
%-----------------------------------------------------------------------
      function [A,B,PIVOT,NPIVOT] = REDH(A,M,N,B,ERRLIM)
%                               USE HOUSEHOLDER TRANSFORMATIONS TO
%                               REDUCE A TO ROW ECHELON FORM
      I = 1;
      for L=1:N
%                               USE PIVOT A(I,L) TO KNOCK OUT ELEMENTS
%                               I+1 TO M IN COLUMN L.
         if (I+1 <= M)
%                               CHOOSE UNIT M-VECTOR W (WHOSE FIRST
%                               I-1 COMPONENTS ARE ZERO) SUCH THAT WHEN
%                               COLUMN L IS PREMULTIPLIED BY
```

```
%                               H = I - 2W*W^T, COMPONENTS I+1 THROUGH
%                               M ARE ZEROED.
            W = CALW(A(1:M,L),M,I);
%                               PREMULTIPLY A BY H = I - 2W*W^T
            for K=L:N
               WTA = 0.0;
               for J=I:M
                  WTA = WTA + W(J)*A(J,K);
               end
               TWOWTA = 2*WTA;
               for J=I:M
                  A(J,K) = A(J,K) - TWOWTA*W(J);
               end
            end
%                               PREMULTIPLY B BY H = I - 2W*W^T
            WTA = 0.0;
            for J=I:M
               WTA = WTA + W(J)*B(J);
            end
            TWOWTA = 2*WTA;
            for J=I:M
               B(J) = B(J) - TWOWTA*W(J);
            end
         end
%                               PIVOT A(I,L) IS NONZERO AFTER PROCESSING
%                               COLUMN L--MOVE DOWN TO NEXT ROW, I+1
         if (abs(A(I,L)) <= ERRLIM)
            A(I,L) = 0.0;
         end
         if (A(I,L) ~= 0.0)
            NPIVOT = I;
            PIVOT(NPIVOT) = L;
            I = I+1;
            if (I > M)
               return
            end
         end
      end
%%%%%%%%%%
      function W = CALW(A,M,I)
%                               FUNCTION CALW CALCULATES A UNIT
%                               M-VECTOR W (WHOSE FIRST I-1 COMPONENTS
%                               ARE ZERO) SUCH THAT PREMULTIPLYING THE
%                               VECTOR A BY H = I - 2W*W^T ZEROES
%                               COMPONENTS I+1 THROUGH M.
      S = 0.0;
      for J=I:M
         S = S + A(J)^2;
         W(J) = A(J);
      end
      if (A(I) >= 0.0)
         BETA = sqrt(S);
      else
         BETA = -sqrt(S);
      end
```

```
      W(I) = A(I) + BETA;
      TWOALP = sqrt(2*BETA*W(I));
%                                TWOALP=0 ONLY IF A(I),...,A(M) ARE ALL
%                                ZERO.  IN THIS CASE, RETURN WITH W=0
      if (TWOALP == 0.0)
         return
      end
%                                NORMALIZE W
      for J=I:M
         W(J) = W(J)/TWOALP;
      end
%-----------------------------------------------------------------------
%------------------------- FIGURE 2.4.1 --------------------------------
%-----------------------------------------------------------------------
      function YOUT = DLSQSP(X,N,XD,YD,M,XOUT,NOUT)
%
%  FUNCTION DLSQSP CALCULATES A NATURAL CUBIC SPLINE WITH KNOTS AT
%    X(1),...,X(N) WHICH IS THE LEAST SQUARES FIT TO THE DATA POINTS
%    (XD(I),YD(I)), I=1,...,M, AND EVALUATES THIS SPLINE AT THE OUTPUT
%    POINTS XOUT(1),...,XOUT(NOUT).
%
%  ARGUMENTS
%
%              ON INPUT                          ON OUTPUT
%              --------                          ---------
%
%    X       - A VECTOR OF LENGTH N CONTAINING
%              THE SPLINE KNOTS.
%
%    N       - THE NUMBER OF KNOTS.
%              (N >= 3).
%
%    XD      - A VECTOR OF LENGTH M CONTAINING
%              THE X-COORDINATES OF THE DATA
%              POINTS.
%
%    YD      - A VECTOR OF LENGTH M CONTAINING
%              THE Y-COORDINATES OF THE DATA
%              POINTS.
%
%    M       - THE NUMBER OF DATA POINTS.
%
%    XOUT    - A VECTOR OF LENGTH NOUT CONTAINING
%              THE X-COORDINATES AT WHICH THE
%              CUBIC SPLINE IS EVALUATED.  THE
%              ELEMENTS OF XOUT MUST BE IN
%              ASCENDING ORDER.
%
%    YOUT    -                                   A VECTOR OF LENGTH NOUT.
%                                                YOUT(I) CONTAINS THE
%                                                VALUE OF THE SPLINE
%                                                AT XOUT(I).
%
%    NOUT    - THE NUMBER OF OUTPUT POINTS.
%
```

```
%-----------------------------------------------------------------------
      ZERO = 0.0;
      for J=1:N
         Y(J) = 0.0;
      end
      for J=1:N
         Y(J) = 1.0;
%                            CALCULATE PHI(J,X), NATURAL CUBIC SPLINE
%                            WHICH IS EQUAL TO ONE AT KNOT X(J) AND
%                            ZERO AT OTHER KNOTS.  THEN SET
%                               A(I,J) = PHI(J,XD(I)), I=1,...,M
         A(1:M,J) = DSPLN(X,Y,N,ZERO,ZERO,XD,M);
         Y(J) = 0.0;
      end
%                            CALL DLLSQR TO MINIMIZE NORM OF A*Y-YD
      Y = DLLSQR(A,M,N,YD);
%                            LEAST SQUARES SPLINE IS
%                             Y(1)*PHI(1,X) + ... + Y(N)*PHI(N,X).
%                            EVALUATE SPLINE AT XOUT(1),...,XOUT(NOUT)
      YOUT = DSPLN(X,Y,N,ZERO,ZERO,XOUT,NOUT);
%-----------------------------------------------------------------------
%-------------------------- FIGURE 3.2.1 ----------------------------
%-----------------------------------------------------------------------
      function [A,X] = DEGSYM(A,N)
%
%  FUNCTION DEGSYM SOLVES THE EIGENVALUE PROBLEM
%
%                A*X = LAMBDA*X
%
%    WHERE A IS A SYMMETRIC MATRIX.
%
%
%  ARGUMENTS
%
%             ON INPUT                          ON OUTPUT
%             --------                          ---------
%
%    A      - THE N BY N SYMMETRIC MATRIX.      A DIAGONAL MATRIX,
%                                               WITH THE EIGENVALUES
%                                               OF A ON THE DIAGONAL.
%
%    N      - THE SIZE OF MATRIX A.
%
%    X      -                                   AN N BY N MATRIX WHICH
%                                               CONTAINS THE EIGEN-
%                                               VECTORS OF A IN ITS
%                                               COLUMNS, IN THE SAME
%                                               ORDER AS THE EIGENVALUES
%                                               APPEAR ON THE DIAGONAL.
%
%-----------------------------------------------------------------------
%                              EPS = MACHINE FLOATING POINT RELATIVE
%                                    PRECISION
% *****************************
      EPS = eps;
```

```
% ****************************
%                              ANORM = SUM OF ALL SQUARES
%                              X INITIALIZED TO IDENTITY
      ANORM = 0.0;
      for I=1:N
         for J=1:N
            ANORM = ANORM + A(I,J)^2;
            X(I,J) = 0.0;
         end
         X(I,I) = 1.0;
      end
      ERRLIM = 1000*EPS*ANORM;
%                              EK = SUM OF OFF-DIAGONAL SQUARES
      EK = 0.0;
      for I=1:N
      for J=1:N
         if (I ~= J)
            EK = EK + A(I,J)^2;
         end
      end
      end
      if (EK <= ERRLIM)
         return
      end
      THRESH = 0.5*EK/N/(N-1);
      while (1 > 0)
         for I=1:N-1
            for J=I+1:N
%                              IF A(J,I)^2 LESS THAN HALF THE
%                              AVERAGE FOR OFF-DIAGONALS, SKIP IT.
               if (A(J,I)^2 <= THRESH)
                  continue
               end
%                              KNOCKING OUT A(J,I) WILL DECREASE OFF-
%                              DIAGONAL SUM OF SQUARES BY 2*A(J,I)^2.
               EK = EK - 2*A(J,I)^2;
%                              CALCULATE NEW THRESHOLD.
               THRESH = 0.5*EK/N/(N-1);
%                              CALCULATE C,S
               BETA = (A(I,I)-A(J,J))/(2.*A(J,I));
               FRACT = 0.5*BETA/sqrt(1.0+BETA^2);
               S = sqrt(max(0.5-FRACT,0.0));
               C = sqrt(max(0.5+FRACT,0.0));
%                              PREMULTIPLY A BY Qij^T
               for K=1:N
                  PIK =  C*A(I,K)+S*A(J,K);
                  PJK = -S*A(I,K)+C*A(J,K);
                  A(I,K) = PIK;
                  A(J,K) = PJK;
               end
%                              POSTMULTIPLY A AND X BY Qij
               for K=1:N
                  BKI =  C*A(K,I)+S*A(K,J);
                  BKJ = -S*A(K,I)+C*A(K,J);
                  A(K,I) = BKI;
```

```
                  A(K,J) = BKJ;
                  XKI =  C*X(K,I)+S*X(K,J);
                  XKJ = -S*X(K,I)+C*X(K,J);
                  X(K,I) = XKI;
                  X(K,J) = XKJ;
               end
%                                CHECK FOR CONVERGENCE
               if (EK <= ERRLIM)
                  return
               end
            end
         end
%                                RETURN TO BEGINNING OF CYCLE
      end
%-----------------------------------------------------------------------
%--------------------------- FIGURE 3.3.4 ----------------------------
%-----------------------------------------------------------------------
      function EIG = DEGNON(A,N)
%
%  FUNCTION DEGNON SOLVES THE EIGENVALUE PROBLEM
%
%                 A*X = LAMBDA*X
%
%    WHERE A IS A GENERAL REAL MATRIX.
%
%
%  ARGUMENTS
%
%              ON INPUT                          ON OUTPUT
%              --------                          ---------
%
%    A       - THE N BY N MATRIX.
%
%    N       - THE SIZE OF MATRIX A.
%
%    EIG     -                                   A COMPLEX N-VECTOR
%                                                CONTAINING THE EIGEN-
%                                                VALUES OF A.
%
%-----------------------------------------------------------------------
%                                EPS = MACHINE FLOATING POINT RELATIVE
%                                      PRECISION
% *****************************
      EPS = eps;
% *****************************
%                                AMAX = MAXIMUM ELEMENT OF A
      AMAX = 0.0;
      for I=1:N
      for J=1:N
         AMAX = max(AMAX,abs(A(I,J)));
      end
      end
      ERRLIM = sqrt(EPS)*AMAX;
%                                REDUCTION TO HESSENBERG FORM
      A = HESSQ(A,N);
```

```
%                               REDUCTION TO QUASI-TRIANGULAR FORM
      A = QR(A,N,ERRLIM);
%                               EXTRACT EIGENVALUES OF QUASI-TRIANGULAR
%                               MATRIX
      I = 1;
      while (I <= N-1)
         if (A(I+1,I) == 0.0)
%                               1 BY 1 BLOCK ON DIAGONAL
            EIG(I) = A(I,I);
            I = I+1;
         else
%                               2 BY 2 BLOCK ON DIAGONAL
            DISC = (A(I,I)-A(I+1,I+1))^2 + 4.0*A(I,I+1)*A(I+1,I);
            TERM =  0.5*(A(I,I)+A(I+1,I+1));
            if (DISC >= 0.0)
               EIG(I)   = TERM + 0.5*sqrt(DISC);
               EIG(I+1)= TERM - 0.5*sqrt(DISC);
            else
               EIG(I)   = TERM + 0.5*sqrt(-DISC)*i;
               EIG(I+1)= TERM - 0.5*sqrt(-DISC)*i;
            end
            I = I+2;
         end
      end
      if (I == N)
         EIG(N) = A(N,N);
      end
%%%%%%%%%
      function A = HESSQ(A,N)
      if (N <= 2)
         return
      end
%                               USE GIVENS ROTATIONS TO REDUCE A
%                               TO UPPER HESSENBERG FORM
      for I=2:N-1
         for J=I+1:N
            if (A(J,I-1) == 0.0)
              continue
            end
            DEN = sqrt(A(I,I-1)^2+A(J,I-1)^2);
            C = A(I,I-1)/DEN;
            S = A(J,I-1)/DEN;
%                               PREMULTIPLY BY Qij^T
            for K=I-1:N
               PIK = C*A(I,K) + S*A(J,K);
               PJK =-S*A(I,K) + C*A(J,K);
               A(I,K) = PIK;
               A(J,K) = PJK;
            end
%                               POSTMULTIPLY BY Qij
            for K=1:N
               BKI = C*A(K,I) + S*A(K,J);
               BKJ =-S*A(K,I) + C*A(K,J);
               A(K,I) = BKI;
               A(K,J) = BKJ;
```

```
            end
         end
      end
%%%%%%%%%
      function A = QR(A,N,ERRLIM)
      if (N <= 2)
         return
      end
%                                  USE QR ITERATION TO REDUCE HESSENBERG
%                                  MATRIX A TO QUASI-TRIANGULAR FORM
      NITER = 1000*N;
      for ITER=1:NITER
%                                  REDUCE A TO UPPER TRIANGULAR FORM USING
%                                  ORTHOGONAL REDUCTION (PREMULTIPLY BY
%                                  Qij^T MATRICES)
         for I=1:N-1
            if (A(I+1,I) == 0.0)
               C = 1.0;
               S = 0.0;
            else
               DEN = sqrt(A(I,I)^2 + A(I+1,I)^2);
               C = A(I,I)/DEN;
               S = A(I+1,I)/DEN;
            end
%                                  USE SAVE TO SAVE C,S FOR POST-
%                                  MULTIPLICATION PHASE
            SAVE(1,I) = C;
            SAVE(2,I) = S;
            if (S == 0.0)
               continue
            end
%                                  IF MATRIX SYMMETRIC, LIMITS ON K
%                                  CAN BE:  K = I : min(I+2,N)
            for K=I:N
               PIK = C*A(I,K) + S*A(I+1,K);
               PJK =-S*A(I,K) + C*A(I+1,K);
               A(I,K)   = PIK;
               A(I+1,K) = PJK;
            end
         end
%                                  NOW POSTMULTIPLY BY Qij MATRICES
         for I=1:N-1
            C = SAVE(1,I);
            S = SAVE(2,I);
            if (S == 0.0)
               continue
            end
%                                  IF MATRIX SYMMETRIC, LIMITS ON K
%                                  CAN BE:  K = max(1,I-1) : I+1
            for K=1:I+1
               BKI = C*A(K,I) + S*A(K,I+1);
               BKJ =-S*A(K,I) + C*A(K,I+1);
               A(K,I)   = BKI;
               A(K,I+1) = BKJ;
            end
```

```
         end
%                                SET NEARLY ZERO SUBDIAGONALS TO ZERO,
%                                TO AVOID UNDERFLOW.
         for I=1:N-1
            if (abs(A(I+1,I)) < ERRLIM)
               A(I+1,I) = 0.0;
            end
         end
%                                CHECK FOR CONVERGENCE TO "QUASI-
%                                TRIANGULAR" FORM.
         ICONV = 1;
         for I=2:N-1
            if (A(I,I-1) ~= 0.0 & A(I+1,I) ~= 0.0)
               ICONV = 0;
            end
         end
         if (ICONV == 1)
            return
         end
      end
%                                HAS NOT CONVERGED IN NITER ITERATIONS
      error('***** QR ITERATION DOES NOT CONVERGE *****')
%-----------------------------------------------------------------------
%------------------------- FIGURE 3.4.2 --------------------------------
%-----------------------------------------------------------------------
      function A = HESSH(A,N)
      if (N <= 2)
         return
      end
%                                USE HOUSEHOLDER TRANSFORMATIONS TO
%                                REDUCE A TO UPPER HESSENBERG FORM
      for I=2:N-1
%                                CHOOSE UNIT N-VECTOR W (WHOSE FIRST
%                                I-1 COMPONENTS ARE ZERO) SUCH THAT WHEN
%                                COLUMN I-1 IS PREMULTIPLIED BY
%                                H = I - 2W*W^T, COMPONENTS I+1 THROUGH
%                                N ARE ZEROED.
         W = CALW(A(1:N,I-1),N,I);
%                                PREMULTIPLY A BY H = I - 2W*W^T
         for K=I-1:N
            WTA = 0.0;
            for J=I:N
               WTA = WTA + W(J)*A(J,K);
            end
            TWOWTA = 2*WTA;
            for J=I:N
               A(J,K) = A(J,K) - TWOWTA*W(J);
            end
         end
%                                POSTMULTIPLY A BY H = I - 2W*W^T
         for K=1:N
            ATW = 0.0;
            for J=I:N
               ATW = ATW + A(K,J)*W(J);
            end
```

```
         TWOATW = 2*ATW;
         for J=I:N
            A(K,J) = A(K,J) - TWOATW*W(J);
         end
      end
   end
%-----------------------------------------------------------------------
%--------------------------- FIGURE 3.4.4 ------------------------------
%-----------------------------------------------------------------------
   function A = HESSM(A,N)
   if (N <= 2)
      return
   end
%                                USE Mij TRANSFORMATIONS TO REDUCE A
%                                TO UPPER HESSENBERG FORM
   for I=2:N-1
%                                SEARCH FROM A(I,I-1) ON DOWN FOR
%                                LARGEST POTENTIAL PIVOT, A(L,I-1)
      BIG = abs(A(I,I-1));
      L = I;
      for J=I+1:N
         if (abs(A(J,I-1)) > BIG)
            BIG = abs(A(J,I-1));
            L = J;
         end
      end
%                                IF ALL SUBDIAGONAL ELEMENTS IN COLUMN
%                                I-1 ALREADY ZERO, GO ON TO NEXT COLUMN
      if (BIG == 0.0)
         continue
      end
%                                PREMULTIPLY BY Pil
%                                (SWITCH ROWS I AND L)
      for K=I-1:N
         TEMP = A(L,K);
         A(L,K) = A(I,K);
         A(I,K) = TEMP;
      end
%                                POSTMULTIPLY BY Pil^(-1) = Pil
%                                (SWITCH COLUMNS I AND L)
      for K=1:N
         TEMP = A(K,L);
         A(K,L) = A(K,I);
         A(K,I) = TEMP;
      end
      for J=I+1:N
         R = A(J,I-1)/A(I,I-1);
         if (R == 0.0)
            continue
         end
%                                PREMULTIPLY BY Mij^(-1)
%                                (SUBTRACT R TIMES ROW I FROM ROW J)
         for K=I-1:N
            A(J,K) = A(J,K) - R*A(I,K);
         end
```

```
%                               POSTMULTIPLY BY Mij
%                               (ADD R TIMES COLUMN J TO COLUMN I)
            for K=1:N
               A(K,I) = A(K,I) + R*A(K,J);
            end
         end
      end
%-----------------------------------------------------------------------
%-------------------------- FIGURE 3.4.5 -----------------------------
%-----------------------------------------------------------------------
      function A = LR(A,N,ERRLIM)
      if (N <= 2)
         return
      end
%                               USE LR ITERATION TO REDUCE HESSENBERG
%                               MATRIX A TO QUASI-TRIANGULAR FORM
%
%                               PIVOT = 'T' IF PIVOTING ALLOWED
      PIVOT = 'T';
      NITER = 1000*N;
      for ITER=1:NITER
%                               REDUCE A TO UPPER TRIANGULAR FORM USING
%                               GAUSSIAN ELIMINATION (PREMULTIPLY BY
%                               Mij^(-1) MATRICES)
         for I=1:N-1
            if (abs(A(I,I)) < ERRLIM & PIVOT == 'T')
%                               SWITCH ROWS I AND I+1 IF NECESSARY
%                               (PREMULTIPLY BY Pi,i+1)
               IPERM(I) = I+1;
               for K=I:N
                  TEMP = A(I+1,K);
                  A(I+1,K) = A(I,K);
                  A(I,K) = TEMP;
               end
            else
               IPERM(I) = I;
            end
            if (abs(A(I+1,I)) < ERRLIM)
               R = 0.0;
            else
               if (abs(A(I,I)) < ERRLIM)
                  error('***** LR ITERATION DOES NOT CONVERGE *****')
               end
               R = A(I+1,I)/A(I,I);
            end
%                               USE SAVE TO SAVE R FOR POST-
%                               MULTIPLICATION PHASE
            SAVE(I) = R;
            if (R == 0.0)
               continue
            end
%                               IF MATRIX TRIDIAGONAL, AND PIVOTING NOT
%                               DONE, LIMITS ON K CAN BE:  K = I , I+1
            for K=I:N
               A(I+1,K) = A(I+1,K) - R*A(I,K);
```

```
            end
         end
%                              NOW POSTMULTIPLY BY Mij MATRICES
         for I=1:N-1
            if (IPERM(I) ~= I)
%                              SWITCH COLUMNS I AND I+1 IF NECESSARY
%                              (POSTMULTIPLY BY Pi,i+1^(-1) = Pi,i+1)
               for K=1:I+1
                  TEMP = A(K,I+1);
                  A(K,I+1) = A(K,I);
                  A(K,I) = TEMP;
               end
            end
            R = SAVE(I);
            if (R == 0.0)
               continue
            end
%                              IF MATRIX TRIDIAGONAL, AND PIVOTING NOT
%                              DONE, LIMITS ON K CAN BE:  K = I , I+1
            for K=1:I+1
               A(K,I) = A(K,I) + R*A(K,I+1);
            end
         end
%                              SET NEARLY ZERO SUBDIAGONALS TO ZERO,
%                              TO AVOID UNDERFLOW.
         for I=1:N-1
            if (abs(A(I+1,I)) < ERRLIM)
               A(I+1,I) = 0.0;
            end
         end
%                              CHECK FOR CONVERGENCE TO "QUASI-
%                              TRIANGULAR" FORM.
         ICONV = 1;
         for I=2:N-1
            if (A(I,I-1) ~= 0.0 & A(I+1,I) ~= 0.0)
               ICONV = 0;
            end
         end
         if (ICONV == 1)
            return
         end
      end
      error('***** LR ITERATION DOES NOT CONVERGE *****')
%-----------------------------------------------------------------------
%--------------------------- FIGURE 3.5.2 ------------------------------
%-----------------------------------------------------------------------
      function [EIG,V] = DPOWER(A,N,EIG,V,IUPDAT)
%
% FUNCTION DPOWER FINDS ONE EIGENVALUE OF A, AND A CORRESPONDING
%   EIGENVECTOR, USING THE SHIFTED INVERSE POWER METHOD.
%
% ARGUMENTS
%
%          ON INPUT                          ON OUTPUT
%          --------                          ---------
```

```
%
%    A      - THE N BY N MATRIX.
%
%    N      - THE SIZE OF MATRIX A.
%
%    EIG    - A (COMPLEX) INITIAL GUESS AT      AN EIGENVALUE OF A,
%             AN EIGENVALUE.                    NORMALLY THE ONE CLOSEST
%                                               TO THE INITIAL GUESS.
%
%    V      - A (COMPLEX) STARTING VECTOR       AN EIGENVECTOR OF A,
%             FOR THE SHIFTED INVERSE POWER     CORRESPONDING TO THE
%             METHOD.  IF ALL COMPONENTS OF     COMPUTED EIGENVALUE.
%             V ARE ZERO ON INPUT, A RANDOM
%             STARTING VECTOR WILL BE USED.
%
%    IUPDAT - THE NUMBER OF SHIFTED INVERSE
%             POWER ITERATIONS TO BE DONE
%             BETWEEN UPDATES OF P.  IF
%             IUPDAT=1, P WILL BE UPDATED EVERY
%             ITERATION.  IF IUPDAT > 1000,
%             P WILL NEVER BE UPDATED.
%
%-----------------------------------------------------------------------
%                                 EPS = MACHINE FLOATING POINT RELATIVE
%                                       PRECISION
% *****************************
      EPS = eps;
% *****************************
%                                IF V = 0, GENERATE A RANDOM STARTING
%                                VECTOR
      if (V == 0)
      SEED = N+10000;
      DEN = 2.0^31-1.0;
      for I=1:N
         SEED = mod(7^5*SEED,DEN);
         V(I) = SEED/(DEN+1.0);
      end
      end
%                                 NORMALIZE V, AND SET VN=V
      VNORM = 0.0;
      for I=1:N
         VNORM = VNORM + abs(V(I))^2;
      end
      VNORM = sqrt(VNORM);
      for I=1:N
         V(I) = V(I)/VNORM;
         VN(I) = V(I);
      end
%                                 BEGIN SHIFTED INVERSE POWER ITERATION
      NITER = 1000;
      for ITER=0:NITER
         if (mod(ITER,IUPDAT) == 0)
%                                 EVERY IUPDAT ITERATIONS, UPDATE PN
%                                 AND SOLVE (A-PN*I)*VNP1 = VN
            PN = EIG;
```

```
            for I=1:N
            for J=1:N
               if (I == J)
                  B(I,J) = A(I,J) - PN;
               else
                  B(I,J) = A(I,J);
               end
            end
            end
            [VNP1,IPERM,B] = DLINEQ(B,N,V);
         else
%                              BETWEEN UPDATES, WE CAN USE THE LU
%                              DECOMPOSITION OF B=A-PN*I CALCULATED
%                              EARLIER, TO SOLVE B*VNP1=VN FASTER
            VNP1 = DRESLV(B,N,V,IPERM);
         end
%                              CALCULATE NEW EIGENVALUE ESTIMATE,
%                              PN + (VN*VN)/(VN*VNP1)
         RNUM = 0.0;
         RDEN = 0.0;
         for I=1:N
            RNUM = RNUM + VN(I)*VN(I);
            RDEN = RDEN + VN(I)*VNP1(I);
         end
         R = RNUM/RDEN;
         EIG = PN + R;
%                              SET V = NORMALIZED VNP1
         VNORM = 0.0;
         for I=1:N
            VNORM = VNORM + abs(VNP1(I))^2;
         end
         VNORM = sqrt(VNORM);
         for I=1:N
            V(I) = VNP1(I)/VNORM;
         end
%                              IF R*VNP1 = VN  (R = (VN*VN)/(VN*VNP1) ),
%                              ITERATION HAS CONVERGED.
         ERRMAX = 0.0;
         for I=1:N
            ERRMAX = max(ERRMAX,abs(R*VNP1(I)-VN(I)));
         end
         if (ERRMAX <= sqrt(EPS))
            return
         end
%                              SET VN = V = NORMALIZED VNP1
         for I=1:N
            VN(I) = V(I);
         end
      end
      error('***** INVERSE POWER METHOD DOES NOT CONVERGE *****')
%-----------------------------------------------------------------------
%-------------------------- FIGURE 3.6.1 ----------------------------
%-----------------------------------------------------------------------
      function POLY = DEGENP(A,B,N,LAM0)
%
```

```
% FUNCTION DEGENP CALLS HESSQZ TO REDUCE THE GENERALIZED EIGENVALUE PROBLEM
%
%              A*X = LAMBDA*B*X
%
% TO A SIMILAR PROBLEM WITH THE SAME EIGENVALUES, WHERE A IS UPPER
% HESSENBERG AND B IS UPPER TRIANGULAR, THEN CALCULATES THE COEFFICIENTS
% OF THE POLYNOMIAL DET(A-LAMBDA*B).  THE ROOTS OF THIS POLYNOMIAL WILL
% BE THE EIGENVALUES OF THE GENERALIZED PROBLEM.
%
% ARGUMENTS
%
%            ON INPUT                          ON OUTPUT
%            --------                          ---------
%
%   A      - THE N BY N A MATRIX.
%
%   B      - THE N BY N B MATRIX.
%
%   N      - THE SIZE OF MATRICES A AND B.
%
%   LAMO   - A SCALAR, SEE POLY. (SET
%            LAMO=0 TO GET THE USUAL
%            POLYNOMIAL COEFFICIENTS)
%
%   POLY   -                                   VECTOR OF LENGTH N+1,
%                                              CONTAINING THE POLYNOMIAL
%                                              COEFFICIENTS. DET(A-LAMBDA*B)=
%                                                 SUM FROM I=0 TO N OF
%                                                 POLY(I+1)*(LAMBDA-LAMO)^I
%
%-----------------------------------------------------------------------
%
%                       CALL HESSQZ TO REDUCE A TO UPPER HESSENBERG AND
%                       B TO UPPER TRIANGULAR FORM
      [A B] = HESSQZ(A,B,N);
%                       DEGENP USES A RECURRENCE RELATION TO CALCULATE
%                       DET(A-LAMBDA*B) AND ALL N DERIVATIVES AT
%                       LAMBDA = LAMO, FROM WHICH THE POLYNOMIAL
%                       COEFFICIENTS CAN BE FOUND.
      DET(1,N+1) = 1.0;
      for I=2:N+1
         DET(I,N+1) = 0.0;
      end
      DET(1,N) = A(N,N)-LAMO*B(N,N);
      DET(2,N) = -B(N,N);
      for I=3:N+1
        DET(I,N) = 0.0;
      end
      for K=N-1:-1:1
         DET(1,K) = (A(K,K)-LAMO*B(K,K))*DET(1,K+1);
         for I=1:N
            DET(I+1,K) = (A(K,K)-LAMO*B(K,K))*DET(I+1,K+1)-B(K,K)*DET(I,K+1);
         end
         FACT = 1.0;
         for J=K+1:N
```

```
            FACT = -FACT*A(J,J-1);
            if (A(K,J)==0.0 & B(K,J)==0.0)
               continue
            end
            DET(1,K) = DET(1,K) + FACT*(A(K,J)-LAM0*B(K,J))*DET(1,J+1);
            for I=1:N
               DET(I+1,K) = DET(I+1,K) + ...
               FACT*((A(K,J)-LAM0*B(K,J))*DET(I+1,J+1)-B(K,J)*DET(I,J+1));
            end
         end
      end
      for I=1:N+1
         POLY(I) = DET(I,1);
      end
%%%%%%%%%%
      function [A,B] = HESSQZ(A,B,N)
%
%  FUNCTION HESSQZ REDUCES THE GENERALIZED EIGENVALUE PROBLEM
%
%                A*X = LAMBDA*B*X
%
%  TO A SIMILAR PROBLEM WITH THE SAME EIGENVALUES, WHERE A IS
%  UPPER HESSENBERG AND B IS UPPER TRIANGULAR.
%
%  ARGUMENTS
%
%             ON INPUT                          ON OUTPUT
%             --------                          ---------
%
%    A      - THE N BY N A MATRIX.              A IS UPPER HESSENBERG.
%
%    B      - THE N BY N B MATRIX.              B IS UPPER TRIANGULAR.
%
%    N      - THE SIZE OF MATRICES A AND B.
%
%-----------------------------------------------------------------------
%                              PREMULTIPLY A AND B BY ORTHOGONAL MATRIX
%                              (PRODUCT OF GIVENS MATRICES) Q, SUCH
%                              THAT QB IS UPPER TRIANGULAR.
      for I=1:N-1
         for J=I+1:N
            if (B(J,I) == 0.0)
               continue
            end
            DEN = sqrt(B(I,I)^2+B(J,I)^2);
            S = -B(J,I)/DEN;
            C =  B(I,I)/DEN;
            for K=I:N
               BIK = C*B(I,K)-S*B(J,K);
               BJK = S*B(I,K)+C*B(J,K);
               B(I,K) = BIK;
               B(J,K) = BJK;
            end
            for K=1:N
               AIK = C*A(I,K)-S*A(J,K);
```

```
            AJK = S*A(I,K)+C*A(J,K);
            A(I,K) = AIK;
            A(J,K) = AJK;
         end
      end
   end
%                               PREMULTIPLY A AND B BY ORTHOGONAL MATRIX
%                               Q, AND POSTMULTIPLY BY ORTHOGONAL MATRIX
%                               Z, SUCH THAT QAZ IS UPPER HESSENBERG AND
%                               QBZ IS STILL UPPER TRIANGULAR
   for I=1:N-2
      for J=N:-1:I+2
         if (A(J,I) == 0.0)
            continue
         end
%                               PREMULTIPLY A TO ZERO A(J,I)
         DEN = sqrt(A(J-1,I)^2+A(J,I)^2);
         S = -A(J,I)/DEN;
         C =  A(J-1,I)/DEN;
         for K=I:N
            A1K = C*A(J-1,K) - S*A(J,K);
            A2K = S*A(J-1,K) + C*A(J,K);
            A(J-1,K) = A1K;
            A(J,K) = A2K;
         end
%                               PREMULTIPLY B BY SAME MATRIX, CREATING
%                               NEW NONZERO B(J,J-1)
         for K=J-1:N
            B1K = C*B(J-1,K) - S*B(J,K);
            B2K = S*B(J-1,K) + C*B(J,K);
            B(J-1,K) = B1K;
            B(J,K) = B2K;
         end
         if (B(J,J-1) == 0.0)
            continue
         end
%                               POSTMULTIPLY B TO ZERO B(J,J-1)
         DEN = sqrt(B(J,J-1)^2+B(J,J)^2);
         S = -B(J,J-1)/DEN;
         C =  B(J,J)/DEN;
         for K=1:J
            BK1 = C*B(K,J-1) + S*B(K,J);
            BK2 = -S*B(K,J-1) + C*B(K,J);
            B(K,J-1) = BK1;
            B(K,J) = BK2;
         end
%                               POSTMULTIPLY A BY SAME MATRIX
         for K=1:N
            AK1 = C*A(K,J-1) + S*A(K,J);
            AK2 = -S*A(K,J-1) + C*A(K,J);
            A(K,J-1) = AK1;
            A(K,J) = AK2;
         end
      end
   end
```

```
%-----------------------------------------------------------------------
%--------------------------- FIGURE 4.5.1 ----------------------------
%-----------------------------------------------------------------------
      function [P,X,Y] = DLPRG(A,B,C,N,M)
%
%  FUNCTION DLPRG USES THE SIMPLEX METHOD TO SOLVE THE PROBLEM
%
%             MAXIMIZE      P = C(1)*X(1) + ... + C(N)*X(N)
%
%   WITH X(1),...,X(N) NONNEGATIVE, AND
%
%          A(1,1)*X(1) + ... + A(1,N)*X(N)  =  B(1)
%            .                   .               .
%            .                   .               .
%          A(M,1)*X(1) + ... + A(M,N)*X(N)  =  B(M)
%
%   WHERE B(1),...,B(M) ARE ASSUMED TO BE NONNEGATIVE.
%
%  ARGUMENTS
%
%             ON INPUT                         ON OUTPUT
%             --------                         ---------
%
%   A       - THE M BY N CONSTRAINT COEFFICIENT
%             MATRIX.
%
%   B       - A VECTOR OF LENGTH M CONTAINING
%             THE RIGHT HAND SIDES OF THE
%             CONSTRAINTS.  THE COMPONENTS OF
%             B MUST ALL BE NONNEGATIVE.
%
%   C       - A VECTOR OF LENGTH N CONTAINING
%             THE COEFFICIENTS OF THE OBJECTIVE
%             FUNCTION.
%
%   N       - THE NUMBER OF UNKNOWNS.
%
%   M       - THE NUMBER OF CONSTRAINTS.
%
%   P       -                                  THE MAXIMUM OF THE
%                                              OBJECTIVE FUNCTION.
%
%   X       -                                  A VECTOR OF LENGTH N
%                                              WHICH CONTAINS THE LP
%                                              SOLUTION.
%
%   Y       -                                  A VECTOR OF LENGTH M
%                                              WHICH CONTAINS THE DUAL
%                                              SOLUTION.
%
%-----------------------------------------------------------------------
%                              EPS = MACHINE FLOATING POINT RELATIVE
%                                    PRECISION
% ****************************
      EPS = eps;
```

```
% ****************************
      P = 0;
      X = zeros(N,1);
      Y = zeros(M,1);
%                               BASIS(1),...,BASIS(M) HOLD NUMBERS OF
%                               BASIS VARIABLES.  INITIAL BASIS CONSISTS
%                               OF ARTIFICIAL VARIABLES ONLY
      for I=1:M
         BASIS(I) = N+I;
         if (B(I) < 0.0)
            disp ('***** ALL B(I) MUST BE NONNEGATIVE *****')
            return
         end
      end
%                               INITIALIZE SIMPLEX TABLEAU
      for I=1:M+2
      for J=1:N+M+1
         TAB(I,J) = 0.0;
      end
      end
%                               LOAD A INTO UPPER LEFT HAND CORNER
%                               OF TABLEAU
      for I=1:M
      for J=1:N
         TAB(I,J) = A(I,J);
      end
      end
%                               LOAD M BY M IDENTITY TO RIGHT OF A
%                               AND LOAD B INTO LAST COLUMN
      for I=1:M
         TAB(I,N+I) = 1.0;
         TAB(I,N+M+1) = B(I);
      end
%                               ROW M+1 CONTAINS -C, INITIALLY
      for J=1:N
         TAB(M+1,J) = -C(J);
      end
%                               ROW M+2 CONTAINS COEFFICIENTS OF
%                               "ALPHA", WHICH IS TREATED AS +INFINITY
      for I=1:M
         TAB(M+2,N+I) = 1.0;
      end
%                               CLEAR "ALPHAS" IN LAST ROW
      for I=1:M
      for J=1:N+M+1
         TAB(M+2,J) = TAB(M+2,J) - TAB(I,J);
      end
      end
%                               SIMPLEX METHOD CONSISTS OF TWO PHASES
      for IPHASE=1:2
         if (IPHASE == 1)
%                               PHASE I:  ROW M+2 (WITH COEFFICIENTS OF
%                               ALPHA) SEARCHED FOR MOST NEGATIVE ENTRY
            MROW = M+2;
            LIM = N+M;
```

```
        else
%                              PHASE II:  FIRST N ELEMENTS OF ROW M+1
%                              SEARCHED FOR MOST NEGATIVE ENTRY
%                              (COEFFICIENTS OF ALPHA NONNEGATIVE NOW)
           MROW = M+1;
           LIM = N;
%                              IF ANY ARTIFICIAL VARIABLES LEFT IN
%                              BASIS AT BEGINNING OF PHASE II, THERE
%                              IS NO FEASIBLE SOLUTION
           for I=1:M
              if (BASIS(I) > LIM)
                 disp ('***** NO FEASIBLE SOLUTION *****')
                 return
              end
           end
        end
%                              THRESH = SMALL NUMBER.  WE ASSUME SCALES
%                              OF A AND C ARE NOT *TOO* DIFFERENT
        THRESH = 0.0;
        for J=1:LIM
           THRESH = max(THRESH,abs(TAB(MROW,J)));
        end
        THRESH = 1000*EPS*THRESH;
%                              BEGINNING OF SIMPLEX STEP
        while (1 > 0)
%                              FIND MOST NEGATIVE ENTRY IN ROW MROW,
%                              IDENTIFYING PIVOT COLUMN JP
           CMIN = -THRESH;
           JP = 0;
           for J=1:LIM
              if (TAB(MROW,J) < CMIN)
                 CMIN = TAB(MROW,J);
                 JP = J;
              end
           end
%                              IF ALL ENTRIES NONNEGATIVE (ACTUALLY,
%                              IF GREATER THAN -THRESH) PHASE ENDS
           if (JP == 0)
              break
           end
%                              FIND SMALLEST POSITIVE RATIO
%                              B(*)/TAB(*,JP), IDENTIFYING PIVOT
%                              ROW IP
           RATMIN = 0.0;
           IP = 0;
           for I=1:M
              if (TAB(I,JP) > THRESH)
                 RATIO = TAB(I,N+M+1)/TAB(I,JP);
                 if (IP == 0 || RATIO < RATMIN)
                    RATMIN = RATIO;
                    IP = I;
                 end
              end
           end
%                              IF ALL RATIOS NONPOSITIVE, MAXIMUM
```

```
%                               IS UNBOUNDED
           if (IP == 0)
              disp ('***** UNBOUNDED MAXIMUM *****')
              return
           end
%                               ADD X(JP) TO BASIS
           BASIS(IP) = JP;
%                               NORMALIZE PIVOT ROW TO MAKE TAB(IP,JP)=1
           AMULT = 1.0/TAB(IP,JP);
           for J=1:N+M+1
              TAB(IP,J) = AMULT*TAB(IP,J);
           end
%                               ADD MULTIPLES OF PIVOT ROW TO OTHER
%                               ROWS, TO KNOCK OUT OTHER ELEMENTS IN
%                               PIVOT COLUMN
           for I=1:MROW
              if (I == IP)
                 continue
              end
              AMULT = TAB(I,JP);
              for J=1:N+M+1
                 TAB(I,J) = TAB(I,J) - AMULT*TAB(IP,J);
              end
           end
        end
%                               END OF SIMPLEX STEP
     end
%                               END OF PHASE II; READ X,P,Y FROM
%                               FINAL TABLEAU
     for J=1:N
        X(J) = 0.0;
     end
     for I=1:M
        K = BASIS(I);
        X(K) = TAB(I,N+M+1);
     end
     P = TAB(M+1,N+M+1);
     for I=1:M
        Y(I) = TAB(M+1,N+I);
     end
%-----------------------------------------------------------------------
%--------------------------- FIGURE 4.6.1 ------------------------------
%-----------------------------------------------------------------------
     function [P,X,Y] = DLPRV(A,IROW,JCOL,NZ,B,C,N,M)
%
%  FUNCTION DLPRV USES THE REVISED SIMPLEX METHOD TO SOLVE THE PROBLEM
%
%             MAXIMIZE     P = C(1)*X(1) + ... + C(N)*X(N)
%
%    WITH X(1),...,X(N) NONNEGATIVE, AND
%
%          A(1,1)*X(1) + ... + A(1,N)*X(N)   = B(1)
%            .                    .              .
%            .                    .              .
%          A(M,1)*X(1) + ... + A(M,N)*X(N)   = B(M)
```

```
%
%    WHERE B(1),...,B(M) ARE ASSUMED TO BE NONNEGATIVE.
%
%  ARGUMENTS
%
%              ON INPUT                          ON OUTPUT
%              --------                          ---------
%
%    A       - A(IZ) IS THE CONSTRAINT MATRIX
%              ELEMENT IN ROW IROW(IZ), COLUMN
%              JCOL(IZ), FOR IZ=1,...,NZ.
%
%    IROW    - (SEE A).
%
%    JCOL    - (SEE A).
%
%    NZ      - NUMBER OF NONZEROS IN A.
%
%    B       - A VECTOR OF LENGTH M CONTAINING
%              THE RIGHT HAND SIDES OF THE
%              CONSTRAINTS.  THE COMPONENTS OF
%              B MUST ALL BE NONNEGATIVE.
%
%    C       - A VECTOR OF LENGTH N CONTAINING
%              THE COEFFICIENTS OF THE OBJECTIVE
%              FUNCTION.
%
%    N       - THE NUMBER OF UNKNOWNS.
%
%    M       - THE NUMBER OF CONSTRAINTS.
%
%    P       -                                   THE MAXIMUM OF THE
%                                                OBJECTIVE FUNCTION.
%
%    X       -                                   A VECTOR OF LENGTH N
%                                                WHICH CONTAINS THE LP
%                                                SOLUTION.
%
%    Y       -                                   A VECTOR OF LENGTH M
%                                                WHICH CONTAINS THE DUAL
%                                                SOLUTION.
%
%-----------------------------------------------------------------------
%
%                              EPS = MACHINE FLOATING POINT RELATIVE
%                                    PRECISION
% *****************************
      EPS = eps;
% *****************************
      P = 0;
      X = zeros(N,1);
      Y = zeros(M,1);
%                              INITIALIZE Ab^(-1) TO IDENTITY
      for I=1:M
      for J=1:M
```

```
            ABINV(I,J) = 0.0;
            if (I == J)
               ABINV(I,J) = 1.0;
            end
         end
      end
%                                OBJECTIVE FUNCTION COEFFICIENTS ARE
%                                CC(I,1) + CC(I,2)*ALPHA, WHERE "ALPHA"
%                                IS TREATED AS INFINITY
      for I=1:N+M
         CC(I,1) = 0.0;
         CC(I,2) = 0.0;
         if (I <= N)
            CC(I,1) = C(I);
         else
            CC(I,2) = -1;
         end
      end
%                                BASIS(1),...,BASIS(M) HOLD NUMBERS OF
%                                BASIS VARIABLES.  INITIAL BASIS CONSISTS
%                                OF ARTIFICIAL VARIABLES ONLY
      for I=1:M
         K = N+I;
         BASIS(I) = K;
%                                INITIALIZE Y TO Ab^(-T)*Cb = Cb
         YY(I,1) = CC(K,1);
         YY(I,2) = CC(K,2);
%                                INITIALIZE Xb TO Ab^(-1)*B = B
         XB(I) = B(I);
         if (B(I) < 0.0)
            disp ('***** ALL B(I) MUST BE NONNEGATIVE *****')
            return
         end
      end
%                                SIMPLEX METHOD CONSISTS OF TWO PHASES
      for IPHASE=1:2
         if (IPHASE == 1)
%                                PHASE I:  ROW 2 OF D (WITH COEFFICIENTS OF
%                                ALPHA) SEARCHED FOR MOST NEGATIVE ENTRY
            MROW = 2;
            LIM = N+M;
         else
%                                PHASE II:  FIRST N ELEMENTS OF ROW 1 OF
%                                D SEARCHED FOR MOST NEGATIVE ENTRY
%                                (COEFFICIENTS OF ALPHA NONNEGATIVE NOW)
            MROW = 1;
            LIM = N;
%                                IF ANY ARTIFICIAL VARIABLES LEFT IN
%                                BASIS AT BEGINNING OF PHASE II, THERE
%                                IS NO FEASIBLE SOLUTION
            for I=1:M
               if (BASIS(I) > LIM)
                  disp (' ***** NO FEASIBLE SOLUTION *****')
                  return
               end
```

```
            end
         end
%                              THRESH = SMALL NUMBER.  WE ASSUME SCALES
%                              OF A AND C ARE NOT *TOO* DIFFERENT
         THRESH = 0.0;
         for J=1:LIM
            THRESH = max(THRESH,abs(CC(J,MROW)));
         end
         THRESH = 1000*EPS*THRESH;
%                              BEGINNING OF SIMPLEX STEP
         while (1 > 0)
%                              D^T = Y^T*A - C^T
            for IR=1:MROW
               for J=1:N+M
                  D(J,IR) = -CC(J,IR);
               end
%                              LAST M COLUMNS OF A FORM IDENTITY MATRIX
               for J=1:M
                  D(N+J,IR) = D(N+J,IR) + YY(J,IR);
               end
%                              FIRST N COLUMNS STORED IN SPARSE A MATRIX
               for IZ=1:NZ
                  I = IROW(IZ);
                  J = JCOL(IZ);
                  D(J,IR) = D(J,IR) + A(IZ)*YY(I,IR);
               end
            end
%                              FIND MOST NEGATIVE ENTRY OF ROW MROW
%                              OF D, IDENTIFYING PIVOT COLUMN JP
            CMIN = -THRESH;
            JP = 0;
            for J=1:LIM
               if (D(J,MROW) < CMIN)
                  CMIN = D(J,MROW);
                  JP = J;
               end
            end
%                              IF ALL ENTRIES NONNEGATIVE (ACTUALLY,
%                              IF GREATER THAN -THRESH) PHASE ENDS
            if (JP == 0)
               break
            end
%                              COPY JP-TH COLUMN OF A ONTO VECTOR AP
            if (JP <= N)
%                              JP-TH COLUMN IS PART OF SPARSE A MATRIX
               for I=1:M
                  AP(I) = 0;
               end
               for IZ=1:NZ
                  J = JCOL(IZ);
                  if (J == JP)
                     I = IROW(IZ);
                     AP(I) = A(IZ);
                  end
               end
```

```
         else
%                             JP-TH COLUMN IS COLUMN OF FINAL IDENTITY
            for I=1:M
               AP(I) = 0;
            end
            AP(JP-N) = 1;
         end
%                             V = Ab^(-1)*AP
         for I=1:M
            V(I) = 0.0;
            for J=1:M
               V(I) = V(I) + ABINV(I,J)*AP(J);
            end
         end
%                             FIND SMALLEST POSITIVE RATIO
%                             Xb(I)/V(I), IDENTIFYING PIVOT ROW IP
         RATMIN = 0.0;
         IP = 0;
         for I=1:M
            if (V(I) > THRESH)
               RATIO = XB(I)/V(I);
               if (IP==0 || RATIO < RATMIN)
                  RATMIN = RATIO;
                  IP = I;
               end
            end
         end
%                             IF ALL RATIOS NONPOSITIVE, MAXIMUM
%                             IS UNBOUNDED
         if (IP == 0)
            disp (' ***** UNBOUNDED MAXIMUM *****')
            return
         end
%                             ADD X(JP) TO BASIS
         BASIS(IP) = JP;
%                             UPDATE Ab^(-1) = E^(-1)*Ab^(-1)
%                             Xb = E^(-1)*Xb
         for J=1:M
            ABINV(IP,J) = ABINV(IP,J)/V(IP);
         end
         XB(IP) = XB(IP)/V(IP);
         for I=1:M
            if (I == IP)
               continue
            end
            for J=1:M
               ABINV(I,J) = ABINV(I,J) - V(I)*ABINV(IP,J);
            end
            XB(I) = XB(I) - V(I)*XB(IP);
         end
%                             CALCULATE Y = Ab^(-T)*Cb
         for IR=1:MROW
            for I=1:M
               YY(I,IR) = 0.0;
               for J=1:M
```

```
                    K = BASIS(J);
                    YY(I,IR) = YY(I,IR) + ABINV(J,I)*CC(K,IR);
                 end
              end
           end
        end
%                                END OF SIMPLEX STEP
     end
%                                END OF PHASE II; CALCULATE X
     for J=1:N
        X(J) = 0.0;
     end
     for I=1:M
        K = BASIS(I);
        X(K) = XB(I);
        Y(I) = YY(I,1);
     end
%                                CALCULATE P
     P = 0.0;
     for I=1:N
        P = P + C(I)*X(I);
     end
%-----------------------------------------------------------------------
%-------------------------- FIGURE 4.6.2 -------------------------------
%-----------------------------------------------------------------------
     function [P,X,Y] = DLPRVS(A,IROW,JCOL,NZ,B,C,N,M)
%
% FUNCTION DLPRVS USES THE REVISED SIMPLEX METHOD TO SOLVE THE PROBLEM
%
%           MAXIMIZE     P = C(1)*X(1) + ... + C(N)*X(N)
%
%   WITH X(1),...,X(N) NONNEGATIVE, AND
%
%          A(1,1)*X(1) + ... + A(1,N)*X(N)   = B(1)
%            .                    .              .
%            .                    .              .
%          A(M,1)*X(1) + ... + A(M,N)*X(N)   = B(M)
%
%   WHERE B(1),...,B(M) ARE ASSUMED TO BE NONNEGATIVE.
%
% ARGUMENTS
%
%           ON INPUT                          ON OUTPUT
%           --------                          ---------
%
%   A      - A(IZ) IS THE CONSTRAINT MATRIX
%            ELEMENT IN ROW IROW(IZ), COLUMN
%            JCOL(IZ), FOR IZ=1,...,NZ.
%
%   IROW   - (SEE A).
%
%   JCOL   - (SEE A).
%
%   NZ     - NUMBER OF NONZEROS IN A.
%
```

```
%    B      - A VECTOR OF LENGTH M CONTAINING
%             THE RIGHT HAND SIDES OF THE
%             CONSTRAINTS.  THE COMPONENTS OF
%             B MUST ALL BE NONNEGATIVE.
%
%    C      - A VECTOR OF LENGTH N CONTAINING
%             THE COEFFICIENTS OF THE OBJECTIVE
%             FUNCTION.
%
%    N      - THE NUMBER OF UNKNOWNS.
%
%    M      - THE NUMBER OF CONSTRAINTS.
%
%    P      -                                 THE MAXIMUM OF THE
%                                             OBJECTIVE FUNCTION.
%
%    X      -                                 A VECTOR OF LENGTH N
%                                             WHICH CONTAINS THE LP
%                                             SOLUTION.
%
%    Y      -                                 A VECTOR OF LENGTH M
%                                             WHICH CONTAINS THE DUAL
%                                             SOLUTION.
%
%  NOTE: DLPRVS CALLS A SPARSE LINEAR SYSTEM SOLVER
%
%          X = SPARSOL(A,IROW,JCOL,NZ,B,N)
%
%  TO SOLVE THE N BY N SPARSE SYSTEM AX=B, WHERE A(IZ), IZ=1,...,NZ,
%  IS THE NONZERO ELEMENT OF A IN ROW IROW(IZ), COLUMN JCOL(IZ), B IS
%  THE RIGHT HAND SIDE VECTOR AND X IS THE SOLUTION.
%
%  THE CALL TO SPARSOL SHOULD BE REPLACED BY A CALL TO A DIRECT LINEAR
%  SYSTEM SOLVER (ITERATIVE SOLVERS NOT RECOMMENDED).
%
%-----------------------------------------------------------------------
%
%                              EPS = MACHINE FLOATING POINT RELATIVE
%                                    PRECISION
% *****************************
      EPS = eps;
% *****************************
      P = 0;
      X = zeros(N,1);
      Y = zeros(M,1);
%                              INITIALIZE Ab TO IDENTITY
      NZB = M;
      for I=1:M
         IROWB(I) = I;
         JCOLB(I) = I;
         AB(I) = 1.0;
      end
%                              OBJECTIVE FUNCTION COEFFICIENTS ARE
%                              CC(I,1) + CC(I,2)*ALPHA, WHERE "ALPHA"
%                              IS TREATED AS INFINITY
```

```
      for I=1:N+M
         CC(I,1) = 0.0;
         CC(I,2) = 0.0;
         if (I <= N)
            CC(I,1) = C(I);
         else
            CC(I,2) = -1;
         end
      end
%                                 BASIS(1),...,BASIS(M) HOLD NUMBERS OF
%                                 BASIS VARIABLES.  INITIAL BASIS CONSISTS
%                                 OF ARTIFICIAL VARIABLES ONLY
      for I=1:M
         K = N+I;
         BASIS(I) = K;
%                                 INITIALIZE Y TO Ab^(-T)*Cb = Cb
         YY(I,1) = CC(K,1);
         YY(I,2) = CC(K,2);
%                                 INITIALIZE Xb TO Ab^(-1)*B = B
         XB(I) = B(I);
         if (B(I) < 0.0)
            disp ('***** ALL B(I) MUST BE NONNEGATIVE *****')
            return
         end
      end
%                                 SIMPLEX METHOD CONSISTS OF TWO PHASES
      for IPHASE=1:2
         if (IPHASE == 1)
%                                 PHASE I:  ROW 2 OF D (WITH COEFFICIENTS OF
%                                 ALPHA) SEARCHED FOR MOST NEGATIVE ENTRY
            MROW = 2;
            LIM = N+M;
         else
%                                 PHASE II:  FIRST N ELEMENTS OF ROW 1 OF
%                                 D SEARCHED FOR MOST NEGATIVE ENTRY
%                                 (COEFFICIENTS OF ALPHA NONNEGATIVE NOW)
            MROW = 1;
            LIM = N;
%                                 IF ANY ARTIFICIAL VARIABLES LEFT IN
%                                 BASIS AT BEGINNING OF PHASE II, THERE
%                                 IS NO FEASIBLE SOLUTION
            for I=1:M
               if (BASIS(I) > LIM)
                  disp (' ***** NO FEASIBLE SOLUTION *****')
                  return
               end
            end
         end
%                                 THRESH = SMALL NUMBER.  WE ASSUME SCALES
%                                 OF A AND C ARE NOT *TOO* DIFFERENT
         THRESH = 0.0;
         for J=1:LIM
            THRESH = max(THRESH,abs(CC(J,MROW)));
         end
         THRESH = 1000*EPS*THRESH;
```

```
%                               BEGINNING OF SIMPLEX STEP
         while (1 > 0)
%                               D^T = Y^T*A - C^T
            for IR=1:MROW
               for J=1:N+M
                  D(J,IR) = -CC(J,IR);
               end
%                               LAST M COLUMNS OF A FORM IDENTITY MATRIX
               for J=1:M
                  D(N+J,IR) = D(N+J,IR) + YY(J,IR);
               end
%                               FIRST N COLUMNS STORED IN SPARSE A MATRIX
               for IZ=1:NZ
                  I = IROW(IZ);
                  J = JCOL(IZ);
                  D(J,IR) = D(J,IR) + A(IZ)*YY(I,IR);
               end
            end
%                               FIND MOST NEGATIVE ENTRY OF ROW MROW
%                               OF D, IDENTIFYING PIVOT COLUMN JP
            CMIN = -THRESH;
            JP = 0;
            for J=1:LIM
               if (D(J,MROW) < CMIN)
                  CMIN = D(J,MROW);
                  JP = J;
               end
            end
%                               IF ALL ENTRIES NONNEGATIVE (ACTUALLY,
%                               IF GREATER THAN -THRESH) PHASE ENDS
            if (JP == 0)
               break
            end
%                               COPY JP-TH COLUMN OF A ONTO VECTOR AP
            if (JP <= N)
%                               JP-TH COLUMN IS PART OF SPARSE A MATRIX
               for I=1:M
                  AP(I) = 0;
               end
               for IZ=1:NZ
                  J = JCOL(IZ);
                  if (J == JP)
                     I = IROW(IZ);
                     AP(I) = A(IZ);
                  end
               end
            else
%                               JP-TH COLUMN IS COLUMN OF FINAL IDENTITY
               for I=1:M
                  AP(I) = 0;
               end
               AP(JP-N) = 1;
            end
%                               SOLVE Ab*V = AP
            V = SPARSOL(AB,IROWB,JCOLB,NZB,AP,M);
```

```
%                               FIND SMALLEST POSITIVE RATIO
%                               Xb(I)/V(I), IDENTIFYING PIVOT ROW IP
         RATMIN = 0.0;
         IP = 0;
         for I=1:M
            if (V(I) > THRESH)
               RATIO = XB(I)/V(I);
               if (IP==0 || RATIO < RATMIN)
                  RATMIN = RATIO;
                  IP = I;
               end
            end
         end
%                               IF ALL RATIOS NONPOSITIVE, MAXIMUM
%                               IS UNBOUNDED
         if (IP == 0)
            disp (' ***** UNBOUNDED MAXIMUM *****')
            return
         end
%                               ADD X(JP) TO BASIS
         BASIS(IP) = JP;
%                               UPDATE Ab.  PUT NONZEROS OF JP-TH
%                               COLUMN OF A (=AP) INTO COLUMN IP
%                               OF SPARSE MATRIX Ab
         NZBOLD = NZB;
         for I=1:M
            if (AP(I) ~= 0.0)
               NZB = NZB+1;
               IROWB(NZB) = I;
               JCOLB(NZB) = IP;
               AB(NZB) = AP(I);
            end
         end
         NZBNEW = NZB;
%                               REMOVE ELEMENTS OF OLD COLUMN IP
         NZB = 0;
         for IZ=1:NZBNEW
            if (JCOLB(IZ) ~= IP || IZ > NZBOLD)
               NZB = NZB+1;
               JCOLB(NZB) = JCOLB(IZ);
               IROWB(NZB) = IROWB(IZ);
               AB(NZB) = AB(IZ);
            end
         end
%                               SOLVE Ab*Xb = B
         XB = SPARSOL(AB,IROWB,JCOLB,NZB,B,M);
%                               SOLVE Ab^T*Y = Cb
         for IR=1:MROW
            for J=1:M
               K = BASIS(J);
               V(J) = CC(K,IR);
            end
            YY(1:M,IR) = SPARSOL(AB,JCOLB,IROWB,NZB,V,M);
         end
      end
```

```
%                               END OF SIMPLEX STEP
      end
%                               END OF PHASE II; CALCULATE X
      for J=1:N
         X(J) = 0.0;
      end
      for I=1:M
         K = BASIS(I);
         X(K) = XB(I);
         Y(I) = YY(I,1);
      end
%                               CALCULATE P
      P = 0.0;
      for I=1:N
         P = P + C(I)*X(I);
      end
%-----------------------------------------------------------------------
%---------------------------- FIGURE 4.7.1 -----------------------------
%-----------------------------------------------------------------------
      function [CMIN,X,Y] = DTRAN(WCAP,SREQ,COST,NW,NS)
%
% FUNCTION DTRAN SOLVES THE TRANSPORTATION PROBLEM
%
%   MINIMIZE    CMIN = COST(1,1)*X(1,1) + ... + COST(NW,NS)*X(NW,NS)
%
%   WITH X(1,1),...,X(NW,NS) NONNEGATIVE, AND
%
%         X(1,1) + ... + X(1,NS)  .LE. WCAP(1)
%            .               .            .
%            .               .            .
%         X(NW,1)+ ... + X(NW,NS) .LE. WCAP(NW)
%         X(1,1) + ... + X(NW,1)    =  SREQ(1)
%            .               .            .
%            .               .            .
%         X(1,NS)+ ... + X(NW,NS)   =  SREQ(NS)
%
% CAUTION: IF TOTAL STORE REQUIREMENTS EXACTLY EQUAL TOTAL WAREHOUSE
%          CAPACITIES, ALTER ONE WCAP(I) OR SREQ(I) SLIGHTLY, SO THAT
%          WAREHOUSE CAPACITIES SLIGHTLY EXCEED STORE REQUIREMENTS.
%
% ARGUMENTS
%
%            ON INPUT                          ON OUTPUT
%            --------                          ---------
%
%   WCAP   - A VECTOR OF LENGTH NW CONTAINING
%            THE WAREHOUSE CAPACITIES.
%
%   SREQ   - A VECTOR OF LENGTH NS CONTAINING
%            THE STORE REQUIREMENTS.
%
%   COST   - THE NW BY NS COST MATRIX. COST(I,J)
%            IS THE PER UNIT COST TO SHIP FROM
%            WAREHOUSE I TO STORE J.
%
```

```
%     NW      - THE NUMBER OF WAREHOUSES.
%
%     NS      - THE NUMBER OF STORES.
%
%     CMIN    -                                     THE TOTAL COST OF THE
%                                                   OPTIMAL ROUTING.
%
%     X       -                                     AN NW BY NS MATRIX
%                                                   CONTAINING THE OPTIMAL
%                                                   ROUTING.  X(I,J) UNITS
%                                                   SHOULD BE SHIPPED FROM
%                                                   WAREHOUSE I TO STORE J.
%
%     Y       -                                     A VECTOR OF LENGTH NW+NS
%                                                   CONTAINING THE DUAL. Y(I)
%                                                   GIVES THE DECREASE IN
%                                                   TOTAL COST PER UNIT
%                                                   INCREASE IN WCAP(I), FOR
%                                                   SMALL INCREASES, AND
%                                                   -Y(NW+J) GIVES THE
%                                                   INCREASE IN TOTAL COST
%                                                   PER UNIT INCREASE IN
%                                                   SREQ(J).
%
%-----------------------------------------------------------------------
      M = NW+NS;
      N = NW*NS+NW;
%                                 SET UP SPARSE CONSTRAINT MATRIX
      NZ = 0;
      for I=1:NW
         for J=1:NS
            NZ = NZ+1;
            IROW(NZ) = I;
            JCOL(NZ) = (I-1)*NS + J;
            A(NZ) = 1.0;
         end
         NZ = NZ+1;
         IROW(NZ) = I;
         JCOL(NZ) = NW*NS+I;
         A(NZ) = 1.0;
%                                 LOAD WAREHOUSE CAPACITIES INTO B
         B(I) = WCAP(I);
      end
      for J=1:NS
         for I=1:NW
            NZ = NZ+1;
            IROW(NZ) = NW+J;
            JCOL(NZ) = J + (I-1)*NS;
            A(NZ) = 1.0;
         end
%                                 LOAD STORE REQUIREMENTS INTO B
         B(NW+J) = SREQ(J);
      end
%                                 FIRST NW*NS ENTRIES IN C ARE
%                                 -COST(I,J).  NEGATIVE SIGN USED
```

```
%                               BECAUSE WE WANT TO MINIMIZE COST
      K = 0;
      for I=1:NW
         for J=1:NS
            K = K+1;
            C(K) = -COST(I,J);
         end
      end
%                               NEXT NW COSTS ARE ZERO, CORRESPONDING
%                               TO WAREHOUSE CAPACITY SLACK VARIABLES
      for I=1:NW
         K = K+1;
         C(K) = 0.0;
      end
%                               USE REVISED SIMPLEX METHOD TO SOLVE
%                               TRANSPORTATION PROBLEM
      [P,XSOL,Y] = DLPRV(A,IROW,JCOL,NZ,B,C,N,M);
%                               FORM OPTIMAL ROUTING MATRIX, X
      CMIN = -P;
      K = 0;
      for I=1:NW
         for J=1:NS
            K = K+1;
            X(I,J) = XSOL(K);
         end
      end
%-----------------------------------------------------------------------
%-------------------------- FIGURE 5.3.1 ---------------------------
%-----------------------------------------------------------------------
      function F = DFFT(F,M)
%
%  FUNCTION DFFT PERFORMS A FAST FOURIER TRANSFORM ON THE COMPLEX
%     VECTOR F, OF LENGTH N=2^M.  THE FOURIER TRANSFORM IS DEFINED BY
%       Y(K) = SUM FROM J=1 TO N OF: EXP[I*2*PI*(K-1)*(J-1)/N]*F(J)
%     WHERE I = SQRT(-1).
%
%  ARGUMENTS
%
%             ON INPUT                          ON OUTPUT
%             --------                          ---------
%
%    F      - THE COMPLEX VECTOR OF LENGTH      THE TRANSFORMED VECTOR
%             2^M TO BE TRANSFORMED.            Y.
%
%    M      - THE LENGTH OF THE VECTOR F
%             IS ASSUMED TO BE 2^M.
%
%-----------------------------------------------------------------------
%                               FOURIER TRANSFORM OF A 1-VECTOR IS
%                               UNCHANGED
      if (M == 0)
         return
      end
      N = 2^M;
      H = cos(2*pi/N) + sin(2*pi/N)*i;
```

```
      N2 = N/2;
%                              COPY ODD COMPONENTS OF F TO FODD
%                              AND EVEN COMPONENTS TO FEVEN
      for K=1:N2
         FODD(K) = F(2*K-1);
         FEVEN(K)= F(2*K);
      end
%                              TRANSFORM N/2-VECTORS FODD AND FEVEN
      FODD  = DFFT(FODD ,M-1);
      FEVEN = DFFT(FEVEN,M-1);
      D = 1.0;
%                              Y = (FODD+D*FEVEN , FODD-D*FEVEN)
      for K=1:N2
         U = D*FEVEN(K);
         F(K)    = FODD(K) + U;
         F(N2+K) = FODD(K) - U;
         D = D*H;
      end
%-----------------------------------------------------------------------
%-------------------------- FIGURE 5.3.4 ---------------------------
%-----------------------------------------------------------------------
      function F = NRFFT(F,M)
%
%  FUNCTION NRFFT PERFORMS A FAST FOURIER TRANSFORM ON THE COMPLEX
%     VECTOR F, OF LENGTH N=2^M.  THE FOURIER TRANSFORM IS DEFINED BY
%       Y(K) = SUM FROM J=1 TO N OF: EXP[I*2*PI*(K-1)*(J-1)/N]*F(J)
%     WHERE I = SQRT(-1).
%
%  ARGUMENTS
%
%              ON INPUT                          ON OUTPUT
%              --------                          ---------
%
%    F      - THE COMPLEX VECTOR OF LENGTH      THE TRANSFORMED VECTOR
%              2^M TO BE TRANSFORMED.            Y.
%
%    M      - THE LENGTH OF THE VECTOR F
%              IS ASSUMED TO BE 2^M.
%
%-----------------------------------------------------------------------
      N = 2^M;
%                              REORDER ELEMENTS OF F AS THEY WILL BE
%                              IN TRANSFORMED VECTOR
      for I=M:-1:2
         NI = 2^I;
         N2 = NI/2;
         for J=0:NI:N-1
            for K=1:N2
               FODD(K) = F(J+2*K-1);
               FEVEN(K) = F(J+2*K);
            end
            for K=1:N2
               F(J+K) = FODD(K);
               F(J+N2+K) = FEVEN(K);
            end
```

```
         end
      end
%                               PROCESS ELEMENTS BY GROUPS OF 2,
%                               THEN BY GROUPS OF 4,8,...,2^M
      for I=1:M
         NI = 2^I;
         N2 = NI/2;
         H = cos(2*pi/NI) + sin(2*pi/NI)*i;
         for J=0:NI:N-1
            D = 1.0;
            for K=1:N2
               FODD(K) = F(J+K);
               FEVEN(K) = F(J+N2+K);
               U = D*FEVEN(K);
               F(J+K) = FODD(K) + U;
               F(J+N2+K) = FODD(K) - U;
               D = D*H;
            end
         end
      end
```

Appendix B—Answers to Selected Exercises

1.1. $\frac{1}{2}N^3$ multiplications (50% more than Gauss elimination).

1.5. Suppress pivoting, change "DO 25 K=I+1,N" to "DO 25 K=J,N" and only one further change is required.

1.6. (a).
$$L = \begin{bmatrix} 1 & 0 & 0 \\ -0.75 & 1 & 0 \\ -0.25 & 0.2 & 1 \end{bmatrix}, U = \begin{bmatrix} -4 & -4 & 4 \\ 0 & -5 & 5 \\ 0 & 0 & 1 \end{bmatrix}$$

1.7. When $N = 20$, $U(\pi/2) = 1.0041157$ for (a),(b), and (c). Here is a periodic tridiagonal solver for part (b):

```
      SUBROUTINE TRIPER(A,B,C,X,F,N)
      IMPLICIT DOUBLE PRECISION(A-H,O-Z)
C         SOLVE LINEAR SYSTEM WITH A PERIODIC, TRI-DIAGONAL MATRIX:
C
C               B(1)  C(1)                          A(1)
C               A(2)  B(2)  C(2)
C                     A(3)  B(3)  C(3)
C                             .     .     .
C
C               C(N)                          A(N)  B(N)
C
C                                A = SUBDIAGONAL OF COEFFICIENT MATRIX
C                                B = MAIN DIAGONAL
C                                C = SUPERDIAGONAL
C                                D = LAST ROW
C                                E = LAST COLUMN
C                                F = RIGHT HAND SIDE VECTOR
C                                X = SOLUTION
      DIMENSION A(N),B(N),C(N),D(N),E(N),X(N),F(N)
C                                COPY F ONTO X
      DO 5 I=1,N
```

```
         X(I) = F(I)
         D(I) = 0.0
         E(I) = 0.0
    5 CONTINUE
      D(1) = C(N)
      E(1) = A(1)
C                                  BEGIN FORWARD ELIMINATION
      DO 10 K=1,N-2
         IF (B(K).EQ.0.0) GO TO 20
         AMULA = -A(K+1)/B(K)
         AMULD = -D(K)/B(K)
         B(K+1) = B(K+1) + AMULA*C(K)
         B(N) = B(N) + AMULD*E(K)
         IF (K.LT.N-2) THEN
            E(K+1) = E(K+1) + AMULA*E(K)
            D(K+1) = D(K+1) + AMULD*C(K)
         ELSE
            C(K+1) = C(K+1) + AMULA*E(K)
            A(N) = A(N) + AMULD*C(K)
         ENDIF
         X(K+1) = X(K+1) + AMULA*X(K)
         X(N) = X(N) + AMULD*X(K)
   10 CONTINUE
      IF (B(N-1).EQ.0.0) GO TO 20
      AMULA = -A(N)/B(N-1)
      B(N) = B(N) + AMULA*C(N-1)
      X(N) = X(N) + AMULA*X(N-1)
      IF (B(N).EQ.0.0) GO TO 20
C                                  BACK SUBSTITUTION
      X(N) = X(N)/B(N)
      X(N-1) = (X(N-1)-C(N-1)*X(N))/B(N-1)
      DO 15 K=N-2,1,-1
         X(K) = (X(K)-C(K)*X(K+1)-E(K)*X(N))/B(K)
   15 CONTINUE
      RETURN
C                                  ZERO PIVOT ENCOUNTERED
   20 PRINT 25
   25 FORMAT (' ZERO PIVOT ENCOUNTERED')
      RETURN
      END
```

1.8. DBRESLV, shown below, will do about $N(N_{UD} + 2N_{LD})$ multiplications.

```
      SUBROUTINE DBRESLV(A,N,NLD,NUD,X,B,IPERM)
      IMPLICIT DOUBLE PRECISION (A-H,O-Z)
C                                  DECLARATIONS FOR ARGUMENTS
      DOUBLE PRECISION A(N,-NLD:NUD+NLD),X(N),B(N)
```

```
      INTEGER N,NLD,NUD,IPERM(N)
C                                 DECLARATIONS FOR LOCAL VARIABLES
      DOUBLE PRECISION B_(N),LJI
      DO 10 K=1,N
C                                 COPY B TO B_, SO B WILL NOT BE ALTERED
         B_(K) = B(K)
   10 CONTINUE
C                                 BEGIN FORWARD ELIMINATION
      DO 35 I=1,N-1
C                                 SWITCH B_(I) AND B_(L)
         L = IPERM(I)
         TEMP = B_(L)
         B_(L) = B_(I)
         B_(I) = TEMP
         DO 30 J=I+1,MIN(I+NLD,N)
C                                 CHOOSE MULTIPLIER TO ZERO AJI
            LJI = A(J,I-J)
C                                 SUBTRACT LJI TIMES B_(I) FROM B_(J)
            B_(J) = B_(J) - LJI*B_(I)
   30    CONTINUE
   35 CONTINUE
C                                 SOLVE U*X = B_ USING BACK SUBSTITUTION
      X(N) = B_(N)/A(N,0)
      DO 45 I=N-1,1,-1
         SUM = 0.0
         DO 40 J=I+1,MIN(I+NUD+NLD,N)
            SUM = SUM + A(I,J-I)*X(J)
   40    CONTINUE
         X(I) = (B_(I)-SUM)/A(I,0)
   45 CONTINUE
      RETURN
      END
```

1.13. $\lambda_{l,m,n} = 6 - 2\cos(\pi l/M) - 2\cos(\pi m/M) - 2\cos(\pi n/M)$

1.14. Number of iterations is about $2M^2\ln(1/\epsilon)/\pi^2$.

1.17. $A_{11}^{-1} = 0.7250783463$.

2.4. Texas A&M = 4.75, Texas = −14 (assuming UTEP is last variable, and thus UTEP = 0).

2.5. $p_3(x) = -3.4 + 6.5x - 2.25x^2 + 0.25x^3$.

2.6. (a).
$$Q = \frac{1}{65}\begin{bmatrix} 39 & 20 & -48 \\ -52 & 15 & -36 \\ 0 & 60 & 25 \end{bmatrix}, R = \begin{bmatrix} 5 & 0 \\ 0 & 13 \\ 0 & 0 \end{bmatrix}, \boldsymbol{x} = \begin{bmatrix} -1/25 \\ -5/169 \end{bmatrix}$$

2.8. (a).
$$Q = \begin{bmatrix} 0.44721 & -0.63246 & 0.53452 & -0.31623 & 0.11952 \\ 0.44721 & -0.31623 & -0.26726 & 0.63246 & -0.47809 \\ 0.44721 & 0.00000 & -0.53452 & 0.00000 & 0.71714 \\ 0.44721 & 0.31623 & -0.26726 & -0.63246 & -0.47809 \\ 0.44721 & 0.63246 & 0.53452 & 0.31623 & 0.11952 \end{bmatrix}$$

(b). Here is the modified REDQ called by DLLSQ2:

```
      SUBROUTINE REDQ2(A,M,N,B,PIVOT,NPIVOT,ERRLIM)
      IMPLICIT DOUBLE PRECISION (A-H,O-Z)
C                                DECLARATIONS FOR ARGUMENTS
      DOUBLE PRECISION A(M,N),B(M),ERRLIM
      INTEGER PIVOT(M),M,N,NPIVOT
C                                USE GIVENS ROTATIONS TO REDUCE
C                                A TO ROW ECHELON FORM
      I = 1
      DO 15 L=1,N
C                                USE PIVOT A(I,L) TO KNOCK OUT
C                                ELEMENTS I+1 TO M IN COLUMN L.
         DO 10 J=I+1,M
            IF (A(J,L).EQ.0.0) GO TO 10
            C = COS(A(J,L))
            S = SIN(A(J,L))
C                                PREMULTIPLY B BY Qij**T
            BI = C*B(I) + S*B(J)
            BJ =-S*B(I) + C*B(J)
            B(I) = BI
            B(J) = BJ
   10    CONTINUE
C                                PIVOT A(I,L) IS NONZERO AFTER
C                                PROCESSING COLUMN L--MOVE DOWN
C                                TO NEXT ROW, I+1
         IF (ABS(A(I,L)).LE.ERRLIM) A(I,L) = 0.0
         IF (A(I,L).NE.0.0) THEN
            NPIVOT = I
            PIVOT(NPIVOT) = L
            I = I+1
            IF (I.GT.M) RETURN
         ENDIF
   15 CONTINUE
      RETURN
      END
```

2.10. $p_5(x) = -4.37752 + 3.71201x + 5.08381x^2 - 4.48259x^3 + 1.16114x^4 - 0.09685x^5$.

3.3. Move "20 CONTINUE" to before "EK = 0.0".

3.9. 39 iterations (101 without shifts).

3.10. (a).
$$\begin{bmatrix} 1 & -1 & 6 & 1 \\ 2 & -1 & 9 & 2 \\ 0 & -1 & 6 & 1 \\ 0 & 0 & -13 & -2 \end{bmatrix}$$

(b).
$$\begin{bmatrix} 1 & 2 & 0 & 0 \\ -1 & 5 & -1 & 0 \\ 0 & 16 & -39/16 & -13 \\ 0 & 0 & 21/256 & 7/16 \end{bmatrix}$$

3.12. (a).
$$A(\text{new}) = \begin{bmatrix} 8.16288 & 0.00000 & 3.24497 \\ 0.00000 & 1.83772 & 2.33884 \\ 3.24497 & 2.33884 & 6.00000 \end{bmatrix}$$

(b).
$$A(\text{new}) = \begin{bmatrix} 7.60000 & 3.00000 & 0.00000 \\ 3.00000 & 7.77600 & 0.83200 \\ 0.00000 & 0.83200 & 0.62400 \end{bmatrix}$$

(c).
$$A(\text{new}) = \begin{bmatrix} 6.25000 & 3.00000 & 0.00000 \\ 2.81250 & 8.01667 & 4.00000 \\ 0.00000 & 1.84889 & 1.73333 \end{bmatrix}$$

3.14. $(\boldsymbol{v}_0^{\mathrm{T}}\boldsymbol{v}_{2n+1})/(\boldsymbol{v}_0^{\mathrm{T}}\boldsymbol{v}_{2n})$, so by doing n iterations with $\boldsymbol{u} = \boldsymbol{v}_n$ you can get the same result as doing $2n$ iterations with $\boldsymbol{u} = \boldsymbol{v}_0$.

3.17. (a). 9 iterations

(b). $>$ 1000 iterations

(c). 19 iterations

3.21. (a). Singular values = 0.16934, 4.41337, 6.29126, 16.10713, 20.13639.

(b).
$$UDV^{\mathrm{T}} = \begin{bmatrix} -4.98654 & 7.00467 & 3.08472 & 3.94083 & -8.00078 \\ 5.00304 & 8.00106 & 3.01916 & 5.98661 & 7.99982 \\ 3.01693 & -6.99412 & -2.89340 & -4.07446 & 4.99902 \\ -3.00037 & -.00013 & 3.99767 & 5.00162 & 3.00002 \\ 6.99995 & 3.99998 & 4.99971 & 9.00020 & 5.00000 \end{bmatrix}$$

3.22. (b).
$$A^{+} = \begin{bmatrix} \frac{1}{6} & \frac{1}{6} & \frac{1}{6} \\ \frac{1}{6} & \frac{1}{6} & \frac{1}{6} \end{bmatrix}$$

4.1. (b). Minimum is $P = 9$, at (3, 0).

4.2. Minimize $\boldsymbol{b}^{\mathrm{T}}\boldsymbol{y}$

with $A^{\mathrm{T}}\boldsymbol{y} \geq \boldsymbol{c}$ and $y_1, ..., y_k \geq 0$.

4.5. (a). $p_2(x) = \frac{10}{9} + \frac{13}{9}x - \frac{1}{9}x^2$.

(b). $p_2(x) = 1 + x + 0x^2$.

(c). $p_2(x) = 1 + 1.3286x - 0.0714x^2$.

Here is a Fortran function which minimizes the L_∞ norm error:

```
      FUNCTION RMININF(A,M,N,x,b)
      IMPLICIT DOUBLE PRECISION (A-H,O-Z)
      DIMENSION A(M,N),x(N),b(M)
      DIMENSION A1(N+1,2*M),B1(N+1),C1(2*M),X1(2*M),Y1(N+1)
      M1 = N+1
      N1 = 2*M
      DO I=1,N
         A1(I,1:M) = -A(1:M,I)
         A1(I,M+1:2*M) = A(1:M,I)
      ENDDO
      A1(N+1,1:2*M) = 1
      B1(1:N) = 0
      B1(N+1) = 1
      C1(1:M) = -b(1:M)
      C1(M+1:2*M) = b(1:M)
      CALL DLPRG(A1,B1,C1,N1,M1,P,X1,Y1)
      x(1:N) = Y1(1:N)
      RMININF = Y1(N+1)
      RETURN
      END
```

4.7. Results will vary depending on the random number generator used, but here is one set of results:

M	Iterations	Possible Bases
10	16	$3*10^7$
20	45	$4*10^{15}$
30	83	$7*10^{23}$
40	138	$1*10^{32}$
50	195	$2*10^{40}$

4.14. If $N_W = N_S = k$, DLPRG requires $O(k^3)$ memory and does $O(k^3)$ operations. DTRAN requires $O(k^2)$ memory and does $O(k^2)$ operations.

5.1. There should be peaks at $f(1)$, $f(16)$ and $f(5462)$.

5.5. (b). When $N = 2^9$, $U(0.5, 0.5) = -0.262116$.

(c). When $N = 2^9$, $U(0.5, 0.5) = -0.262116$.

(d). When $N = 2^8$, maximum error $= 1.12 * 10^{-4}$.

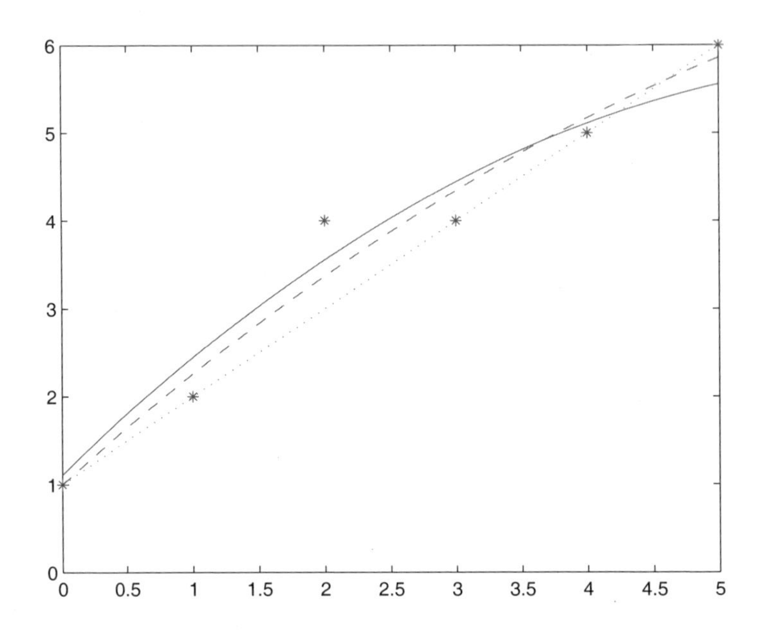

Problem 4.5
L_∞ fit (solid line), L_1 fit (dotted line) and L_2 fit (dashed line)

5.7. (a). About 3N COMPLEX*16 words are used.

(b). Make the user pass FODD and FEVEN in as workarrays (a little extra work for the user, that is why DFFT doesn't do this):

```
SUBROUTINE DFFT(F,N,FODD,FEVEN)
```

then use the array F as workspace for the lower level DFFT calls:

```
CALL DFFT(FODD,M-1,F(1),F(N2+1))
CALL DFFT(FEVEN,M-1,F(1),F(N2+1))
```

Now no automatic arrays are used, so the only memory used is the 2 N words allocated by the user in the main calling program.

5.8. Plot after removing high frequencies is shown on next page.

6.2. With $M = 16$, U at the midpoint should be 0.055881.

6.3. Here is a parallelized version of REDQ, where each processor stores only its own columns of A:

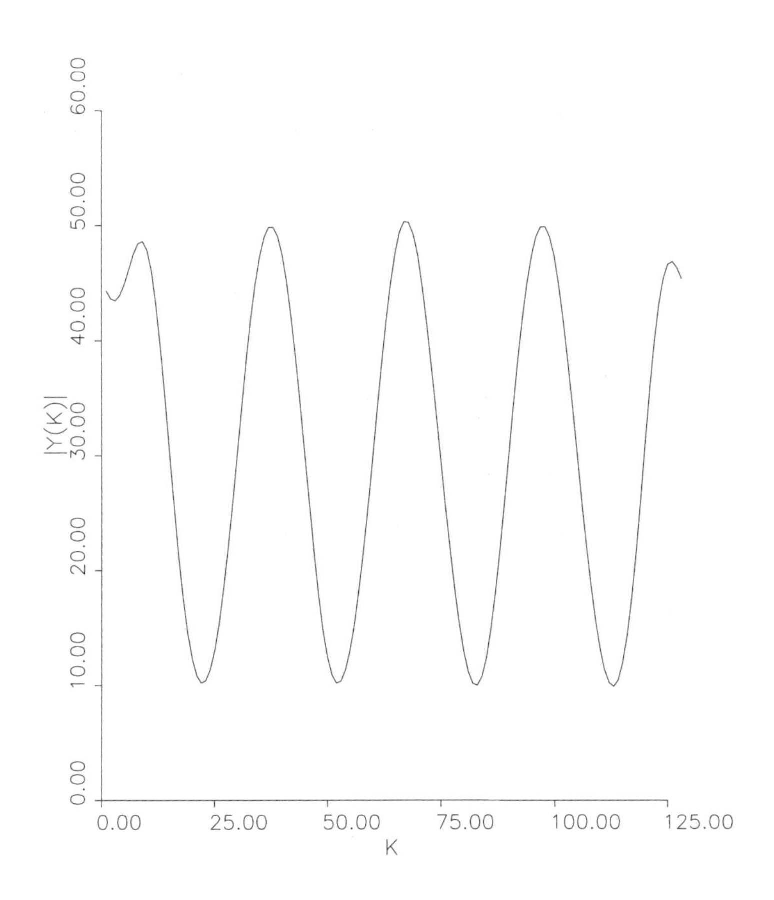

Problem 5.8
Figure 5.3.2 Without High Frequencies

```
      SUBROUTINE PREDQ(A,M,N,B,PIVOT,NPIVOT,ERRLIM)
      IMPLICIT DOUBLE PRECISION (A-H,O-Z)
C                                   DECLARATIONS FOR ARGUMENTS
      DOUBLE PRECISION A(M,*),B(M),ERRLIM
      INTEGER PIVOT(M),M,N,NPIVOT
C                                   DECLARATIONS FOR LOCAL VARIABLES
      DOUBLE PRECISION COLUMNL(M)
      include 'mpif.h'
```

```
C                               NPES = NUMBER OF PROCESSORS
      CALL MPI_COMM_SIZE (MPI_COMM_WORLD,NPES,IERR)
C                               ITASK = MY PROCESSOR NUMBER
      CALL MPI_COMM_RANK (MPI_COMM_WORLD,ITASK,IERR)
C                               USE GIVENS ROTATIONS TO REDUCE A
C                               TO ROW ECHELON FORM
      I = 1
      DO 15 L=1,N
C                               JTASK OWNS ACTIVE COLUMN L
         JTASK = MOD(L-1,NPES)
         IF (ITASK.EQ.JTASK) THEN
C                               IF JTASK IS ME, SAVE ACTIVE COLUMN IN
C                               VECTOR COLUMNL
            LME = (L-1)/NPES+1
            DO 1 J=I,M
               COLUMNL(J) = A(J,LME)
    1       CONTINUE
         ENDIF
C                               RECEIVE COLUMNL FROM PROCESSOR JTASK
         CALL MPI_BCAST(COLUMNL(I),M-I+1,MPI_DOUBLE_PRECISION,
     &   JTASK,MPI_COMM_WORLD,IERR)
C                               L0 = FIRST COLUMN >= L BELONGING TO ME
         LL0 = (L-1+NPES-(ITASK+1))/NPES
         L0 = ITASK+1+LL0*NPES
C                               USE PIVOT A(I,L) TO KNOCK OUT ELEMENTS
C                               I+1 TO M IN COLUMN L.
         DO 10 J=I+1,M
            IF (COLUMNL(J).EQ.0.0) GO TO 10
            DEN = SQRT(COLUMNL(I)**2+COLUMNL(J)**2)
            C = COLUMNL(I)/DEN
            S = COLUMNL(J)/DEN
C                               PREMULTIPLY A BY Qij**T
            DO 5 K=L0,N,NPES
               KME = (K-1)/NPES+1
               BIK = C*A(I,KME) + S*A(J,KME)
               BJK =-S*A(I,KME) + C*A(J,KME)
               A(I,KME) = BIK
               A(J,KME) = BJK
    5       CONTINUE
            BIL = C*COLUMNL(I) + S*COLUMNL(J)
            BJL =-S*COLUMNL(I) + C*COLUMNL(J)
            COLUMNL(I) = BIL
            COLUMNL(J) = BJL
C                               PREMULTIPLY B BY Qij**T
            BI = C*B(I) + S*B(J)
            BJ =-S*B(I) + C*B(J)
            B(I) = BI
            B(J) = BJ
```

```
  10    CONTINUE
C                                 PIVOT A(I,L) NONZERO AFTER PROCESSING
C                                 COLUMN L--MOVE DOWN TO NEXT ROW, I+1
        IF (ABS(COLUMNL(I)).LE.ERRLIM) COLUMNL(I) = 0.0
        IF (COLUMNL(I).NE.0.0) THEN
           NPIVOT = I
           PIVOT(NPIVOT) = L
           I = I+1
           IF (I.GT.M) RETURN
        ENDIF
  15 CONTINUE
     RETURN
     END
```

6.7. Here is a parallelized version of HESSH:

```
      SUBROUTINE PHESSH(A,N)
      IMPLICIT DOUBLE PRECISION (A-H,O-Z)
C                                 DECLARATIONS FOR ARGUMENTS
      DOUBLE PRECISION A(N,N)
C                                 DECLARATIONS FOR LOCAL VARIABLES
      DOUBLE PRECISION W(N),ATWI(N),ATW(N)
      include 'mpif.h'
      IF (N.LE.2) RETURN
C                                 INITIALIZE MPI
      CALL MPI_INIT (IERR)
C                                 NPES = NUMBER OF PROCESSORS
      CALL MPI_COMM_SIZE (MPI_COMM_WORLD,NPES,IERR)
C                                 ITASK = MY PROCESSOR NUMBER (0,...,NPES-1).
C                                 I WILL NEVER TOUCH ANY COLUMNS OF A EXCEPT
C                                 MY COLUMNS, ITASK+1+ K*NPES, K=0,1,2,...
      CALL MPI_COMM_RANK (MPI_COMM_WORLD,ITASK,IERR)
C                                 USE HOUSEHOLDER TRANSFORMATIONS TO
C                                 REDUCE A TO UPPER HESSENBERG FORM
      DO 40 I=2,N-1
C                                 JTASK IS PROCESSOR THAT OWNS COLUMN I-1
         JTASK = MOD(I-2,NPES)
         IF (ITASK.EQ.JTASK) THEN
C                                 CHOOSE UNIT N-VECTOR W (WHOSE FIRST
C                                 I-1 COMPONENTS ARE ZERO) SUCH THAT WHEN
C                                 COLUMN I-1 IS PREMULTIPLIED BY
C                                 H = I - 2W*W**T, COMPONENTS I+1 THROUGH
C                                 N ARE ZEROED.
            CALL CALW(A(1,I-1),N,W,I)
         ENDIF
C                                 RECEIVE VECTOR W FROM PROCESSOR JTASK
         CALL MPI_BCAST(W(I),N-I+1,MPI_DOUBLE_PRECISION,JTASK,
```

```
     &    MPI_COMM_WORLD,IERR)
C                                    PREMULTIPLY A BY H = I - 2W*W**T
         L0 = (I-2+NPES-(ITASK+1))/NPES
         I0 = ITASK+1+L0*NPES
         DO 15 K=I0,N,NPES
            WTA = 0.0
            DO 5 J=I,N
               WTA = WTA + W(J)*A(J,K)
    5       CONTINUE
            TWOWTA = 2*WTA
            DO 10 J=I,N
               A(J,K) = A(J,K) - TWOWTA*W(J)
   10       CONTINUE
   15    CONTINUE
C                                    POSTMULTIPLY A BY H = I - 2W*W**T
         L0 = (I-1+NPES-(ITASK+1))/NPES
         I0 = ITASK+1+L0*NPES
         DO 25 K=1,N
            ATWI(K) = 0.0
            DO 20 J=I0,N,NPES
               ATWI(K) = ATWI(K) + A(K,J)*W(J)
   20       CONTINUE
   25    CONTINUE
         CALL MPI_ALLREDUCE(ATWI,ATW,N,MPI_DOUBLE_PRECISION,
     &    MPI_SUM,MPI_COMM_WORLD,IERR)
         DO 35 K=1,N
            TWOATW = 2*ATW(K)
            DO 30 J=I0,N,NPES
               A(K,J) = A(K,J) - TWOATW*W(J)
   30       CONTINUE
   35    CONTINUE
   40 CONTINUE
      IF (NPES.GT.1) THEN
C                                    REDISTRIBUTE ENTIRE A MATRIX TO ALL
C                                      PROCESSORS
         DO 45 J=1,N
            JTASK = MOD(J-1,NPES)
C                                    RECEIVE NON-ZERO PORTION OF COLUMN J
C                                      FROM PROCESSOR JTASK WHICH OWNS IT
            CALL MPI_BCAST(A(1,J),MIN(J+1,N),MPI_DOUBLE_PRECISION,
     &        JTASK,MPI_COMM_WORLD,IERR)
   45    CONTINUE
      ENDIF
      CALL MPI_FINALIZE (IERR)
      RETURN
      END
```

References

Bisseling, R.H. (2004), *Parallel Scientific Computing*, Oxford University Press.

Brigham, E.O. (1974), *The Fast Fourier Transform*, Prentice Hall.

Brigham, E.O. (1988), *The Fast Fourier Transform and Its Applications*, Prentice Hall.

Buchanan, J.L. and P.R. Turner (1992), *Numerical Methods and Analysis*, McGraw-Hill.

Cooley, J.W. and J.W. Tukey (1965), "An algorithm for the machine calculations of complex Fourier series," *Mathematics of Computation* **19**, 297–301.

Dantzig, G.B. (1951), "Maximization of a linear function of variables subject to linear inequalities," in *Activity Analysis of Production and Allocation*, John Wiley & Sons. pp. 339–347.

Dantzig, G.B. (1963), *Linear Programming and Extensions*, Princeton University Press.

de Boor, C. (1978), *A Practical Guide to Splines*, Springer.

Dongarra, J.J., J.R. Bunch, C.B. Moler, and G.W. Stewart (1979), *LINPACK User's Guide*, SIAM.

Dongarra, J.J., J. Du Croz, S. Hammarling and R.J. Hanson (1988), "An extended set of FORTRAN basic linear algebra subprograms," *ACM Transactions on Mathematical Software* **14**, 1–32.

Duff, I. and J. Reid (1983), "The Multifrontal Solution of Indefinite Sparse Symmetric Linear Equations," *ACM Transactions on Mathematical Software* **9**, 302–325.

Duff, I., A. Erisman, and J. Reid (1986), *Direct Methods for Sparse Matrices*, Oxford University Press.

Forsythe, G. and P. Henrici (1960), "The cyclic Jacobi method for computing the principal values of a complex matrix," *Transactions of the American Mathematical Society* **94**, 1–23.

Forsythe, G., M. Malcolm, and C.B. Moler (1977), *Computer Methods for Mathematical Computations*, Prentice Hall.

Francis, J.G.F. (1961), "The *Q-R* transformation, parts I and II," *Computer Journal* **4**, 265–271, 332–345.

George, A. and J.W.H. Liu (1981), *Computer Solution of Large Sparse Positive Definite Systems*, Prentice Hall.

Gill, P., W. Murray, and M. Wright (1991), *Numerical Linear Algebra and Optimization, Volume I*, Addison-Wesley.

Hagar, W.W. (1988), *Applied Numerical Linear Algebra*, Prentice Hall.

Hanselman, D. and B. Littlefield (2001), *Mastering MATLAB 6*, Prentice Hall.

Heath, M.T. (2002), *Scientific Computing, An Introductory Survey*, McGraw-Hill.

Higham, D.J. and N.J. Higham (2000), *MATLAB Guide*, SIAM.

Kahaner, D, C. Moler, and S. Nash (1989), *Numerical Methods and Software*, Prentice Hall.

Kincaid, D. and W. Cheney (2004), *Numerical Mathematics and Computing, Fifth Edition*, Brooks and Cole.

Leiss, E. (1995), *Parallel and Vector Computing*, McGraw-Hill.

Pacheco, P. (1996), *Parallel Programming with MPI*, Morgan Kaufmann.

Rutishauser, H. (1958), "Solution of eigenvalue problems with the *L-R* transformation," *U.S. Bureau of Standards Applied Mathematics Series* **49**, 47–81.

Schendel, U. (1984), *Introduction to Numerical Methods for Parallel Computers*, Ellis Horwood.

Schrijver, A. (1986), *Theory of Linear and Integer Programming*, John Wiley & Sons.

Sewell, G. (2005), *The Numerical Solution of Ordinary and Partial Differential Equations, Second Edition*, John Wiley & Sons.

Sewell, G. (2010), "Solving PDEs in non-Rectangular 3D Regions using a Collocation Finite Element Method," *Advances in Engineering Software* **41**, 748–753.

Singleton, R. (1967), "On computing the fast Fourier transform," *Communications of the ACM* **10**(10), 647–654.

Smith, B.T., J.M. Boyle, J.J. Dongarra, B.S. Garbow, U. Ikebe, V.C. Klema, and C.B. Moler (1976), *Matrix Eigensystem Routines—EISPACK Guide*, Springer.

Stewart, G.W. (1973), *Introduction to Matrix Computations*, Academic Press.

Trefethen, N. (2002), "A Hundred-dollar, Hundred-digit Challenge," *SIAM News* **35**(1), 1.

Wilkinson, J.H. (1965), *The Algebraic Eigenvalue Problem*, Oxford University Press.

Young, D.H. (1971), *Iterative Solution of Large Linear Systems*, Academic Press.